Transition, Turbulence and Combustion

ICASE/LaRC Interdisciplinary Series in Science and Engineering

Volume 3

Transition, Turbulence and Combustion

Volume II

Turbulence and Combustion

edited by

M. Y. Hussaini
Institute of Computer Applications in Science and Engineering (ICASE), NASA Langley Research Center, Hampton, Virginia, U.S.A.

Thomas B. Gatski
NASA Langley Research Center, Hampton, Virginia, U.S.A.

and

T. L. Jackson
Institute of Computer Applications in Science and Engineering (ICASE), NASA Langley Research Center, Hampton, Virginia, U.S.A.

KLUWER ACADEMIC PUBLISHERS
DORDRECHT / BOSTON / LONDON

Library of Congress Cataloging-in-Publication Data

Transition, turbulence and combustion / edited by Thomas G. Gatski, M.Y. Hussaini, T.L. Jackson.
p. cm. -- (ICASE/LaRC interdisciplinary series in science and engineering ; v. 3)
Proceedings of the Workshop on Transition, Turbulence and Combustion, held June 7-July 2, 1993.
Includes index.
Contents: v. 1. Transition -- v. 2. Turbulence and combustion.
ISBN 0-7923-3085-4(acid-free paper). -- ISBN 0-7923-3086-2 (set)
1. Fluid dynamics--Congresses. 2. Turbulence--Congresses. 3. Combustion--Congresses. I. Gatski, T. B. II. Hussaini, M. Yousuff. III. Jackson, Thomas L. (Thomas Luther), 1957- . IV. Workshop on Transition, Turbulence and Combustion (1993) V. Series.
TA357.T713 1994
532'.0526--dc20 94-32719

ISBN 0-7923-3085-4 (Vol.II)
ISBN 0-7923-3086-2 (Set)

Published by Kluwer Academic Publishers,
P.O. Box 17, 3300 AA Dordrecht, The Netherlands.

Kluwer Academic Publishers incorporates
the publishing programmes of
D. Reidel, Martinus Nijhoff, Dr W. Junk and MTP Press.

Sold and distributed in the U.S.A. and Canada
by Kluwer Academic Publishers,
101 Philip Drive, Norwell, MA 02061, U.S.A.

In all other countries, sold and distributed
by Kluwer Academic Publishers Group,
P.O. Box 322, 3300 AH Dordrecht, The Netherlands.

Printed on acid-free paper

Printed in the Netherlands

CONTENTS

Volume II

PREFACE

These two volumes contain the proceedings of the Workshop on Transition, Turbulence and Combustion, sponsored by the Institute for Computer Applications in Science and Engineering (ICASE) and the NASA Langley Research Center (LaRC), during June 7 to July 2, 1993. Volume I contains the contributions from the transition research, and Volume II contains the contributions from both the turbulence and combustion research. This is the third workshop in the series on the subject. The first was held in 1989, the second in 1991, and their proceedings were published by Springer-Verlag under the titles "Instability and Transition" (edited by M. Y. Hussaini and R. G. Voigt) and "Instability, Transition and Turbulence" (edited by M. Y. Hussaini, A. Kumar and C. L. Streett) respectively.

The objectives of these workshops are to expose the academic community to current technologically important issues of transition, turbulence and combustion, and to acquaint the academic community with the unique combination of theoretical, computational and experimental capabilities at LaRC. It is hoped these will foster continued interactions, and accelerate progress in elucidating the fundamental phenomena of transition, turbulence and combustion.

The research areas of interest in transition covered the full range of the subject: linear and nonlinear stability, direct and large-eddy simulation and phenomenological modeling of the transition zone.

In turbulence, the interest was focussed on second-order closures and the formulation of near-wall corrections to existing high Reynolds number models, and closure model development based on turbulent flow structures and RNG theory.

Interest in combustion research was focussed on three areas: structure and stability of counterflow diffusion flames, development of novel mixing enhancement techniques for non-premixed combustion, and methods of modeling the interaction between turbulence and chemical kinetics.

There were approximately a hundred participants in the Workshop, one-third of which were LaRC and ICASE scientists and contractors; the rest were mainly from universities. There were also some participants from US aircraft industries. These participants were divided into eleven groups of five to ten members each. The group leaders were: S. P. Wilkinson (Experiment), M. R. Malik (Stability theory and applications); P. Hall (Advanced asymptotics), M. G. Macaraeg (Receptivity), C. L. Streett (Simulations I) and

G. Erlebacher (Simulations II), T. B. Gatski (Quasi-equilibrium turbulent flows), Ye Zhou (RNG theory and turbulence modeling), J. P. Drummond (Counter-jet diffusion flames), T. L. Jackson (Turbulent Combustion) and A. Kumar (Shock-induced combustion/detonation).

These proceedings contain research papers resulting from projects which, for the most part, were begun during the Workshop and completed shortly thereafter, or resulting from earlier research reexamined during the Workshop in light of interactions and discussions. The papers are organized, reviewed and summarized by the group leaders. In addition to these research articles, the proceedings also contain the presentations by D. M. Bushnell and M. Goldstein, two of the many talks on special topics delivered during the Workshop. Also included is the summary of the panel discussion on renormalization group methods in turbulence modeling.

It is our pleasure to acknowledge the assistance and cooperation of many LaRC scientists and other logistical staff who contributed to the smooth operation and success of the Workshop. In particular, we thank Douglas Dwoyer and Jerry Hefner for their support and encouragement, and Michael Card and Edwin Prior for their continuous assistance; we thank Dennis Bushnell, Marvin Goldstein and Eli Reshotko for their tutorials which discussed the research needs and opportunities as well as the state-of-the-art tools; and also the group leaders for their efforts in supervising, reviewing and summarizing the research output of their respective groups. Finally, we thank all the participants who made the Workshop a success.

It is indeed a pleasure to acknowledge the assistance of Linda Johnson who took care of the arrangements for the Workshop, Emily Todd who coordinated the preliminary correspondence for the Workshop as well as the collection and format editing of the manuscripts, Shannon Fedors who did an excellent job as the Workshop secretary, Barbara Stewart who typed or reformatted the manuscripts, Shelly Millen who made editing changes and formatted figures, and Etta Blair who handled reimbursements for the participants. We are pleased to express our sincere appreciation to all of them. Thanks are also due to the the staff of Kluwer Academic Publishers for their cooperation in bringing out this volume.

TBG, MYH and TLJ

The research of the following authors who contributed to this volume was supported by the National Aeronautics and Space Administration under NASA Contract No. NAS1-19480 while they were in residence at the Transition, Turbulence and Combustion Workshop.

H. Aksoy
S. Balachandar
David C. Banks
Gal Berkooz
S. Biringen
Nicholas D. Blackaby
Gregory Blaisdell
Thomas J. Bridges
T. Brown
Thomas C. Corke
Adrian V. Coward
Andrew H. Dando
A. O. Demuren
James P. Denier
M. R. Dhanak
Gordon Erlebacher
Hermann F. Fasel
David A. Fuciarelli
Sharath S. Girimaji
Philip Hall
Carlos Hartel
Dan S. Henningson
K. M. Isaac
T. L. Jackson
A. K. Kapila
Keith F. Knasiak
F. Kozukso
Linda D. Kral
J. T. Lachowicz
S. H. Lam
M. Matalon
David McComb
Hubert L. Meitz
S. R. Otto
Demetrios T. Papageorgiou
Ugo Piomelli
R. W. Pitz
James J. Quirk
Satish C. Reddy
Helen L. Reed
D. N. Riahi
J. R. Ristorcelli
R. Rubinstein
E. M. Saiki
Peter J. Schmid
Sharon O. Seddougui
K. Seshadri
Charles G. Speziale
Evangelos T. Spyropoulous
Siva Thangam
George Vahala
Robert Wilson
Bassam A. Younis
Ye Zhou
Y. H. Zurigat

CONTRIBUTORS

Ridha Abid
High Technology Corporation
NASA Langley Research Center
Hampton, VA 23681-0001

H. Aksoy
Mechanical and Aerospace Engineering
Arizona State University
Tempe, AZ 85287-6106

Harold L. Atkins
NASA Langley Research Center
Hampton, VA 23681-0001

S. Balachandar
University of Illinois
Urbana, IL 61801

David C. Banks
Institute for Computer Applications in Science and Engineering
NASA Langley Research Center
Hampton, VA 23681-0001

Gal Berkooz
Center for Applied Math. and Dept. of Mechanical and Aerospace Engineering
Cornell University
Ithaca, NY 14853

S. Biringen
Department of Aerospace Engineering Sciences
University of Colorado
Campus Box 429
Boulder, CO 80309

Nicholas D. Blackaby
University of Manchester
Manchester, M13 9PL
UNITED KINGDOM

Gregory A. Blaisdell
Purdue University
West Lafayette, IN 47907-1282

Thomas J. Bridges
Mathematics Institute
University of Warwick
Coventry CV4 7AL
GREAT BRITAIN

T. M. Brown
Department of Mechanical Engineering
Vanderbilt University
Nashville, TN 37235

Dennis M. Bushnell
NASA Langley Research Center
Hampton, VA 23681-0001

Mark H. Carpenter
NASA Langley Research Center
Hampton, VA 23681-0001

D. Cavalieri
Mechanical and Aerospace
Engineering Department
Illinois Institute of Technology
Chicago, IL 60616

Chau-Lyan Chang
High Technology Corporation
Hampton, VA 23666

Meelan Choudhari
High Technology Corporation
Hampton, VA 23666

Alison J. Cooper
Mathematics Institute
University of Warwick
Coventry CV4 7AL
GREAT BRITAIN

Thomas C. Corke
Fluid Dynamics Research
Center
Mechanical and Aerospace
Engineering Department
Illinois Institute of Technology
Chicago, IL 60616

Adrian V. Coward
University of Manchester
Manchester M13 9PL
UNITED KINGDOM

Andrew H. Dando
Department of Mathematics
Oxford Road
University of Manchester
Manchester M13 9PL
UNITED KINGDOM

A. O. Demuren
Old Dominion University
Norfolk, VA 23529

James P. Denier
School of Mathematics
University of New South Wales
P. O. Box 1
Kensington, NSW 2033
AUSTRALIA

M. R. Dhanak
Florida Atlantic University
Boca Raton, FL 33431

S. P. G. Dinavahi
Analytical Services &
Materials, Inc.
NASA Langley Research Center
Hampton, VA 23681-0001

Phil Drummond
NASA Langley Research Center
Hampton, VA 23681-0001

Juan Elezgaray
CRPP/CNRS
Bordeaux
FRANCE

Nabil M. El-Hady
Analytical Services &
Materials, Inc.
107 Research Drive
Hampton, VA 23666

Gordon E. Erlebacher
Institute for Computer
Applications in Science
and Engineering
NASA Langley Research Center
Hampton, VA 23681-0001

Hermann F. Fasel
University of Arizona
Tucson, AZ 85721

David A. Fuciarelli
Arizona State University
Tempe, AZ 85287-6106

T. B. Gatski
NASA Langley Research Center
Hampton, VA 23681-0001

Sharath S. Girimaji
Institute for Computer
Applications in Science
and Engineering
NASA Langley Research Center
Hampton, VA 23681-0001

Peyman Givi
State University of New York
Buffalo, NY 14260

M. E. Goldstein
NASA Lewis Research Center
Cleveland, OH 44135

Philip Hall
University of Manchester
Manchester M13 9PL
UNITED KINGDOM

Carlos Härtel
DLR, Institute for Fluid
Mechanics
Bunsenstr. 10, D-37073
Göttingen
GERMANY

Dan S. Henningson
Aeronautical Research Institute
of Sweden
Bromma
SWEDEN
and
Department of Mechanics
Royal Institute of Technology
Stockholm
SWEDEN

F. Q. Hu
Department of Mathematics
and Statistics
Old Dominion University
Norfolk, VA 23529

Xiaoli Huai
Department of Mechanical Engineering
University of Maryland
College Park, MD 20742

K. M. Isaac
The University of Missouri-Rolla
Mechanical & Aerospace Engineering Department
Rolla, MO 65401

Farhad A. Jaberi
State University of New York
Buffalo, NY 14260

T. L. Jackson
Institute for Computer Applications in Science and Engineering
NASA Langley Research Center
Hampton, VA 23681-0001

Ronald D. Joslin
NASA Langley Research Center
Hampton, VA 23681-0001

A. K. Kapila
Rensselaer Polytechnic Institute
Troy, NY 12180-3590

Keith F. Knasiak
Fluid Dynamics Research Center
Mechanical and Aerospace Engineering Department
Illinois Institute of Technology
Chicago, IL 60616

F. Kozusko
Department of Mathematics and Statistics
Old Dominion University
Norfolk, VA 23529

Linda D. Kral
McDonnell Douglas Corporation
P. O. Box 516
Mail Code 106-4126
St. Louis, MO 63166-0516

Ajay Kumar
NASA Langley Research Center
Hampton, VA 23681-0001

J. T. Lachowicz
Mechanical and Aerospace Engineering Department
North Carolina State University
Raleigh, NC 27695

Sau-hai Lam
Department of Mechanical and Aerospace Engineering
Princeton University
Princeton, NJ 08544

F. Li
High Technology Corporation
Hampton, VA 23666

R.-S. Lin
High Technology Corporation
Hampton, VA 23666

Anders Lundbladh
Royal Institute of Technology
Department of Mechanics
SWEDEN
and
Aeronautical Research Institute of Sweden
Bromma
SWEDEN

Michéle G. Macaraeg
NASA Langley Research Center
Hampton, VA 23681-0001

M. R. Malik
High Technology Corporation
Hampton, VA 23666

Jamal A. Masad
High Technology Corporation
Hampton, VA 23666

M. Matalon
Engineering Sciences and Applied Mathematics
McCormick School of Engineering and Applied Science
Northwestern University
Evanston, IL 60208

David McComb
Department of Physics
University of Edinburgh
UNITED KINGDOM

Hubert L. Meitz
University of Arizona
Tucson, AZ 85721

S. P. Nandula
Department of Mechanical Engineering
Vanderbilt University
Nashville, TN 37235

1st Lt John Nelson
United States Air Force Wright Laboratories
Wright-Patterson Air Force Base, OH 45433-7913

S. R. Otto
Institute for Computer Applications in Science and Engineering
NASA Langley Research Center
Hampton, VA 23681-0001

Demetrios T. Papageorgiou
Department of Mathematics and Center for Applied Mathematics and Statistics
New Jersey Institute of Technology
Newark, NJ 07102

Ugo Piomelli
Department of Mechanical Engineering
University of Maryland
College Park, MD 20742

R. W. Pitz
Department of Mechanical Engineering
Vanderbilt University
Nashville, TN 37235

C. David Pruett
Analytical Services and Materials, Inc.
Hampton, VA 23666

James J. Quirk
Institute for Computer Applications in Science and Engineering
NASA Langley Research Center
Hampton, VA 23681-0001

Satish C. Reddy
New York University
New York, NY 10012

Helen L. Reed
Arizona State University
Tempe, AZ 85287-6106

D. N. Riahi
Department of Theoretical & Applied Mechanics
University of Illinois
Urbana, IL 61801

J. R. Ristorcelli
Institute for Computer Applications in Science and Engineering
NASA Langley Research Center
Hampton, VA 23681-0001

R. Rubinstein
Institute for Computational Mechanics in Propulsion
NASA Lewis Research Center
Cleveland, OH 44135

E. M. Saiki
Department of Aerospace Engineering Sciences
University of Colorado
Campus Box 429
Boulder, CO 80309

Peter J. Schmid
Department of Mathematics
Massachusetts Institute of Technology
Cambridge, MA 02139

Sharon O. Seddougui
University of Birmingham
Birmingham, B15 2TT
UNITED KINGDOM

K. Seshadri
Center for Energy and Combustion Science
University of California, San Diego
La Jolla, CA 92093-0310

Bart A. Singer
High Technology Corporation
NASA Langley Research Center
Hampton, VA 23681-0001

D. J. Singh
Analytical Services & Materials, Inc.
Hampton, VA 23666

Janice M. Smith
NASA Langley Research Center
Hampton, VA 23681-0001

R. M. C. So
Mechanical and Aerospace Engineering
Arizona State University
Tempe, AZ 85287-6106

Charles G. Speziale
Department of Aerospace and Mechanical Engineering
Boston University
Boston, MA 02215

Evangelos T. Spyropoulos
Purdue University
West Lafayette, IN 47907-1282

Craig L. Streett
NASA Langley Research Center
Hampton, VA 23681-0001

J. T. Stuart
Imperial College
London SW7 2BZ
UNITED KINGDOM

Siva Thangam
Stevens Institute of Technology
Hoboken, NJ 07030

George Vahala
College of William and Mary
Williamsburg, VA 23185

M. P. Vonderwell
Department of Theoretical & Applied Mechanics
University of Illinois
Urbana, IL 61801

Stephen P. Wilkinson
NASA Langley Research Center
Hampton, VA 23681-0001

R. V. Wilson
Old Dominion University
Norfolk, VA 23529

Richard W. Wlezien
High Technology Corporation
Hampton, VA 23666

Bassam A. Younis
Department of Civil Engineering
City University
London EC1V OHB
UNITED KINGDOM

Ye Zhou
Institute for Computer
Applications in Science
and Engineering
NASA Langley Research Center
Hampton, VA 23681-0001

Y. H. Zurigat
Mechanical Engineering
Department
University of Jordan
Amman
JORDAN

TURBULENCE

QUASI-EQUILIBRIUM TURBULENT FLOWS

GROUP SUMMARY: QUASI-EQUILIBRIUM TURBULENT FLOWS

T. B. Gatski

NASA Langley Research Center
Hampton, Virginia 23681-0001

The research conducted in this topical area focused on the evaluation and development of turbulent models in both compressible and incompressible turbulent flows. In the following paragraphs, the various research topics are outlined and significant results highlighted. Also included is a summary of the near-wall panel discussion held during the workshop.

S. Dinavahi, J. Ristorcelli and G. Erlebacher used the turbulent results from the DNS of a temporal supersonic boundary layer on a cylinder to study several terms of the Reynolds stress equations which relate to compressibility. New terms, specifically associated with compressibility are identified and their structure in the boundary layer is studied. Although the effects of the spatial growth of the boundary layer were neglected in the DNS, results are still useful because the subsequent Reynolds stress models are based on rational asymptotic expansions of the Navier-Stokes equations, which only bring orders of magnitude into play.

With the development of models for the compressible dissipation and the pressure dilatation, attention needs to be focused on the remaining correlations which appear in the compressible turbulent stress transport equations. Included in this category is the turbulent mass flux, which represents the difference between the Reynolds averaged and Favre averaged mean velocity fields. J. Ristorcelli and S. Dinavahi have developed an algebraic-type model which is proportional to the mean density gradients with a tensor eddy viscosity. Based on temporal DNS data of the early stages of a compressible boundary-layer flow just discussed, the model was found to represent the simulated mass flux quite well. These tests have also shown that the usual assumption of equating the Reynolds averaged and Favre averaged velocities in the viscous terms is erroneous.

Direct simulation data can also be used to validate existing turbulence models. R. Abid has used the DNS results of Blaisdell to evaluate three pressure-strain correlation models. In addition, mod-

M. Y. Hussaini et al. (eds.), Transition, Turbulence and Combustion, Vol. II, 5–8.

els for the dilatation dissipation and pressure-dilatation are added and the performance of the total modeled set evaluated. Each turbulence model is evaluated in its ability to predict the equilibrium states accurately. The results showed that the models using variable density extensions to the compressible regime were unable to correctly predict the equilibrium states. Additional model development is needed in order to correctly account for the compressibility effects.

Due to the ever increasing desire to perform simulations on more practical compressible turbulent flows, the adaptation of large-eddy simulations, that is, the development of subgrid scale models, to spatially evolving flows is necessary. The study of Spyropoulos and Blaisdell has dealt with the evaluation of several dynamic subgrid-scale models for inhomogeneous flows. In the study, both implicit and explicit filters were tested and compared to the traditional sharp cut-off filter as well as experimental data.

It is also becoming increasingly important to establish the predictive capabilities of new closure models, especially Reynolds stress models, due to the increased complexity and cost of computation associated with them. In this regard, B. Younis, T. Gatski and C. Speziale evaluated the pressure-strain model of Speziale, Sarkar, and Gatski in three complex flows involving either rotational body forces or longitudinal curvature in lateral and longitudinal planes. In comparisons with experiments and another pressure-strain model, which is only linear in the stress anisotropies, the quadratic SSG model was found to perform quite well.

The proper subgrid scale (SGS) modeling of the near-wall region in a large eddy simulation is essential in order to correctly predict the turbulent physics. Härtel observed that in the near-wall region of a turbulent channel, there is a buffer layer with a reverse transfer of energy from the subgrid-scale to the large scales. This effect cannot be accounted for by standard subgrid-scale models. After identifying its source, a new algebraic model is proposed with a qualitatively correct energy budget.

Coupled with the choice of an appropriate Reynolds stress closure in the solution of wall-bounded flows is the appropriate means of integrating the model equation to the wall. Following a proposal made by Durbin, A. Demuren and R. Wilson have incorporated near-wall models based on elliptic relaxation equations into two high Reynolds number second-moment closures. While the results show reasonable predictions of the mean flow and Reynolds stress anisotropy, they fail to show the change from a 2-component state to a nearly one-component state observed in close proximity to the wall.

A more traditional approach to developing near-wall corrections to existing high Reynolds number closures is to construct models which yield the correct asymptotic behavior of the turbulence in the near-wall region. H. Aksoy and R. So have modified the high Reynolds number SSG model for application to wall-bounded flows. The new model is tested against experimental fully developed turbulent flow in a square duct. The results are also compared against the predictions of the near-wall model of Lai and So. In addition, recent DNS calculations of turbulent flow in a square duct provide an additional database from which validation studies are made. In the tests conducted to date, the modified SSG model is found to yield fairly good agreement with experiments suggesting its applicability to predicting complex turbulent flows.

Finally, during the workshop an informal panel discussion on near-wall modeling was held which attempted to identify and address some problem areas inherent in using either wall-function approaches or closure schemes integrable directly to the wall. The discussion session included participants of the workshop as well as Prof. R. M. C. So from Arizona State University. In order to keep the discussion focused, only the issue of near-wall models was discussed, while the related though separate issue of low-Reynolds number modeling was excluded.

At the outset, several cases were presented where the absence of a well defined log-layer did not preclude the use of a wall function formulation coupled with a high-Reynolds number turbulence model. These cases include separated flows, flows with additional rates of strain, turbulence driven secondary motions, three-dimensional flows over infinite swept wings and rotating bodies, and even periodic flows. Counter examples of where the wall-function approach is clearly inappropriate include the cases of strongly curved channel flow, some adverse pressure gradient flows, flows with suction or injection, and flows over riblets. However, since no well-posed comparative study has been conducted into the performance of either wall-functions or near-wall models on designated test flows, it was not possible to reach any definitive conclusions on the merits of the approaches.

Several issues related to the use of either approach were discussed. In the wall-function formulation, those included whether to specify the stress anisotropies or their flux at the matching point. It appears that the latter specification of the flux is more appropriate. In the near-wall formulation, more points are needed than the wall-function formulation due to the obvious resolution requirements near the wall, but this does not appear to be overly restrictive.

At this time, several issues are unresolved concerning the development of near-wall closures. Among the questions that need to be examined more closely is the role of asymptotic consistency in the construction of the closure models. While incorporating such behavior into the models will necessarily fix the correct behavior for attached wall-bounded flows, it is not as yet certain whether this will restrict the applicability of such models to other more complex flows. These restrictions could take the form of either poor predictive behavior or diminished robustness in the solution procedure. These points are further complicated by the uncertainty of the other closures in the model equations. In incompressible flows, there is the view that the defect may lie in the dissipation rate while the pressure-strain rate correlation may not require as extensive modification. Another view suggests that the blocking effect of the wall may produce strong anisotropies which may have to be modeled by way of the pressure-strain. An approach which uses elliptic relaxation equations to simulate such an effect would then be worthy of investigation. In the compressible regime, other correlations, such as pressure dilatation, compressible dissipation and mass flux, arise and these complicate the picture even further.

From this brief summary of the near-wall closure discussions, it is clear that a focused effort is required to identify the key deficiencies of turbulent closure models in more complex turbulent flows. The attached flat plate boundary-layer flows can be effectively handled by either wall-function approaches or near-wall closures with the degree of accuracy the only outstanding question. For flows involving non-planar geometries, additional strain rates, or pressure gradient flows, the issue is unclear due to the lack of any controlled studies on model choice, accuracy, and numerical formulation and implementation. These issues can be resolved with comparisons to direct simulations of more complex flows and the behavior of the various unknown correlations in the near-wall region.

ON PREDICTION OF EQUILIBRIUM STATES IN HOMOGENEOUS COMPRESSIBLE TURBULENCE

Ridha Abid[1]

High Technology Corporation
NASA Langley Research Center
Hampton, VA 23681-0001

ABSTRACT

Direct numerical simulations of compressible, homogeneous, turbulent shear flows are used to evaluate Reynolds stress models. Three pressure-strain models, which are either linear, quadratic, or cubic in the anisotropy tensor, are considered. Dilatational dissipation and pressure-dilatation models are inserted into the Reynolds stress closure. Results show that variable-density extensions of incompressible pressure-strain correlation models do not correctly capture the compressibility effects seen in the direct simulations. In particular, the increase in the anisotropy of normal stresses and the reduction in the shear stress are not reproduced by any of the models. Also, the use of the incompressible form of the dissipation-rate equation to determine the solenoidal part of the dissipation is found to be questionable.

1. Introduction

A resurgence of interest in hypersonics has emerged, which is driven by advanced new applications such as high-speed civil transport aircraft, supersonic combustion ramjets and transatmospheric vehicles. These new applications bring into prominence some critical items that have limited the effectiveness of computational fluid dynamics codes used as tools for hypersonic system design. Chief among these is compressible turbulence modeling. Several turbulence models of varying degrees of complexity have been developed that range from the simplest algebraic or zero-equation model to the full Reynolds stress closure. Most of these models are simple extensions of their incompressible counterparts, where compressibility

[1]Research was supported by the National Aeronautics and Space Administration under NASA contract NAS1-19299, NASA Langley Research Center, Hampton, VA 23681-0001.

M. Y. Hussaini et al. (eds.), Transition, Turbulence and Combustion, Vol. II, 9–21.

effects are incorporated into the models through changes in the mean density. However, many studies have shown that this type of model is unable to reproduce some features of flows that depend on compressibility, such as the reduction in the spreading rate of the compressible mixing layer as the Mach number increases. Thus, a better understanding of the effects of compressibility on flow turbulence is needed to improve current turbulence models.

Recently, direct numerical simulation (DNS) of compressible, homogeneous turbulent shear flow by Blaisdell (1990) has shown that the growth of turbulent kinetic energy decreases as the Mach number increases. The reduction is due to two compressibility terms: the dilatational dissipation and pressure-dilatation correlation, which explicitly appear in the turbulent kinetic energy and mean temperature equations. Zeman (1990, 1991) and Sarkar (1991, 1992) have modeled the additional terms; the inclusion of the two terms in two-equation turbulence models leads to a significant improvement in predicting the reduction in spreading rate with increasing Mach number.

In this paper, an assessment of Reynolds stress models in predicting compressible homogeneous shear flow is conducted with the DNS of Blaisdell (1990). Three pressure-strain correlation models, which are variable-density extensions of their incompressible counterparts, are considered. These models are either linear, quadratic, or cubic in the anisotropy stress tensor. The dilatational dissipation and pressure-dilatation correlation, which are the only explicit compressibility terms, are inserted into the Reynolds stress closure. Particular attention will be paid to the ability of each turbulence model to predict equilibrium states accurately.

2. Mathematical Formulation

In compressible, turbulent flows, two averaging techniques are commonly used to define the mean and fluctuating parts of a turbulent variable. Usually, conventional Reynolds averages are used for the pressure p and the density ρ; Favre averages are used for the velocity u_i and the temperature T. Thus, any dependent variable f can be decomposed into mean and fluctuating parts in two ways:

$$f = \overline{f} + f', \quad f = \tilde{f} + f'' \tag{1}$$

where the overbar represents the Reynolds average, the tilde denotes the Favre average, and the primes and double primes are, respectively, the deviations from the Reynolds average and the mass-weighted average. The Favre average $\tilde{f}$ is a density-weighted Reynolds average

$$\tilde{f} = \frac{\overline{\rho f}}{\overline{\rho}} \tag{2}$$

We will consider the problem of compressible, homogeneous shear flow. In this problem, an initially decaying, compressible turbulence is subjected to a uniform shear S with the corresponding mean velocity gradients

$$\tilde{u}_{i,j} = S\delta_{1i}\delta_{2j} \tag{3}$$

The flow is assumed to be an ideal gas, which satisfies the equation of state $\overline{p} = \overline{\rho}R\tilde{T}$. The evolution equation of the mean temperature, derived from the energy equation, is given by

$$c_v \frac{D\tilde{T}}{Dt} = \varepsilon - \pi_d \tag{4}$$

where ε is the total dissipation rate, $\pi_d = \overline{p'd'}/\overline{\rho}$ is the specific pressure-dilatation, and c_v is the specific heat at constant volume.

For a compressible homogeneous shear flow at high Reynolds numbers, the Favre-averaged Reynolds stress tensor $\tau_{ij} = \overline{u_i''u_j''}$ is a solution of the transport equation

$$\frac{D\tau_{ij}}{Dt} = P_{ij} + \Pi_{ij} - \frac{2}{3}\varepsilon\delta_{ij} + \frac{2}{3}\pi_d\delta_{ij} \tag{5}$$

where Π_{ij} is the deviatoric part of the pressure-strain correlation and $P_{ij} = -\tau_{ik}\tilde{u}_{j,k} - \tau_{jk}\tilde{u}_{i,k}$ is the production term. Here, the Kolmogorov hypothesis of isotropy is invoked to model the dissipation rate tensor.

The transport equation for the turbulent kinetic energy $K = \overline{u_i''u_i''}/2$ is obtained by contracting indices in Eq. (5) as

$$\frac{DK}{Dt} = \mathcal{P} - \varepsilon + \pi_d \tag{6}$$

where $\mathcal{P} = -\tau_{ij}\tilde{u}_{ij}$ is the turbulence production.

As shown by Zeman and Sarkar, the dissipation rate of compressible homogeneous flow at high Reynolds numbers can be decomposed into a solenoidal or incompressible part, ε_s, and a compressible part, ε_d as

$$\varepsilon = \varepsilon_s + \varepsilon_d \tag{7}$$

where $\varepsilon_s = \overline{\nu}\overline{\omega_i'\omega_i'}$ and $\varepsilon_d = \frac{4}{3}\overline{\nu}\overline{(u_{i,i}')^2}$ given that $\overline{\nu}$ is the kinematic viscosity and ω_i' is the fluctuating vorticity. Direct simulations of compressible, homogeneous shear flows show that ε_s is largely independent of compressibility and that the growth of turbulent kinetic energy decreases as the Mach number increases because of the augmented contribution of the compressible dissipation. Therefore, ε_d is modeled and ε_s is obtained by solving the incompressible form of the dissipation equation

$$\frac{D\varepsilon_s}{Dt} = -C_{\varepsilon 1}\frac{\varepsilon_s}{K}\tau_{ij}\tilde{u}_{i,j} - C_{\varepsilon 2}\frac{{\varepsilon_s}^2}{K} \tag{8}$$

where $C_{\varepsilon 1}$ and $C_{\varepsilon 2}$ are closure coefficients that are model dependent. The models of the dilatational dissipation and pressure-dilatation correlation considered are

Sarkar

$$\varepsilon_c = 0.5M_t^2\varepsilon_s \tag{9}$$

$$\overline{p'd'} = 0.15\overline{\rho}\mathcal{P}M_t + 0.2\overline{\rho}\varepsilon_s M_t^2 \tag{10}$$

where $M_t = \sqrt{2K/\gamma R\tilde{T}}$ is the turbulent Mach number and γ is the ratio of specific heats ($= 1.4$).

Zeman

$$\varepsilon_c = 0.75(1 - exp(-((M_t - 0.1)/0.6)^2))\varepsilon_s \tag{11}$$

$$\overline{p'd'} = (\overline{p}\gamma)^{-1}(\frac{\overline{p^2} - \overline{p_e^2}}{\tau_f}) = -0.5\frac{D}{Dt}(\frac{\overline{p^2}}{\gamma\overline{p}}) \tag{12}$$

with

$$\tau_f = 0.4\frac{K}{\varepsilon_s}M_t \tag{13}$$

$$\overline{p_e^2} = 2\overline{\rho}^2 K\gamma R\tilde{T}(\frac{M_t^2 + M_t^4}{1 + M_t^2 + M_t^4}) \tag{14}$$

Finally, we consider incompressible pressure-strain models, which are modeled as linear functions of the mean velocity gradients with coefficients that depend algebraically on the anisotropy tensor and the turbulent dissipation rate. Compressibility effects are incorporated through changes in density. Three models will be analyzed: the Launder, Reece, and Rodi (LRR) model (1975); the Fu, Launder, and Tselepidakis (FLT) model (1987); and the Speziale, Sarkar, and Gatski (SSG) model (1991). The detailed form of these models is provided in Abid and Speziale (1993). These models are either linear (LRR), quadratic (SSG), or cubic (FLT) in the anisotropy tensor

$$b_{ij} = \frac{1}{2K}(\tau_{ij} - \frac{2}{3}K\delta_{ij}) \tag{15}$$

Calculations of compressible, homogeneous shear flows show that the Reynolds stresses, the dissipation rate, and the mean temperature grow exponentially, so that the anisotropy tensor b_{ij}, the shear parameter SK/ε_s and the turbulent Mach number M_t achieve equilibrium values that are independent of the initial conditions. Therefore, the system of equations for τ_{ij}, ε, and $\tilde{T}$ is nondimensionalized and is recast into an equivalent set of equations for ε_s/SK, b_{ij}, K, and M_t as follows:

$$\frac{DK^*}{Dt^*} = -(2b_{12} + (\frac{\varepsilon - \pi_d}{SK}))K^* \tag{16}$$

$$\frac{DM_t^2}{Dt^*} = M_t^2(\frac{\varepsilon_s}{SK})(\frac{\mathcal{P}}{\varepsilon_s} - (\frac{\varepsilon - \pi_d}{\varepsilon_s})(1 + \sigma M_t^2)) \tag{17}$$

$$\frac{D}{Dt^*}(\frac{\varepsilon_s}{SK}) = (\frac{\varepsilon_s}{SK})^2((C_{\varepsilon 1} - 1)(\frac{\mathcal{P}}{\varepsilon_s}) - (C_{\varepsilon 2} - (\frac{\varepsilon - \pi_d}{\varepsilon_s}))) \tag{18}$$

$$\frac{Db_{ij}}{Dt^*} = \frac{\Pi_{ij}}{2SK} + (\frac{P_{ij}}{\varepsilon_s} - \frac{2}{3}\frac{\mathcal{P}}{\varepsilon_s}\delta_{ij})(\frac{\varepsilon_s}{2SK}) - (\frac{\mathcal{P}}{\varepsilon_s} - \frac{\varepsilon}{\varepsilon_s} + \frac{\pi_d}{\varepsilon_s})(\frac{\varepsilon_s}{SK})b_{ij} \tag{19}$$

where $t^* = St$ is the dimensionless time, $K^* = K/K_0$, K_0 is the initial value of turbulent kinetic energy, and $\sigma = \gamma(\gamma - 1)/2$. An additional equation, Eq. (12), is needed when Zeman's model is used.

This system of nonlinear ordinary differential equations associated with each Reynolds stress model is solved subject to the initial conditions

$$K^* = 1, \; \frac{\varepsilon_s}{SK} = 0.45, \; M_t^2 = 0.094,$$

$$b_{11} = 0.124, \; b_{22} = -0.106, \; b_{12} = -0.187$$

at time $t^* = 0$, which is taken from the DNS of Blaisdell (run Sha192) at $t^* = 2$. The reason for this is that the initial conditions of the DNS at $t^* = 0$ are nonphysical. See Speziale et al. (1992). The system was integrated with a fourth-order Runge-Kutta scheme. The equilibrium states for each turbulence model are obtained numerically. A comparison of the predictions of the models with the DNS data will be made in the next section.

3. Results and Discussion

The predictions of the SSG, LRR, and FLT models with Sarkar's compressibility corrections will be compared with the DNS results of Blaisdell (run Sha192).

Figures 1 and 2 show the time evolution of the turbulent kinetic energy and the dissipation rate, predicted by the SSG model for two initial turbulent Mach numbers $M_{t0} = 0$ and $M_{t0} = 0.307$. These figures clearly show that a variable-density extension of the SSG model is not capable of capturing the decrease of the growth rate of turbulent kinetic energy as the turbulent Mach number increases. Explicit compressibility corrections are needed to predict the trends of the DNS. Also, the differences in the predictions between turbulence models are shown (Figs. 3 and 4)). The SSG and FLT models reproduce the results fairly well. However, the LRR model performs poorly because it was not well calibrated in incompressible, homogeneous shear flow. See Abid and Speziale (1993).

In Fig. 5, the model predictions for the time evolution of Reynolds stress anisotropies are shown. All of the models substantially underpredict the Reynolds stress anisotropies in comparison with the results of the DNS. Also, the relaxation of b_{ij} to their equilibrium states has not been captured by all the models. As pointed out by Speziale, Gatski, and Sarkar (1992), the long time behavior of turbulence models is tied to their ability to predict the equilibrium values.

In fact, all the models predict exponential long time behavior, i.e., K, ε, $\tilde{T}$, and τ_{ij} proportional to $exp(\lambda_\infty t^*)$, where λ_∞ is the equilibrium growth rate given by

$$\lambda_\infty = -2(b_{12})_\infty - (\frac{\varepsilon - \pi_d}{SK})_\infty \tag{20}$$

The equilibrium states obtained from the various turbulence models are compared with the DNS data of Rogers et al. (1986) for incompressible homogeneous shear flow (Table 1) and the DNS data of Blaisdell (1990) for compressible homogeneous shear flow with Sarkar's model (Table 2). Several observations in regard to these results are noteworthy: the Reynolds stress anisotropies $(b_{ij})_\infty$ and the shear parameter $(SK/\varepsilon_s)_\infty$ are underpredicted by all models, particularly $(b_{11})_\infty$; the shear stress $(b_{12})_\infty$ is erroneously predicted to be insensitive to the compressibility effects; and the SSG and LRR models predict the observed trend that $(b_{11})_\infty$ increases as a function of the turbulent Mach number M_t (however the FLT model erroneously predicts the opposite trend of the DNS data). Similar conclusions are drawn when Zeman's model is used. See Table 3.

All of the turbulence models are clearly incapable of properly accounting for the effects of compressibility on turbulent shear flow. The failure of the Reynolds stresses to predict the increase in the anisotropy of normal stresses and the reduction in the shear stress is partly due to the use of the incompressible pressure-strain correlation models. As shown by Blaisdell and Sarkar (1993), the contribution of the compressibility to the pressure-strain correlation is large and must be taken into account.

Another deficiency of the turbulence models considered in this study lies in the use of the incompressible form of the dissipation equation to obtain the solenoidal part of the dissipation rate. The results presented in Tables 1 and 2 show clearly that $(\mathcal{P}/\varepsilon_s)_\infty$ is erroneously predicted to be sensitive to the compressibility effects. When an equilibrium state is achieved, Eq. (18) gives

$$(\frac{\mathcal{P}}{\varepsilon_s})_\infty = \frac{C_{\varepsilon 2}-1}{C_{\varepsilon 1}-1} - (\frac{1}{C_{\varepsilon 1}-1})(\frac{\varepsilon_d - \pi_d}{\varepsilon_s}) \tag{21}$$

where the first term in the right side represents the equilibrium value of $\mathcal{P}/\varepsilon_s$ in incompressible, homogeneous shear flow. Equation (21) clearly shows that the erroneous prediction of $(\mathcal{P}/\varepsilon_s)_\infty$ is directly tied to the solenoidal dissipation equation. This conclusion can be reached in other ways.

The value of $(\mathcal{P}/\varepsilon_s)_\infty$ can be determined from Eq. (17) as

$$(\frac{\mathcal{P}}{\varepsilon_s})_\infty = (1 + (\frac{\varepsilon_d - \pi_d}{\varepsilon_s}))(1 + \sigma M_{t\infty}^2) \tag{22}$$

The combination of Eqs. (21) and (22) leads to

$$(\frac{\mathcal{P}}{\varepsilon_s})_\infty = \frac{C_{\varepsilon 2}}{(C_{\varepsilon 1} - 1 + \frac{1}{1+\sigma M_{t\infty}^2})} \tag{23}$$

The above equation, which is independent of the dilatational dissipation and pressure-dilatation models, shows that $(\mathcal{P}/\varepsilon_s)_\infty$ cannot be insensitive to the compressibility effects as seen in the direct simulations, otherwise Eq. (23) will predict nonphysical values for M_t ($M_{t\infty} = 1.7$ when $(\mathcal{P}/\varepsilon_s)_\infty = 1.84$). Note that the system of Eqs. (16)-(19) will predict nonphysical results for $M_{t\infty}$ if the explicit compressibility equations are not included.

4. Concluding Remarks

Three Reynolds stress models have been evaluated for the problem of compressible, homogeneous shear flow. The dilatational dissipation and pressure-dilatation correlation models, formulated by Zeman and Sarkar, have been inserted in the Reynolds stress closure. Comparisons between the predictions of the various models and the direct numerical simulation of Blaisdell have been made.

All three Reynolds stress models fail to capture the compressibility effects seen in the direct simulations. In particular, the increase in anisotropy of the normal stresses and the reduction in the shear stress are not reproduced by any of the models. This result is partly due to the use of variable density extensions of incompressible pressure-strain models. Thus, dilatational effects on the pressure-strain correlation must be identified and accounted for in compressible turbulence modeling.

An analysis of the predictions of the equilibrium states has shown that use of the incompressible form of the dissipation to determine the solenoidal dissipation rate is another source of inaccuracy in the predictions. In particular, the erroneous prediction of $(\mathcal{P}/\varepsilon_s)_\infty$ is directly tied to the solenoidal dissipation equation.

The present study indicates the need for further direct numerical simulations of compressible, homogeneous shear flows. Such simulations could be used to distinguish between compressibility and low Reynolds number effects and to provide more information on the equilibrium states.

References

Abid, R. and Speziale, C. G., 1993, "Predicting equilibrium states with Reynolds stress closures in channel flow and homogeneous shear flow," *Phys. Fluids A* **5** , pp. 1151-1158.

Blaisdell, G. A., 1990, "Numerical simulations of compressible homogeneous turbulence," *Ph.D. Thesis*, Mechanical Engineering Dept., Stanford University, Stanford, California.

Blaisdell, G. A. and Sarkar, S., 1993, "Investigation of the pressure strain correlation in compressible homogeneous turbulent shear flow," in *FED* **151**, Transitional and Turbulent Compressible Flows.

Fu, S., Launder, B. E. and Tselepedikas, D. P., 1987, "Accommodating the effect of high strain rates in modeling the pressure-strain correlation," UMIST Mechanical Engineering Department Report TFD/87/5.

Launder, B. E., Reece, G. and Rodi, W., 1975, "Progress in the development of a Reynolds stress turbulence closure," *J. Fluid Mech.* **68**, p. 637.

Rogers, M. M., Moin, P., and Reynolds, W. C., 1986, "The structure and modeling of the hydrodynamic and passive scalar fields in homogeneous turbulent shear flows," *Technical Report* TF-25, Stanford university, Stanford, California.

Sarkar, S., Erlebacher, G. and Hussaini, M. Y., 1991, "Direct simulations of compressible turbulence in a shear flow," *Theoret. Comput. Fluid Dynamics* **2**, pp. 291-305.

Sarkar, S., 1992, "The pressure-dilatation correlation in compressible flows," *Phys. Fluids A* **4**, pp. 1251-1258.

Speziale, C. G., Sarkar, S. and Gatski, T. B., 1991. "Modeling the pressure-strain correlation of turbulence: An invariant dynamical systems approach," *J. Fluid Mech.* **227**, p. 245.

Speziale, G. S., Gatski, T. B. and Sarkar, S., 1992, "On testing models for the pressure-strain correlation of turbulence using direct simulations," *Phys. Fluids A* **4**, pp. 2887-2899.

Zeman, O., 1990, "Dilatation dissipation: The concept and application in modeling compressible mixing layers, " *Phys. Fluids A* **2**, p. 178.

Zeman, O. and Blaisdell, G.A., 1991, "New physics and models for compressible turbulent flows," in *Advances in Turbulence 3*, A. V. Johanson and P. H. Alfredson eds., Springer Verlag, Berlin, pp. 445-454.

Equilibrium Values	LRR Model	SSG Model	FLT Model	DNS Data
b_{11}	0.155	0.219	0.208	0.215
b_{12}	-0.187	-0.164	-0.146	-0.158
b_{22}	-0.121	-0.146	-0.144	-0.153
b_{33}	-0.004	-0.073	-0.064	-0.062
SK/ε_s	5.34	5.77	6.84	5.70
P/ε_s	2.0	1.88	2.0	1.80

Table 1. Predicted Equilibrium Values for Incompressible Homogeneous Shear Flow.

Equilibrium Values	LRR Model	SSG Model	FLT Model	DNS Data
b_{11}	0.166	0.230	0.189	0.424
b_{12}	-0.187	-0.165	-0.148	-0.118
b_{22}	-0.130	-0.148	-0.138	-0.236
b_{33}	-0.036	-0.082	-0.051	-0.188
SK/ε_s	3.77	4.11	4.77	7.82
M_t	0.65	0.60	0.65	0.51
P/ε_s	1.41	1.36	1.41	1.84

Table 2. Predicted Equilibrium Values for Compressible Homogeneous Shear Flow with Sarkar's Model.

Equilibrium Values	LRR Model	SSG Model	FLT Model	DNS Data
b_{11}	0.167	0.231	0.187	0.424
b_{12}	-0.191	-0.167	-0.148	-0.118
b_{22}	-0.131	-0.148	-0.137	-0.236
b_{33}	-0.036	-0.083	-0.050	-0.188
SK/ε_s	3.59	3.95	4.61	7.82
M_t	0.48	0.45	0.48	0.51
P/ε_s	1.41	1.36	1.41	1.84

Table 3. Predicted Equilibrium Values for Compressible Homogeneous Shear Flow with Zeman's Model.

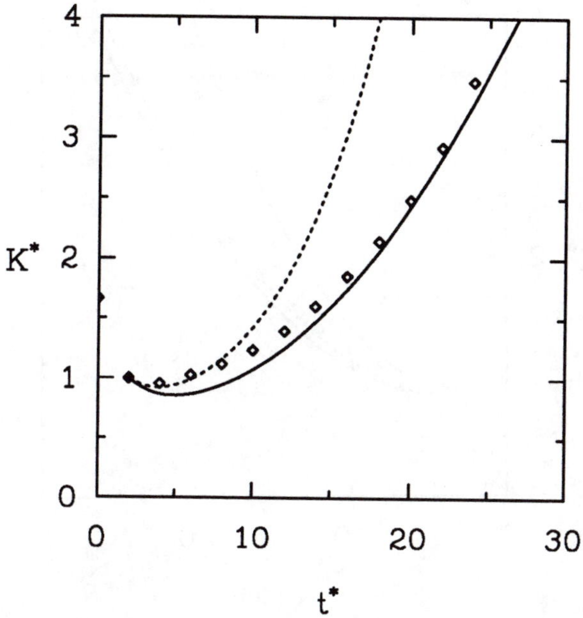

Figure 1. Time evolution of the turbulent kinetic energy using SSG model; — $M_{t0} = 0.307$; - - $M_{t0} = 0$; ◇ DNS of Blaisdell.

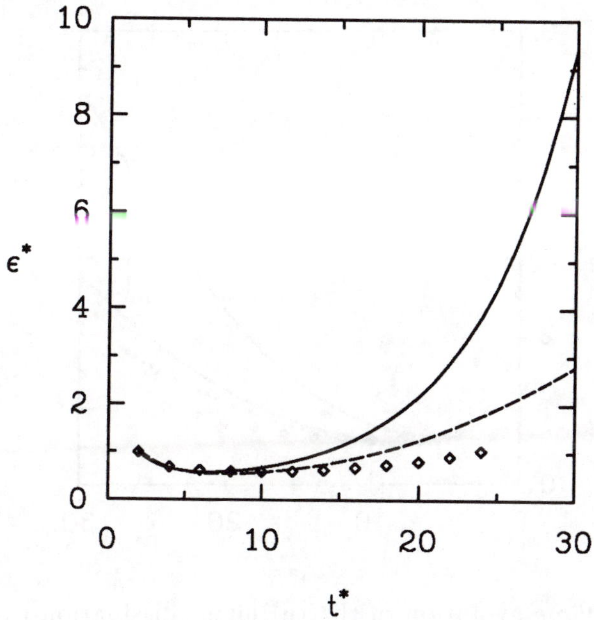

Figure 2. Time evolution of the turbulent dissipation rate using SSG model; — $M_{t0} = 0.307$; - - $M_{t0} = 0$; ◇ DNS of Blaisdell.

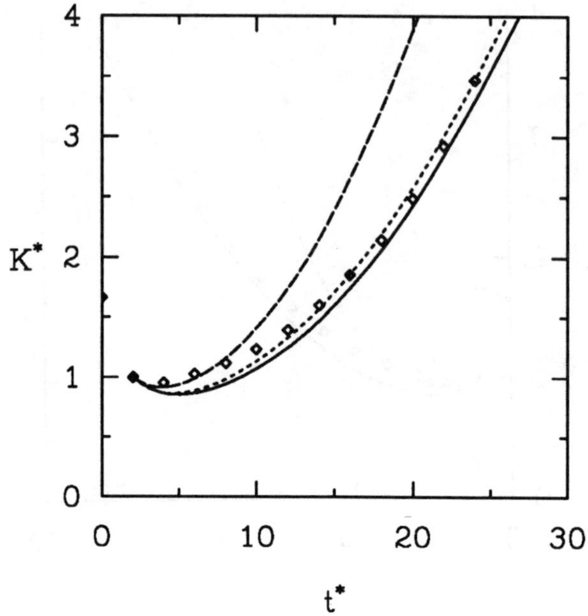

Figure 3. Time evolution of the turbulent kinetic energy : $M_{t0} = 0.307$, — SSG model; - - - - FLT model; – – LRR model; and ◇ DNS of Blaisdell.

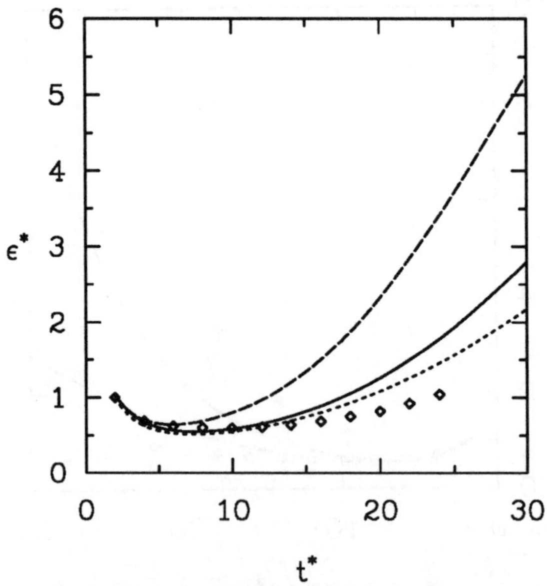

Figure 4. Time evolution of the turbulent dissipation rate : $M_{t0} = 0.307$, — SSG model; - - - - FLT model; – – LRR model; and ◇ DNS of Blaisdell.

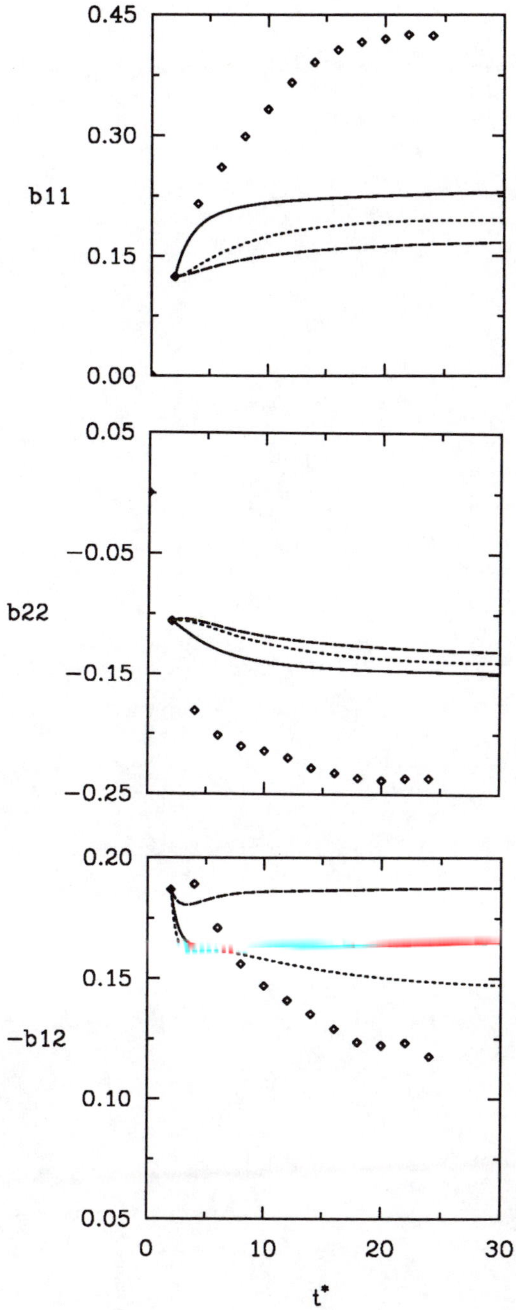

Figure 5. Time Evolution of the turbulent Reynolds stress anisotropies b_{ij}: M_{t0} =0.307 — SSG model; - - - FLT model; – – LRR model; ◇ DNS of Blaisdell.

NEAR-WALL TURBULENCE MODELS AND THEIR APPLICATION TO FLOW IN A SQUARE DUCT

H. Aksoy and R. M. C. So

Mechanical and Aerospace Engineering
Arizona State University
Tempe, Arizona 85287-6106

ABSTRACT

A near-wall correction to the Reynolds-stress model formulated by Speziale et al. (1991) is proposed. Initial tests of this model and another model by Lai and So (1990) with incompressible and compressible boundary-layer flows proved to be encouraging. In this paper, the models are further tested against fully developed turbulent flow in a square duct at a bulk Reynolds number of 50,000 and the results are compared with experimental measurements (Lund, 1977; Eppich, 1982; Gessner et al., 1993). Preliminary calculations give reasonable agreement with measured data and show that near-wall models hold promise for complex flows.

1. Introduction

Almost all practical flows have complexities associated with them and they render the assumption of equilibrium turbulence invalid. The wall-function approximations are based on the assumptions of equilibrium turbulence and constant shear stress near a wall; therefore, there is a need to develop near-wall Reynolds-stress closures that can effectively account for the complexities present in most practical turbulent flows. In order to properly account for these complexities, there is a need to integrate directly to the wall. This means that, in order to satisfy the wall boundary conditions, existing turbulence models have to be modified to include non-equilibrium behavior and viscous effects near a wall.

The approach adopted in the present study is based on an asymptotic analysis of the exact and the high-Reynolds-number (high-Re) modeled turbulence equations. Near-wall correcting functions are sought by requiring the modeled equations to match the behavior of the exact equations near a wall, at least to order y, where y is the coordinate normal to the wall. A thorough description of the near-

M. Y. Hussaini et al. (eds.), Transition, Turbulence and Combustion, Vol. II, 23–37.

wall asymptotic analysis of second order closures is given by Lai and So (1990) and So et al. (1991).

Near-wall correcting functions are formulated for the dynamical-systems based pressure-strain model (SSG) of Speziale et al. (1991). Initial tests of the modified model against incompressible and compressible boundary layers show fairly good agreement with data. In the present study, this model is tested further by using it to calculate an incompressible turbulent flow in a square duct. The reason for choosing the SSG model is due to the fact that it has been calibrated using more recent and reliable data on homogeneous shear flows and shown to give better agreement with data (Speziale et al., 1991; Demuren and Sarkar, 1993) compared to the pressure-strain model (LRR) of Launder et al. (1975). Demuren and Sarkar (1993) have evaluated several pressure-strain models against fully developed turbulent channel flows and showed that the SSG model is the only model that does not require wall reflection terms to give good comparisons with data. The presence of wall reflection terms in the pressure-strain models are undesirable because they incorporate an effect that is difficult to specify in the presence of complex geometries such as corners in the case of a flow in a square duct. In addition, the effects of the wall reflection terms persist even far from the wall and there is uncertainty as to how fast they should decay. On the other hand, the LRR model requires wall-reflection terms to predict the Reynolds-stress correctly.

The turbulent developing flow in a square duct is characterized by the presence of secondary motions (about 3% of the axial bulk velocity) in the plane perpendicular to the streamwise direction. The secondary cells are generated by the presence of Reynolds-stress gradients in the corner region as a result of the anisotropy of the turbulence field. Therefore, this problem provides a very good test case for near-wall Reynolds-stress models. Accurate prediction of the secondary flow has been a challenging task for many turbulence models.

Computations of turbulent developing flows in square and/or rectangular ducts using different turbulence models have been attempted by numerous researchers, e.g. Gessner and Emery (1981) and Sotiropoulos and Patel (1993). However, most calculations invoked the wall function approximations and/or algebraic equations for the Reynolds-stresses. It is believed that an asymptotically correct near-wall correction to a well-calibrated high-Re Reynolds-stress model should help capture the turbulence anisotropy near the corners better than models which utilize wall-functions and algebraic Reynolds-stress equations.

The measurements of Lund (1977), Eppich (1982) and Gessner et al. (1993) in a square duct at a bulk Reynolds number of 50,000 (based on the duct width and bulk velocity U_b) are used to evaluate the model calculations. In addition to the near-wall SSG model, the near-wall model of Lai and So (1990), which is derived based on the LRR model, is also used to calculate the flows and the results compared with measurements and the other calculations.

2. Near-Wall Turbulence Model

The square-duct flow problem is governed by the three components of the mean momentum equation, the continuity equation, the six components of the Reynolds-stress equation and an equation for the dissipation-rate of the turbulent kinetic energy. For an incompressible flow, the Reynolds-stress ($\overline{u_i\, u_j}$), transport equations can be symbolically written as

$$C_{ij} = D_{ij}^v + D_{ij}^t + P_{ij} + \Phi_{ij}^* - \varepsilon_{ij} \tag{1}$$

where the terms from left to right represent convection, molecular diffusion, turbulent diffusion, production by mean strain, velocity-pressure-gradient correlation, and dissipation, respectively. Among these, only the turbulent diffusion, the velocity-pressure-gradient correlation, and the dissipation terms need modeling. The models used in this study for turbulent diffusion and dissipation are the same as the ones proposed by Lai and So (1990) and are not given here. The following proposal is made for the velocity-pressure-gradient correlation term,

$$\Phi_{ij}^* = \Phi_{ij} + \Phi_{ij,w} \tag{2}$$

where Φ_{ij} is given by the high-Re pressure-strain model (SSG) of Speziale et al. (1991) and $\Phi_{ij,w}$ is introduced to compensate for the incorrect behavior of Φ_{ij} while providing additional terms to balance $(\varepsilon_{ij} - D_{ij}^v)$ near the wall. After rearranging some of the terms in the original model, the SSG pressure-strain model can be rewritten as,

$$\Phi_{ij} = -(2C_1\varepsilon + C_1^*\tilde{P})b_{ij} + C_2\varepsilon\left(b_{ik}b_{kj} - \frac{1}{3}\Pi\delta_{ij}\right) -$$
$$\alpha\left(P_{ij} - \frac{2}{3}\delta_{ij}\tilde{P}\right) - \beta\left(D_{ij} - \frac{2}{3}\delta_{ij}\tilde{P}\right) - \gamma k\left(\frac{\partial U_i}{\partial x_j} + \frac{\partial U_j}{\partial x_i}\right) -$$
$$\frac{C_3^*}{2}\Pi^{1/2}k\left(\frac{\partial U_i}{\partial x_j} + \frac{\partial U_j}{\partial x_i}\right) \tag{3}$$

where k is the turbulent kinetic energy and ε its dissipation rate, $U_i = (U, V, W)$ is the mean velocity vector and $x_i = (x, y, z)$ are the physical coordinates. The other variables are defined as

$$b_{ij} = \frac{(\overline{u_i u_j} - \frac{2}{3}\delta_{ij}k)}{2k}, \quad \Pi = b_{kl}b_{kl}, \quad \tilde{P} = \frac{P_{kk}}{2},$$

$$P_{ij} = -\left\{\overline{u_i u_k}\frac{\partial U_j}{\partial x_k} + \overline{u_j u_k}\frac{\partial U_i}{\partial x_k}\right\}, \quad D_{ij} = -\left\{\overline{u_i u_k}\frac{\partial U_k}{\partial x_j} + \overline{u_j u_k}\frac{\partial U_k}{\partial x_i}\right\}$$

$$C_1 = 1.7,\ C_1^* = 1.8,\ C_2 = 4.2,\ C_3^* = 1.3,\ \alpha = \frac{C_4 + C_5}{4},\ \beta = \frac{C_4 - C_5}{4},$$

$$\gamma = \frac{C_4}{3} - \frac{C_3}{2},\ C_3 = 0.8,\ C_4 = 1.25,\ C_5 = 0.4.$$

The constants C_3, C_4, and C_5 appear in the original SSG model (Speziale et al., 1991) only and are collected in the α, β, and γ terms in the present study. It can be seen that the LRR model can be recovered in the linear limit as C_1^*, C_2, and C_3^* go to zero.

Following the near-wall asymptotic analysis of Lai and So (1990) and comparing the near-wall behavior of the exact Reynolds-stress equation with the modeled equation as outlined above, one can see that the main unbalanced terms up to order y^0 are the C_1 and C_2 terms. As for the γ and C_3^* terms, the order near a wall is y^2. Besides the C_1 and C_2 terms, the γ and C_3^* terms may produce an undesirable influence on the shear stress equation. Consequently, this influence could affect the mean velocity calculation through the momentum equation, giving rise to an incorrect mean velocity distribution in the buffer region. Therefore, these problems should be taken into consideration if the governing equations are to be integrated all the way to the wall. Following the general approach to develop a near-wall model as discussed in Lai and So (1990), the near wall correction for the SSG model should compensate for the incorrect behavior of the C_1 and C_2 terms, while providing additional damping in the C_3^* term to ensure that this term will not have an undesirable effect on the shear stress equation in the near-wall region. Taking these considerations into account, a reasonable proposal for the near-wall correction of the SSG model is given as

$$\begin{aligned}\Phi_{ij,w} &= f_{w,1}\left[(2C_1\varepsilon + C_1^*\tilde{P})b_{ij} - C_2\varepsilon(b_{ik}b_{kj} - \frac{1}{3}\Pi\delta_{ij})\right] + \\ & f_{w,3}\frac{C_3^*}{2}\Pi^{1/2}k\left(\frac{\partial U_i}{\partial x_j} + \frac{\partial U_j}{\partial x_i}\right) + \Phi_{ij}^p\end{aligned} \tag{4}$$

where

$$f_{w,1} = \exp[-(R_t/200)^2],\ R_t = \frac{k^2}{v\varepsilon}$$

$$f_{w,3} = 0.79A^{1/2}\exp[-(R_t/180)^{1.5}]$$

$$A = 1 - \frac{9}{2}(b_{ij}b_{ij} - 2b_{ij}b_{jk}b_{ki}),\ n_k = (0,1,0)$$

$$\begin{aligned}\Phi_{ij}^p &= -\frac{1}{3}\left[\frac{\partial}{\partial x_1}\left(\nu\frac{\partial\overline{u_i u_k}}{\partial x_1}\right)n_k n_j + \frac{\partial}{\partial x_1}\left(\nu\frac{\partial\overline{u_j u_k}}{\partial x_1}\right)n_k n_i\right] + \\ & \frac{1}{3}\frac{\partial}{\partial x_m}\left(\nu\frac{\partial\overline{u_1 u_k}}{\partial x_m}\right)\delta_{ij}n_k n_1.\end{aligned} \tag{5}$$

In the expressions given above, R_t is the turbulent Reynolds number, ν is kinematic viscosity, A is the stress invariant, and n_k is the unit vector normal to the wall. Since pressure-diffusion term is explicitly neglected in formulating the near-wall correction to the SSG model, Eq. (5) is proposed to balance the behavior of the modeled equation up to order y^2 in the near-wall region by following the proposal of Tselepidakis (1991). The model for Φ_{ij}^p balances 12, 23 components of the Reynolds-stress equations up to order y^2, 22 component up to order y^3, and has negligible contribution to the other components near the wall. This form of the pressure-diffusion model assumes a viscous-like pressure diffusion.

The near-wall model of the dissipation-rate equation is exactly the same as the one given by So et al. (1991) with the exception of the constant in the damping function $f_{w,2}$ which is modified from 40 to 32 in the present study. This near-wall model for the dissipation-rate equation yields a maximum at the wall in agreement with the direct numerical simulation data.

3. Numerical Procedure and Boundary Conditions

Fully-developed turbulent flow in a square duct is two-dimensional, therefore, can be solved by iterating in the cross-stream plane alone. However, in anticipation of the need to calculate developing flows later, a three-dimensional TEACH-type code is used to solve the governing equations in the square duct. The coordinate system and a quadrant of the square duct geometry is shown in Figure 1. An iterative forward marching solution procedure is used to solve the parabolized equations. In particular, the three-dimensional parabolic finite difference procedure of Patankar and Spalding (1972) is employed. The convergence rate is improved by implementing the SIMPLEC algorithm (Van Doormal and Raithby, 1984) into the code rather than using the original SIMPLE algorithm (Patankar, 1980).

At the boundaries, all the mean velocity components and the Reynolds stresses are set equal to zero, while the dissipation rate is specified as $2v(\partial\sqrt{k}/\partial y)^2$. At the inlet, essentially all variables are specified to be uniform except near the wall. Near the wall, all variables are given as obtained from the results of a boundary layer code at a specified boundary layer thickness of approximately 1 percent of the duct width.

Special attention is given to the placement of grids near the wall. A minimum of 5 grids are placed within $y^+ < 5$ with the first grid point at approximately $y^+ = 1$, and 15-25 grids are located in the region $5 < y^+ < 65$, depending on the total number of grids used. Calculations with 81×81, 121×121 and 161×161 grids have been carried out. There were no significant differences between the calculations obtained with the 121×121 and 161×161 grids. Therefore, only the results from the 121×121 grids are presented here. The forward step size is varied progressively from 0.4 percent of the hydraulic diameter near the inlet to 4 percent until the flow becomes fully developed at 100 hydraulic diameters downstream. The iterations are carried out until the normalized residual sources reduce to 10^{-5} for each variable.

4. Results and Discussions

Sample results are presented in Figures 2 - 7. Figure 2 shows the secondary flow structure at a location where the flow is fully developed, while Figures 3 - 7 show the velocity statistics comparisons of two near-wall models with measurements. In addition to the present near-wall proposal, denoted as modified SSG, the near-wall model of Lai and So (1990), designated as LS, is also included in the plots.

Figure 2 shows the secondary flow structure inside the square duct obtained using the modified SSG model. As a consequence of the Reynolds-stress gradients in the corners, eight symmetrical cells (two in each quadrant) appear in the cross-stream plane. When the original LS model (Lai and So, 1990) is used, two very small cells are also observed next to the corner in each quadrant, one on each side of the corner bisector and extending in size to about 3% of the duct width. Experimental measurements of the secondary flow this close to the corner are not available. Therefore, the existence of these cells cannot be verified. On the other hand, the wall shear stress data along z (Fig. 3) do not exhibit a sharp fall in the shear stress as a result of the secondary flow separation close to the corner. The two small cells disappeared when the Reynolds number was increased to 250,000. Also, when the pressure-diffusion model in the original LS model was replaced by Eq. 5 which assumes a more viscous like diffusion of the Reynolds-stresses rather than a turbulence dominated diffusion as in the original model, the two small cells close to the corner were not present. The present results indicate that the modeling of the pressure-diffusion term in the near-wall region has a significant effect on the calculated secondary flow close to the corner.

In Fig. 3, the calculated wall shear stress along z is compared with experimental data. The modified SSG model predicts a decreasing shear stress towards the wall bisector in accordance with observation. Both models predict an inflexion point at about $y = 0.05a$, somewhat closer to the corner than what the experimental data shows. This inflexion point disappears as the Reynolds number increases (Gessner and Emery, 1981). Plots of shear stress profiles obtained from several high-Re model calculations using wall functions fail to predict this inflexion point at Reynolds numbers of 50,000 and 34,000 as shown by Gessner and Emery (1981).

Figures 4a and 4b compare the calculated and measured axial velocity (U) distributions along the wall and corner bisectors, respectively. Both models overpredict the axial velocity somewhat along the wall bisector. The modified SSG predictions are better near the wall but in the center region the LS model result is closer to the measurements. Along the corner bisector and close to the wall, the LS model calculation is in fairly good agreement with data while the modified SSG model underpredicts U.

Secondary velocity (W) profiles at three different locations along the y-direction are compared with measurements in Fig. 5. The scatter noted in the experimental data arises as a result of the difficulty in measuring the relatively small secondary velocities. Both models underpredict the measurements. In order to successfully predict the secondary velocity profiles, both the Reynolds-stresses and their gradients need to be captured accurately since they are responsible for the production of the axial vorticity in non-circular straight ducts (Speziale, 1986). The difficulty encountered in the accurate prediction of the secondary velocities stems from the deficiencies of both the high-Re and near-wall models in capturing the correct anisotropy levels near the corners. As discussed by Speziale (1986), several High-Re models of higher-order statistics have been formulated in an ad hoc manner to account for the contributions these higher-order statistics make to the generation of secondary flows.

Comparisons of the Reynolds-stresses and turbulent kinetic energy along the wall and corner bisectors are given in Figs. 6 and 7, respectively. The agreement with measurements along the wall bisector is fairly good for both models. However, along the corner bisector, the LS model calculations seem to be in better agreement with data compared to the modified SSG model results.

5. Conclusions

A near-wall correction is formulated for the SSG pressure-strain model and the modified SSG model is used to calculate turbulent flow in a square duct at a Reynolds number of 50,000. These results are compared with measurements and with the LS model calculations to assess the suitability of the SSG model for turbulent complex flow calculations. The results are very encouraging and tend to indicate that a suitably modified SSG model can be made to yield fairly good near-wall results even when applied to calculate complex flows such as encountered in a square duct. The recent DNS calculations of Huser and Biringen (1993) at a bulk Reynolds number of 10,320 should provide a benchmark set of data to further validate the models. It is intended to test the models further against the data of Huser and Biringen (1993) and with three-dimensional developing square duct flows at a Reynolds number of 250,000.

Acknowledgement

Funding support from NASA Langley Research Center under Grant No. NAG-1-1080, monitored by Dr. T. B. Gatski, is gratefully acknowledged.

References

Demuren, A. O. and Sarkar, S., 1993. "Perspective: Systematic study of Reynolds stress closure models in the computations of plane channel flows," *Journal of Fluids Engineering* **115**, pp. 5-12.

Eppich, H. M., 1982. "An experimental investigation of the turbulence structure in the inner region of a three-dimensional streamwise corner flow," MS Thesis, Dept. Mech. Engin., Univ. of Washington.

Gessner, F. B. and Emery, A. F., 1981. "The numerical prediction of developing turbulent flow in rectangular ducts," *J. Fluids Engin.* **103**, pp. 445-455.

Gessner, F. B., Eppich, H. M., and Lund, E. G., 1993. "Reynolds number effects on the near-wall structure of turbulent flow along a streamwise corner," *Near-Wall Turbulent Flows*, R. M. C. So, C. G. Speziale, and B. E. Launder (Editors), pp. 965-975.

Huser, A. and Biringen, S., 1993. "Direct numerical simulation of turbulent flow in a square duct," AIAA-93-0198.

Lai, Y. G. and So, R. M. C., 1990. "On near-wall turbulent flow modeling," *Journal of Fluid Mechanics* **221**, pp. 641-673.

Launder, B. E., Reece, G. J., Rodi, W., 1975. "Progress in the development of a Reynolds-stress turbulence closure," *J. Fluid Mechanics* **68**, pp. 537-566.

Lund, E. G., 1977. "Mean flow and turbulence characteristics in the near corner region of a square duct," MS Thesis, Dept. Mech. Engin., Univ. of Washington.

Patankar, S. V., 1980. *Numerical Heat Transfer and Fluid Flow*, Hemisphere, New York.

Patankar, S. V. and Spalding, D. B., 1972. "A calculation procedure for heat, mass and momentum transfer in 3-D parabolic flows," *International Journal of Heat and Mass Transfer* **15**, pp. 1787-1806.

So, R. M. C., Lai, Y. G., Zhang, H. S., and Hwang, B. C., 1991. "Second-order near-wall turbulence closures: A review," *AIAA Journal* **29**(11), pp. 1819-1835.

So, R. M. C., Zhang, H. S., and Speziale, C. G., 1991. "Near-wall modeling of the dissipation rate equation," *AIAA Journal* **29**, pp. 2069-2076.

Sotiropoulos, F. and Patel, V. C., 1993. "Evaluation of some near-wall models for the Reynolds-stress transport equations in a complex 3-D shear flow," *Near-Wall Turbulent Flows*, R. M. C. So, C. G. Speziale and B. E. Launder (Editors), pp. 987-997.

Speziale, C. G., 1986. "The dissipation rate correlation and turbulent secondary flows in noncircular ducts," *J. Fluids Engin.* **108**, pp. 118-120.

Speziale, C. G., Sarkar, S., and Gatski, T. B., 1991. "Modelling the pressure-strain correlation of turbulence: An invariant dynamical systems approach," *Journal of Fluid Mechanics* **227**, pp. 245-272.

Tselepidakis, D. P., 1991. "Development and application of a new second-order closure for turbulent flows near walls," Ph. D. Thesis, Dept. of Mech. Engin., Univ. of Manchester, Inst. Science and Tech., TFD/92/2.

Van Doormal, J. P. and Raithby, G. D., 1984. "Enhancements of the SIMPLE method for predicting incompressible fluid flows," *Numerical Heat Transfer* **7**, pp. 147-163.

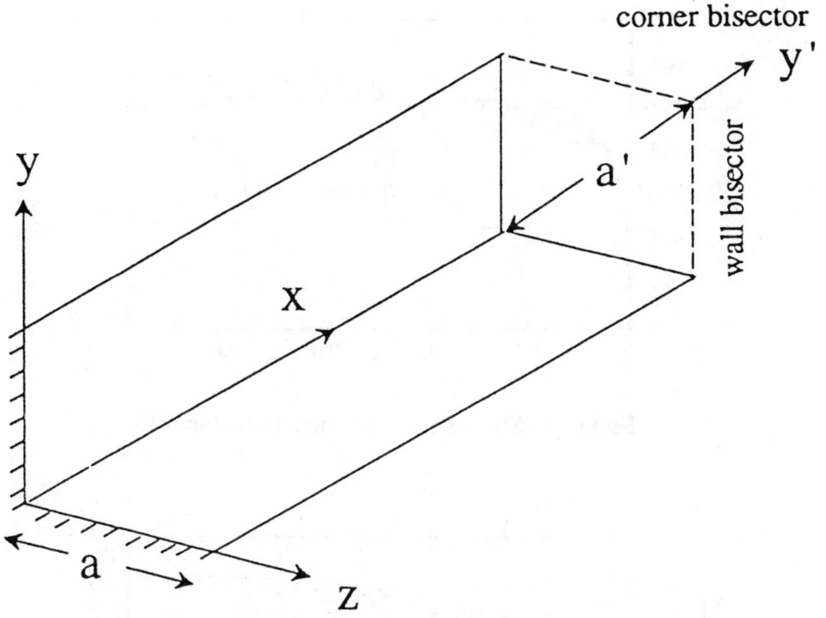

Figure 1. Coordinate system and the square duct geometry.

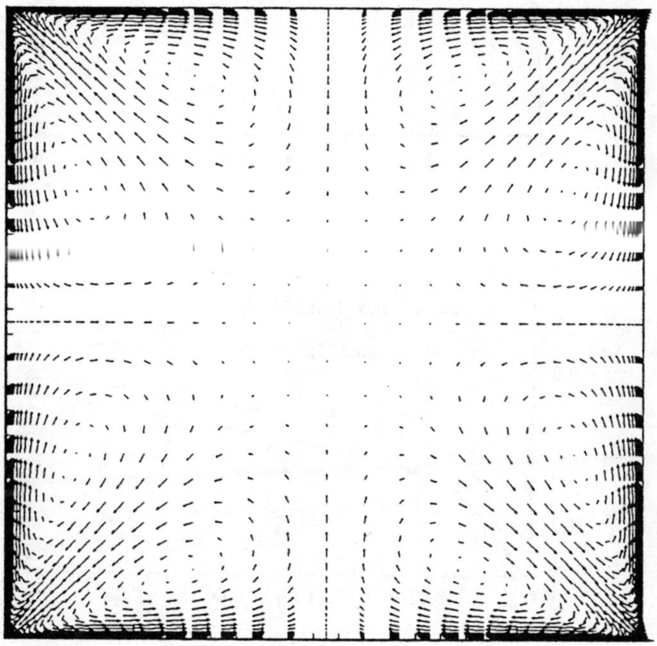

Figure 2. Secondary velocity vectors.

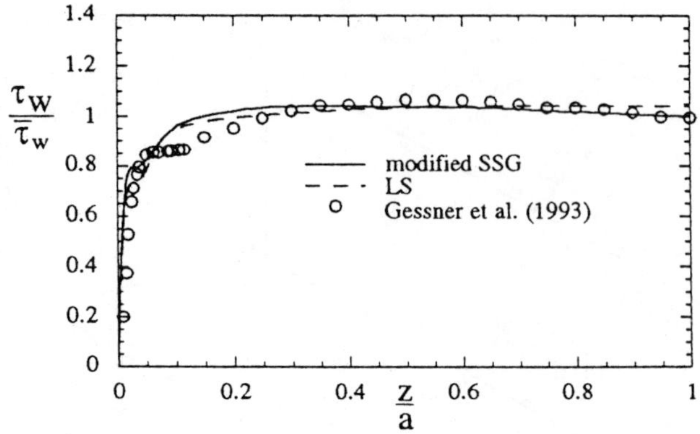

Figure 3. Shear stress variation along the wall.

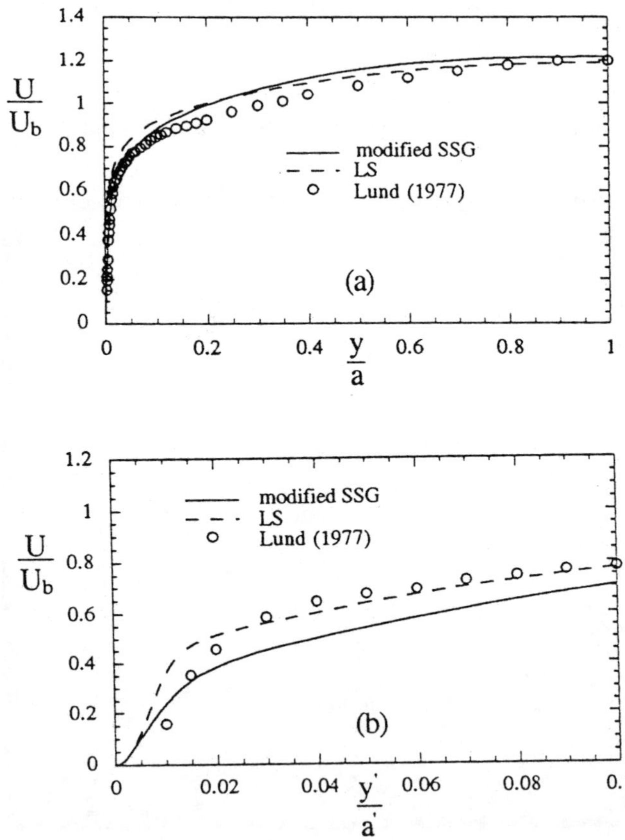

Figure 4. Axial velocity along (a) wall bisector; (b) corner bisector.

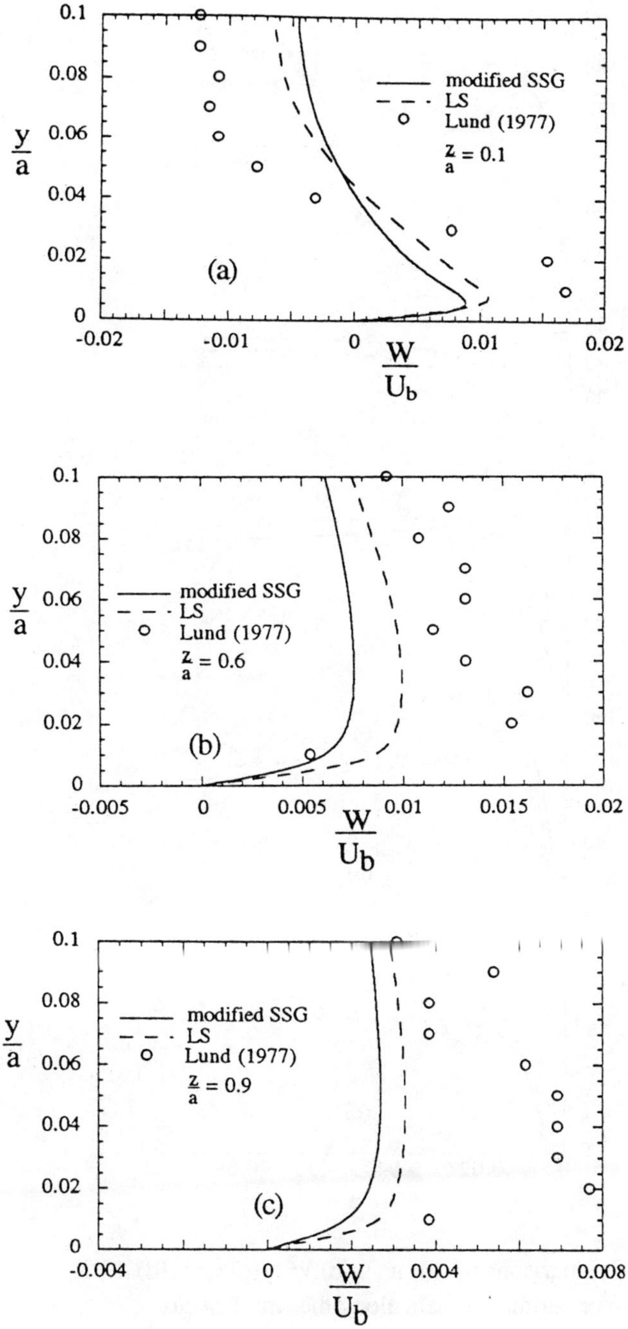

Figure 5. Comparison of secondary velocities : (a) z/a = 0.1; (b) z/a = 0.6; (c) z/a = 0.9.

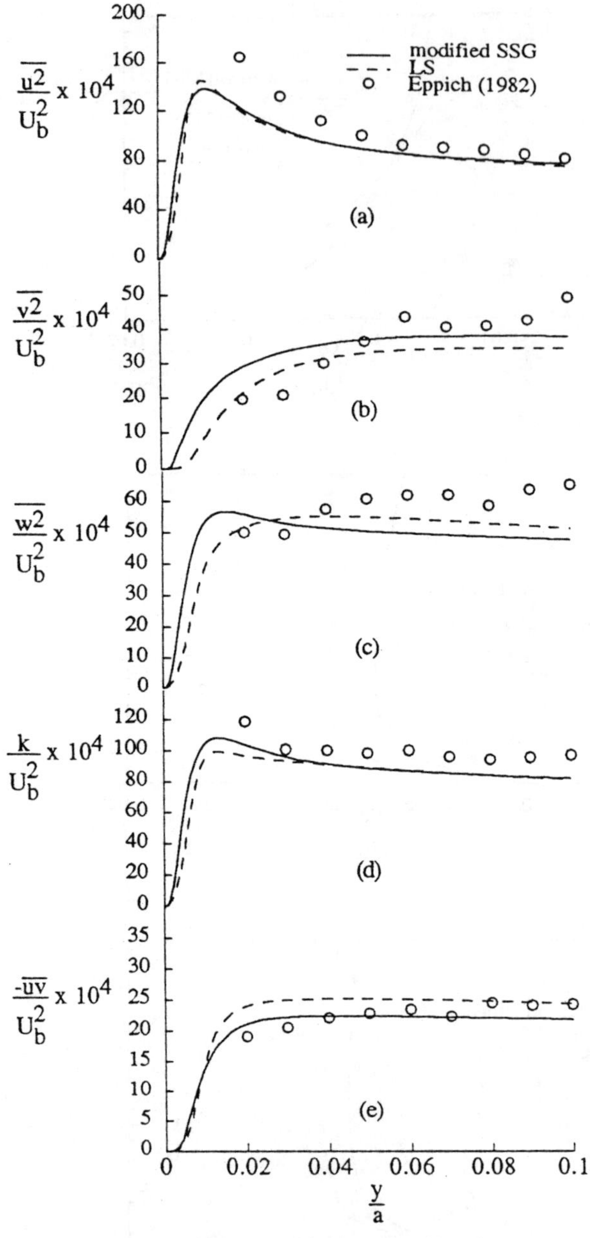

Figure 6. Comparisons of (a) $\overline{u^2}$; (b) $\overline{v^2}$; (c) $\overline{w^2}$; (d) k; (e) - $\overline{uv}$ with experimental data along the wall bisector.

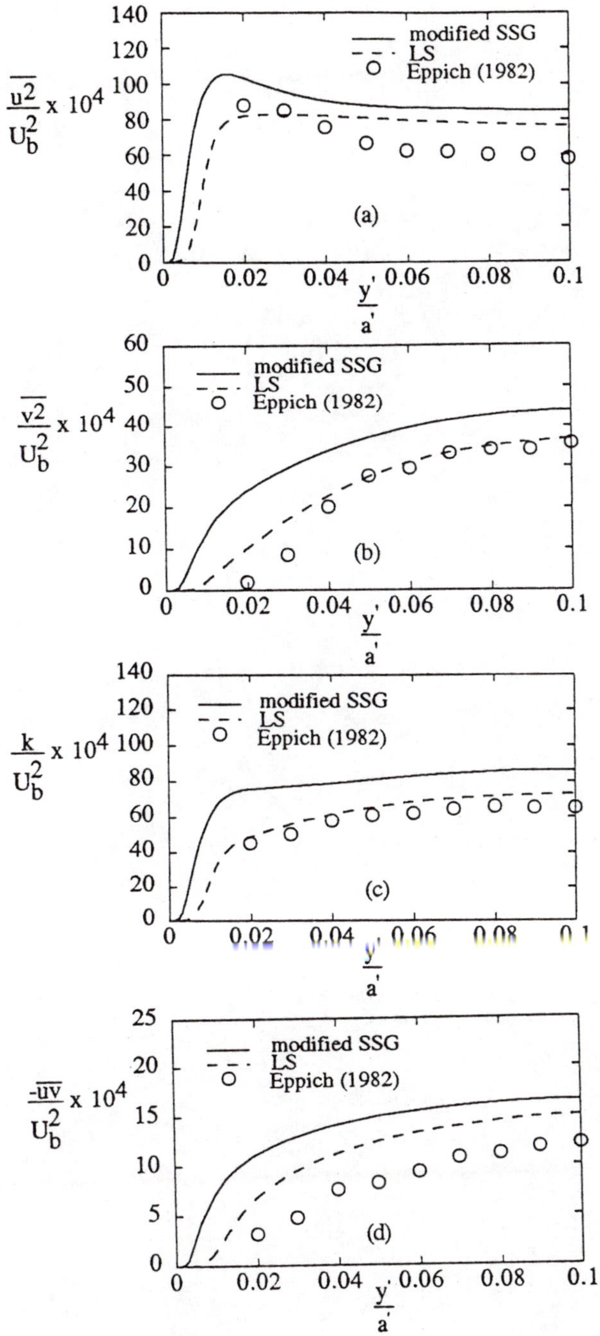

Figure 7. Comparisons of (a) $\overline{u^2}$; (b) $\overline{v^2}$; (c) k ; (d) - $\overline{uv}$ with experimental data along the corner bisector.

SGS MODELING IN THE NEAR-WALL REGION OF TURBULENT FLOWS

Carlos Härtel[1]

DLR, Institute for Fluid Mechanics
Bunsenstr. 10, D-37073 Göttingen, Germany

ABSTRACT

The present paper is concerned with subgrid-scale (SGS) modeling in the near-wall region of turbulent flows, with special emphasis on the energy exchange between resolved and unresolved turbulent motions. A reverse transfer of energy occurring in the buffer layer is discussed in terms of the transport equation for the dominant component of the SGS stress tensor. The results can prove useful if e.g. second-order closures (SOC) are considered. In addition an algebraic Reynolds-stress model suitable for application in large-eddy simulation (LES) is introduced. It is shown that this model can improve SGS modeling as far as the energy balance between resolved and unresolved turbulent motions is concerned.

1. Introduction

In turbulent flows of engineering and geophysical interest the range of excited scales usually is much too large to be described by means of direct numerical simulation (DNS). Therefore the LES technique, where only a fraction of all relevant scales is computed explicitly, has become a wide-spread tool in turbulence research. LES is based on a splitting of all flow variables into a resolvable grid-scale (GS) and an unresolvable SGS part, the latter of which has to be accounted for by a model. Probably the most important effect of SGS models is to provide a proper exchange of energy between GS and SGS motions, as has repeatedly been pointed out (see e.g. Rogallo and Moin (1984)). SGS motions are usually assumed to be dissipative in nature, i.e. to take energy from the resolved scales in the mean. For most free flows, in particular at high Reynolds numbers, this is typically the case. However, in the near-wall region of turbulent flows a "reverse" transfer of energy from SGS to GS turbulent motions can be observed. Currently, the immediate vicinity

[1]Research was partly supported by the DFG.

M. Y. Hussaini et al. (eds.), Transition, Turbulence and Combustion, Vol. II, 39–49.

of solid walls is not computed explicitly in most LES of flows at higher Reynolds numbers, but rather modeled by artificial boundary conditions. Still, a refined treatment of this zone remains desirable whenever possible, particularly in cases where near-wall phenomena can have a decisive influence on the whole flow field. This is to be expected e.g. in many flows exhibiting separation.

For a detailed investigation of near-wall turbulence, DNS databases can be efficiently exploited. This has been done in the past primarily for turbulent channel flow (Piomelli et al., 1987; Härtel and Kleiser, 1992) and for turbulent pipe flow (Härtel et al., 1993). Using DNS data SGS models can be tested in an "a-priori" way, where all SGS quantities of interest are computed directly from the DNS data and then compared with modeled values for the same GS flow field. A-priori tests can give clear indications whether a model in principle is able to represent the physics properly.

Since the reverse flux of kinetic energy in the near-wall flow constitutes a fundamental problem for simple eddy-viscosity modeling, it should be investigated how other closures can cope with this effect.

2. Governing Equations

In LES any dependent variable f of the flow field is split into a grid-scale part $\overline{f}$ and a SGS part f', i.e.

$$f = \overline{f} + f' \quad . \tag{1}$$

The definition of $\overline{f}$ can be based on a filtering operation (Leonard, 1974) applied either in spectral or in real space. A further splitting of the GS flow variables can be performed by using a Reynolds decomposition, where a quantity g is split into a statistical mean value $< g >$ and a fluctuation $\tilde{g}$, i.e.

$$g = < g > + \tilde{g} \quad . \tag{2}$$

Whenever the operator $< \cdot >$ is used here, it symbolizes an averaging over wall-parallel planes and time.

Applying the filtering procedure to the Navier-Stokes equations yields the equations of motion for the grid-scale velocity field. The effect of the small scales is represented by the tensor Q_{ij}, which in LES has to be modeled. It will subsequently be used in the form

$$Q_{ij} = -\left(\overline{u_i' u_j'} + \overline{\overline{u}_i u_j'} + \overline{u_i' \overline{u}_j}\right) \quad . \tag{3}$$

For the temporal evolution of the SGS stress Q_{ij} the following budget equation can be derived

$$\frac{\partial Q_{ij}}{\partial t} = \sum_{l=1}^{7} T_l^{ij} \quad , \tag{4}$$

where the individual terms are given by

$$\begin{aligned}
T_1^{ij} &= -\overline{\overline{u}_k \frac{\partial Q_{ij}}{\partial x_k}} \\
T_2^{ij} &= -\left(\overline{Q_{ik}\frac{\partial \overline{u}_j}{\partial x_k}} + \overline{Q_{jk}\frac{\partial \overline{u}_i}{\partial x_k}}\right) \\
T_3^{ij} &= \frac{\partial}{\partial x_k}(\overline{u_i u_j u_k} - \overline{\overline{u}_i\overline{u_j u_k}} - \overline{\overline{u}_j\overline{u_i u_k}} - \overline{\overline{u}_k\overline{u_i u_j}} + 2\,\overline{\overline{u}_i\overline{u}_j\overline{u}_k}\,) \\
T_4^{ij} &= \frac{\partial}{\partial x_i}(\overline{p u_j} - \overline{\overline{p}\,\overline{u}_j}\,) + \frac{\partial}{\partial x_j}(\overline{p u_i} - \overline{\overline{p}\,\overline{u}_i}\,) \\
T_5^{ij} &= -\overline{p\left(\frac{\partial u_i}{\partial x_j} + \frac{\partial u_j}{\partial x_i}\right)} + \overline{\overline{p}\left(\frac{\partial \overline{u}_i}{\partial x_j} + \frac{\partial \overline{u}_j}{\partial x_i}\right)} \\
T_6^{ij} &= \nu \frac{\partial^2 Q_{ij}}{\partial x_k \partial x_k} \\
T_7^{ij} &= 2\nu\left(\overline{\frac{\partial u_i}{\partial x_k}\frac{\partial u_j}{\partial x_k}} - \overline{\frac{\partial \overline{u}_i}{\partial x_k}\frac{\partial \overline{u}_j}{\partial x_k}}\right) \quad .
\end{aligned}$$

Multiplication of the filtered Navier-Stokes equations by the GS velocity $\overline{u}_i$ gives rise to the following term representing the effect of SGS motions on the energy budget of the resolved flow field

$$\overline{u}_i \frac{\partial Q_{ij}}{\partial x_j} = \frac{\partial}{\partial x_j}(\overline{u}_i Q_{ij}) - Q_{ij}S_{ij} \quad , \tag{5}$$

where $S_{ij} = (\partial\overline{u}_i/\partial x_j + \partial\overline{u}_j/\partial x_i)/2$ is the GS rate-of-strain tensor. The term $Q_{ij}S_{ij}$ in equation (5), which governs the exchange of energy between GS and SGS motions, is frequently called "SGS dissipation". In this paper we will concentrate on its average $\varepsilon := < Q_{ij}S_{ij} >$. By application of (2) ε can further be split into two terms, which should be considered separately

$$\varepsilon = < Q_{ij} >< S_{ij} > + < \tilde{Q}_{ij}\tilde{S}_{ij} > \quad . \tag{6}$$

In equation (6) $\varepsilon^{MS} := < S_{ij} >< Q_{ij} >$ means an enhancement of SGS turbulence in the presence of mean shear. This term is part of the total turbulence production $\mathcal{P}$ considered in statistical turbulence theory. The second term $\varepsilon^{FS} :=< \tilde{Q}_{ij}\tilde{S}_{ij} >$ is the covariance of fluctuating GS rates of strain and fluctuating SGS stresses and accounts for a redistribution of energy within the turbulence spectrum, without affecting the mean flow directly. The interaction between mean flow ($< \overline{u}_i >$), GS turbulence ($\tilde{\overline{u}}_i$) and SGS turbulence (u'_i) is illustrated in Fig. 1.

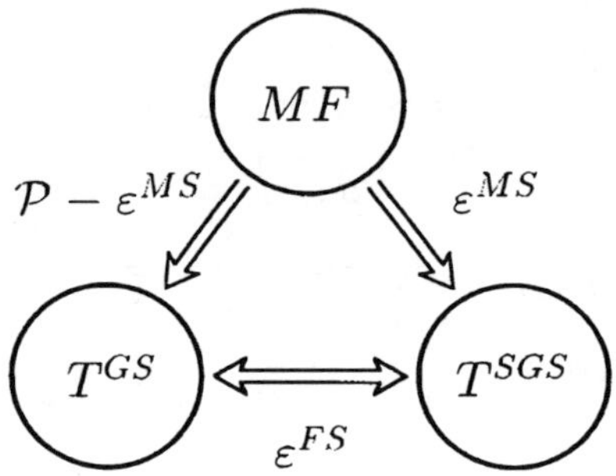

Figure 1: Components of energy transfer between mean flow (MF), grid-scale turbulence (T^{GS}) and subgrid-scale turbulence (T^{SGS}).

3. SGS Dissipation

DNS results for ε^{MS} and ε^{FS} are given in Fig. 2, where x_3 indicates the wall-normal direction. All DNS data are taken from the same turbulent channel flow database at $Re_\tau = 210$ as used in Härtel and Kleiser (1992). For simplicity results will be given for the upper channel half ($0 \leq x_3 \leq 1$) only. While ε^{MS} apparently is nonnegative throughout, ε^{FS} is positive in the viscous sublayer and the outer flow, but exhibits a negative ("reverse") transfer in a region corresponding approximately to the buffer layer. Results for two eddy-viscosity SGS models are added in the figure for comparison. One is the classical Smagorinsky model (Smagorinsky, 1963), the other is an artificial "reference-model" using the true DNS subgrid-scale kinetic energy to evaluate the proper SGS velocity scale (for details see Härtel and Kleiser, 1992). Obviously the reverse energy transfer cannot be represented by either of the models. Looking into the energy balance in the near-wall region more closely (Härtel and Kleiser, 1992; Härtel et al., 1993) one finds that the reverse transfer of energy within the turbulence spectrum is primarily caused by the stress Q_{13}, i.e. the

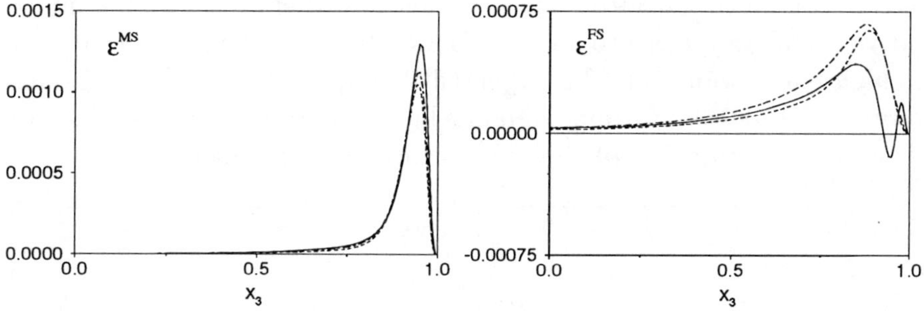

Figure 2: Comparison of DNS data (solid line), Smagorinsky model (dashed line) and reference model (dot-dashed line). Dissipation due to mean (left) and fluctuating (right) rates of strain.

stress aligned with the mean rate of strain. The according component of the SGS dissipation $\varepsilon_{13}^{FS} := < \tilde{Q}_{13}\tilde{S}_{13} >$ is shown in Fig. 3.

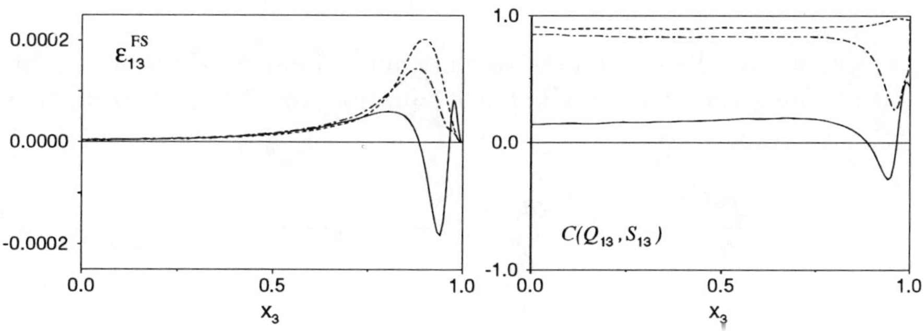

Figure 3: Comparison of DNS data (solid line), Smagorinsky model (dashed line) and reference model (dot-dashed line). Left: Component ε_{13}^{FS} of ε^{FS}. Right: Correlation coefficient between Q_{13} and S_{13}.

Rewriting the covariance ε_{13}^{FS} with the aid of the correlation coefficient C between Q_{13} and S_{13} gives

$$\varepsilon_{13}^{FS} = C(Q_{13}, S_{13}) \ \mathrm{rms}(Q_{13}) \ \mathrm{rms}(S_{13}) \quad . \tag{7}$$

From equation (7) it is clear that changes of sign in ε_{13}^{FS} are due to qualitative changes in the correlation between stress and rate of strain. The corresponding correlation coefficients are also given in Fig. 3. Clearly the negative correlation between stress and rate of strain cannot be captured by the models. Rather a very high nonnegative correlation close to unity is found. It should be noted that such a high correlation must be expected if simple eddy-viscosity models are applied, since a proportionality between stress and rate of strain is a basic assumption of the gradient diffusion hypothesis.

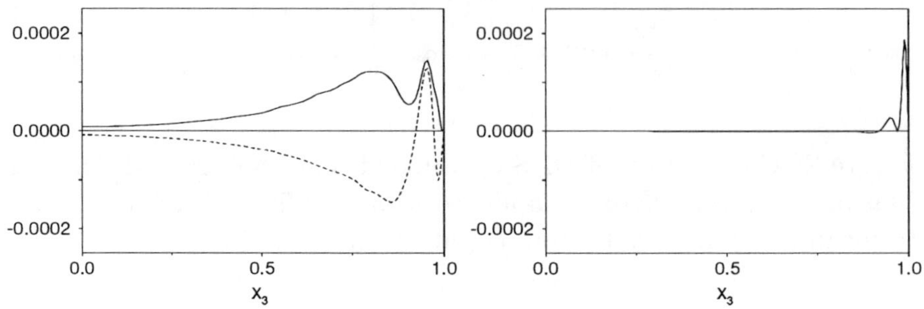

Figure 4: Selected terms from the budget equation (8). Left: $< \tilde{S}_{13}\, \tilde{T}_3^{13} >$ (solid line) and $< \tilde{S}_{13}\, (\tilde{T}_4^{13} + \tilde{T}_5^{13}) >$. Right: $< \tilde{S}_{13}\, \tilde{T}_7^{13} >$

A more detailed insight into the energy budget of the near-wall flow can be gained from the balance equation for ε_{13}^{FS}. Differentiating ε_{13}^{FS} with respect to t gives

$$\frac{\partial\, \varepsilon_{13}^{FS}}{\partial t} = < \tilde{S}_{13} \frac{\partial \tilde{Q}_{13}}{\partial t} > + < \tilde{Q}_{13} \frac{\partial \tilde{S}_{13}}{\partial t} > \quad . \tag{8}$$

In equation (8) the first term on the *r.h.s.* is of particular interest e.g. in the context of SOC, since it directly depends on modeling assumptions for the unknown terms in the balance equation for Q_{13}. By equation (4) this term can be written as

$$< \tilde{S}_{13} \frac{\partial \tilde{Q}_{13}}{\partial t} > = \sum_{l=1}^{7} < \tilde{S}_{13}\, \tilde{T}_l^{13} > \quad . \tag{9}$$

For brevity we will restrict the discussion here to those terms in the *r.h.s.* of equation (9) which involve modeling of unknown SGS correlations, i.e. the terms with $l = 3, 4, 5$ and 7. The corresponding results are shown in Fig.4. Note that for the pressure terms only their sum is given. The viscous term appears to be important in the immediate vicinity of the wall only, while the others contribute considerably to the budget throughout. Apparently the sum of the pressure terms is most significant in the buffer layer, where a strong positive kink can be observed. One hence may conclude that in SOC the modeling of this term will need particular attention, if the near-wall physics is to be captured properly.

4. The Algebraic SSG Model

Algebraic Reynolds-stress relations can be viewed as a compromise between the simple eddy-viscosity models and the complex SOC. In this section we will consider an algebraic model derived by Gatski and Speziale (1992), which is based on the well-known "SSG" second-order closure (Speziale, Sarkar and Gatski, 1991). In this model the SGS stress Q_{13} is computed as

$$Q_{13} = \frac{7.76(1+\eta^2)\,k^{SGS}}{3+\eta^2+6\zeta^2\eta^2+6\zeta^2}\left[S^*_{13} + (S^*_{1k}W^*_{k3} + S^*_{3k}W^*_{k1}) - 2S^*_{1k}S^*_{k3}\right] \tag{10}$$

where

$$\eta = (S^*_{ij}S^*_{ij})^{1/2} \quad , \quad \zeta = (W^*_{ij}W^*_{ij})^{1/2} \tag{11}$$

and

$$S^*_{ij} = 0.0874\,\vartheta^{SGS}\,S_{ij} \quad , \quad W^*_{ij} = 0.1864\,\vartheta^{SGS}\,W_{ij} \quad . \tag{12}$$

Here $W_{ij} = (\partial\overline{u}_i/\partial x_j - \partial\overline{u}_j/\partial x_i)/2$ represents the GS rotation tensor. The SGS time scale ϑ^{SGS}, used for normalizing S_{ij} and W_{ij} in equation (12), is given by

$$\vartheta^{SGS} = \frac{k^{SGS}}{\varepsilon_\nu^{SGS}} \quad , \tag{13}$$

where k^{SGS} and ε_ν^{SGS} symbolize the SGS kinetic energy and SGS viscous dissipation, respectively. While in the original (statistical) form k and ε_ν are taken as statistical mean values, in a LES version these quantities could also be evaluated locally. However, from numerical experimentation we found that ε_ν^{SGS} frequently becomes very small, resulting in much too large values for ϑ^{SGS}. Consequently averages for k^{SGS} and ε_ν^{SGS} will be used here.

To avoid additional transport equations for k^{SGS} and ε_ν^{SGS} in a LES, one has to express these quantities in terms of common LES length and time scales such as the grid size Δ and $(S_{ij}S_{ij})^{1/2}$, i.e.

$$k^{SGS,mod} = F_D \cdot C_k\,\Delta^2\,(S_{ij}S_{ij}) \tag{14a}$$

$$\varepsilon_\nu^{SGS,mod} = C_\varepsilon\,\Delta^2\,(S_{ij}S_{ij})^{3/2} \tag{14b}$$

where F_D denotes a van Driest type damping function, applied to ensure the correct asymptotic behaviour of $k^{SGS,mod}$ at the wall. The constants in equation (14) can be determined theoretically for homogeneous, isotropic turbulence that exhibits an infinitely extended $\kappa^{-5/3}$ inertial range. The procedure is similar to the one used by Lilly (1967), and we will omit the details here. One finds

$$C_k = 0.203 \quad , \quad C_\varepsilon = 0.085 \quad . \tag{15}$$

However, if inhomogeneous wall-bounded turbulent flows are considered, the modeling becomes more complicated. It is not clear in advance, e.g., whether a mean rate of strain (MS) has to be included in the SGS modeling. Very close to solid walls, where the least SGS turbulence activity is found, the highest values of the MS are encountered, suggesting that here excluding the MS might be justified. On the other hand, the MS directly enhances SGS turbulence (see Fig. 1) and hence cannot be neglected generally.

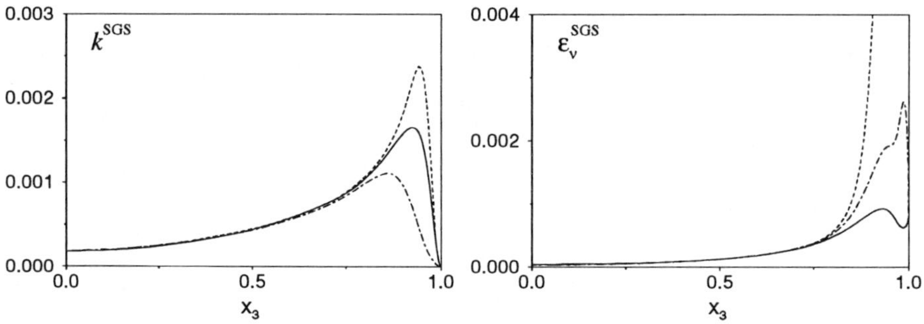

Figure 5: k^{SGS} (left) and ε_ν^{SGS} (right). Comparison of DNS data (solid line) and modeling including (dashed line) and excluding (dot-dashed line) the mean rate of strain.

In Fig. 5 DNS data for k^{SGS} and ε_ν^{SGS} are compared with values modeled after equation (14), including or excluding the mean rate of strain. The model constants here were set to the empirically determined values of $C_k = 0.05$ and $C_\varepsilon = 0.02$. In the core flow both forms approximate the real distribution of k^{SGS} and ε_ν^{SGS} very well, but near the wall deviations become visible in particular for the dissipation rate. Based on these results we decided to include the MS for the

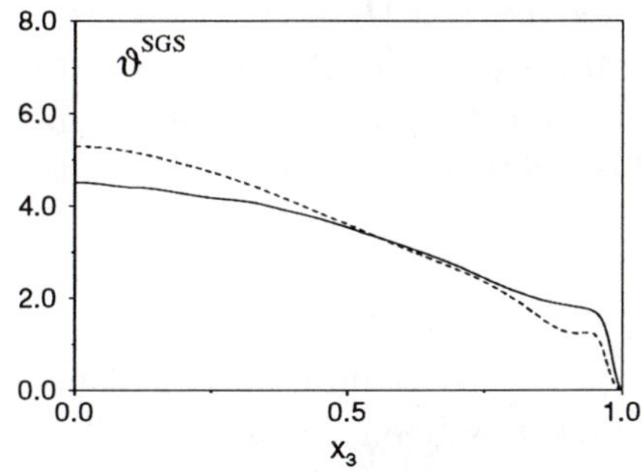

Figure 6: SGS timescale ϑ^{SGS}. Comparison of DNS data (solid line) and modeled values.

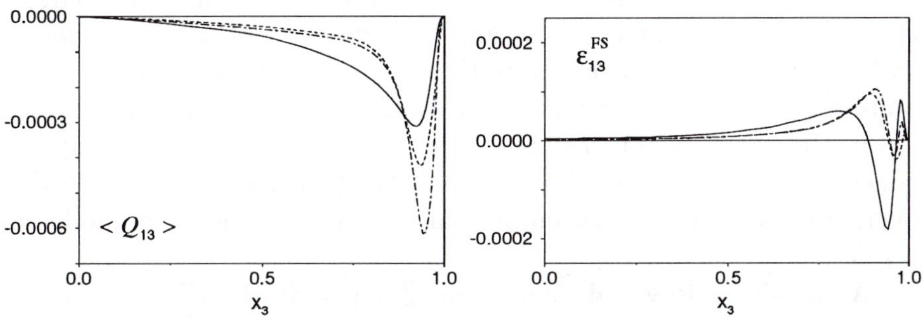

Figure 7: SGS quantities $< Q_{13} >$ (left) and ε_{13}^{FS} (right). Comparison of DNS data (solid line), SSG model with k^{SGS} and ϑ^{SGS} (dashed line), SSG model with $k^{SGS,mod}$ and $\vartheta^{SGS,mod}$ (dot-dashed line).

computation of $k^{SGS,mod}$ but excluded it for $\varepsilon_\nu^{SGS,mod}$, which leads to a fairly good approximation for the time scale $\vartheta^{SGS,mod}$ (Fig. 6).

In Fig. 7 DNS results for the mean stress $< Q_{13} >$ and ε_{13}^{FS} are compared with results from the algebraic SSG models based on k^{SGS} and ε_ν^{SGS} taken from the DNS, and on $k^{SGS,mod}$ and $\varepsilon_\nu^{SGS,mod}$. With respect to $< Q_{13} >$ both versions perform slightly worse, than e.g. the standard Smagorinsky model (see Härtel and Kleiser (1992)), but for ε_{13}^{FS} obviously a considerable improvement can be observed. Though quantitatively deviating from the DNS data, a very good qualitative agreement can be found with a positive energy transfer in the viscous sublayer and the core flow and a region of reverse transfer at almost the correct location. This remarkable improvement can be better understood by looking at equation (10) in more detail. The peculiar decoupling of stresses and rates of strain near the wall is primarily achieved by the fact that the shear $\partial \overline{u}_1 / \partial x_3$ appears in the denominator of the fraction in front of the square brackets to a higher power than in the numerator. Whether this principle can be used to improve simple eddy-viscosity models is currently investigated.

5. Conclusions

In the present paper SGS modeling in the near-wall region of turbulent channel flow has been considered. The work is motivated by the observation that within the buffer layer a reverse transfer of energy from subgrid-scale to grid-scale turbulent motions occurs, which common SGS models cannot account for.

To gain more insight into the energy budget, the transport equation of the dominant contribution to the reverse transfer of energy was considered. It turned out that the sum of pressure-diffusion and pressure-strain correlation is very significant in the buffer layer, i.e. in the region where the reverse transfer is observed. It may be concluded that in SOC this term needs particular attention if the energetic interaction between SGS and GS motions is to be modeled properly.

An algebraic Reynolds-stress model, derived from the well known SSG turbulence model, has been considered as an alternative to common eddy-viscosity models. This model was found to be clearly superior with respect to the energy budget of the near-wall flow. Although it quantitatively differs from the DNS data, a remarkable qualitative agreement is achieved.

References

Gatski, T. B. and Speziale, C. G., 1992. "On explicit algebraic stress models for complex turbulent flows," *J. Fluid Mech.* **254**, pp. 59-77.

Härtel, C. and Kleiser, L., 1992. "Comparative testing of subgrid-scale models for wall-bounded turbulent flows," *Proceedings, First European Computational Fluid Dynamics Conference*, Ch. Hirsch et al., ed., Elsevier Science Publishers, Amsterdam, The Netherlands, Vol. 1, pp. 215-222.

Härtel, C., Kleiser, L., Unger, F., and Friedrich, R., 1993. "Subgrid-scale energy transfer in the near-wall region of turbulent flows," *submitted to Phys. Fluids. A.*

Leonard, A., 1974. "On the energy cascade in large-eddy simulations of turbulent fluid flows," *Adv. Geophys.* **18A**, pp. 237-248.

Lilly, D. K., 1967. "The representation of small-scale turbulence in numerical simulation experiments," *Proceedings, IBM Scientific Computing Symposium on Environmental Sciences*, Yorktown Heights, N.Y., pp. 195-210.

Piomelli, U., Ferziger, J. H., and Moin, P., 1987. "Models for large eddy simulations of turbulent channel flows including transpiration," Dept. Mech. Engng., Stanford Univ., Rep. TF-32.

Rogallo, R. S. and Moin, P., 1984. "Numerical simulation of turbulent flows," *Ann. Rev. Fluid Mech.* **16**, pp. 99-137.

Speziale, C. G., Sarkar, S., and Gatski, T. B., 1991. "Modeling the pressure-strain correlation of turbulence: an invariant dynamical systems approach," *J. Fluid Mech.* **227**, pp. 245-272.

Smagorinsky, J., 1963. "General circulation experiments with the primitive equations. I. The basic experiment," *Monthly Weather Review* **91**, pp. 99-164.

EVALUATION OF INHOMOGENEOUS FORMULATIONS OF THE DYNAMIC SUBGRID-SCALE MODEL

Evangelos T. Spyropoulos and Gregory A. Blaisdell [1]

Purdue University
West Lafayette, Indiana 47907-1282

ABSTRACT

Use of the dynamic subgrid-scale model in large eddy simulations of inhomogeneous flows requires filtering of the flow field in physical space rather than Fourier wave space. The use of spatial filters is investigated by conducting simulations of homogeneous turbulence. Results using various spatial filters are compared to those using the traditional sharp cut-off filter and to experimental data. Several implicit filters are found to perform extremely well. In contrast, explicit filter formulations are found to be very dissipative at higher wavenumbers, while some are potentially unstable.

1. Introduction

There has been much interest in using the dynamic subgrid-scale model to perform large eddy simulations (LES). The dynamic model was first introduced by Germano et al. (1991) and was extended for use in compressible flows by Moin et al. (1991). Since then, further refinements to the model have been proposed (Lilly, 1992; Ghosal et al., 1992).

The dynamic model requires filtering the resolved large scale field. So far it has been mostly implemented in turbulent flows that are homogeneous in at least two directions (e.g., channel flows or turbulence passing through a plane shock) where the filtering can be performed efficiently (and exactly) in wave space using fast Fourier transforms. In more complex inhomogeneous flows this is not possible and some kind of discrete filtering has to be applied in physical

[1] Research was sponsored by the National Aeronautics and Space Administration under NASA Contract No. NAS1-19480 and Grant No. NAG-1-1509 while the authors were in residence at the Institute for Computer Applications in Science and Engineering (ICASE), NASA Langley Research Center, Hampton, VA 23681-0001.

M. Y. Hussaini et al. (eds.), Transition, Turbulence and Combustion, Vol. II, 51–60.

space. The purpose of this study is to examine the performance of spatial filters in actual simulations. A homogeneous flow test case is chosen, so that the effects of such filters can be compared to that of the sharp cut-off filter, defined in wave space, as well as against available experimental data.

2. Mathematical Formulation

2.1. Governing equations and SGS modeling

In LES one computes the motion of the large scale structures, while modeling the nonlinear interactions with the small scales. The governing equations for the large eddies in compressible flows are obtained after filtering the continuity, momentum and energy equations, and recasting in terms of Favre averages. The filtering operation (denoted by an overbar) maintains only the large scales and can be written in terms of a convolution integral as follows

$$\overline{f}(\mathbf{x}) = \int_D G(\mathbf{x} - \mathbf{x}')f(\mathbf{x}')d\mathbf{x}' \tag{1}$$

where f is a turbulent field, G is some spatial filter (usually a sharp cut-off) of width equal to the grid spacing and D is the flow domain.

The effects of the small scales are present in the equations of motion through two terms: the subgrid-scale (SGS) stress $\tau_{ij} = \overline{\rho}(\widetilde{u_i u_j} - \tilde{u}_i\tilde{u}_j)$ and the SGS heat flux $q_i = \overline{\rho}(\widetilde{u_i T} - \tilde{u}_i\tilde{T})$. These terms are modeled here using a compressible flow version of the dynamic SGS model of Germano et al. (1991). (For details see Moin et al. (1991).)

The model involves three coefficients, C, C_I and Pr_t. They are automatically adjusted, as time progresses, based on the resolved flow field information with the aid of a "test" filter (as mentioned in Section 1) that has a filter width coarser than the grid used to perform the computations. This is indeed the main advantage of the dynamic model over other SGS models used in the past; it requires little prior experience on the flow type being considered. In addition, it is more accurate, since it adjusts by itself to the flow conditions.

2.2. Formulation of spatial filters

Many different formulations of spatial filters were examined to determine their behavior in Fourier space. This was done because we believe it would be desirable for the spatial filter to have properties similar to the sharp cut-off filter.

The spatial filters are formulated as follows

$$\overline{f}_j + \sum_{n=1}^{N_i} \alpha_n \left(\frac{\overline{f}_{j-n} + \overline{f}_{j+n}}{2} \right) = \sum_{n=0}^{N_e-1} a_n \left(\frac{f_{j-n} + f_{j+n}}{2} \right) \tag{2}$$

where f_j is the function value and $\overline{f}_j$ is the filtered function value. N_i is the number of implicit coefficients, which are given by $\{\alpha_n\}$, and N_e is the number of explicit coefficients, which are $\{a_n\}$.

The Fourier transform of the filtered function is given by

$$\widehat{\overline{f}} = \widehat{g}\widehat{f} \tag{3}$$

where $\widehat{f}$ is the Fourier transform of the function and $\widehat{g}$ is the filter transfer function. For the filter formulation given above the filter transfer function is

$$\widehat{g}_k = \frac{\sum_{n=0}^{N_e-1} a_n \cos\left(\frac{2\pi kn}{N}\right)}{1 + \sum_{m=1}^{N_i} \alpha_m \cos\left(\frac{2\pi km}{N}\right)} \tag{4}$$

where k is the wavenumber and N is the number of grid points.

The first filters to be considered are approximations to the top hat filter in physical space. The top hat filter is defined by

$$G(x - x') = \begin{cases} 1 & \text{if } x - \Delta/2 \leq x' \leq x + \Delta/2 \\ 0 & \text{otherwise} \end{cases} \tag{5}$$

where Δ, the filter width, is assumed here equal to $2\Delta x$. The convolution integral in equation (1) can be approximated in various ways. Using three collocation points and the trapezoid rule gives $a_0 = 1/2$, $a_1 = 1/4$. With three collocation points one can improve the accuracy of the integration by using Simpson's rule, which gives $a_0 = 2/3$, $a_1 = 1/6$. If the integration is done analytically, then a spectrally accurate formula is obtained.

The filter transfer functions for these schemes are shown in Figure 1(a) against that of the sharp cut-off filter. The top hat filter behaves very differently than the sharp cut-off filter and is more like the Gaussian filter in that some of the low wavenumber modes are reduced in amplitude while some of the high wavenumber modes are retained. The trapezoid rule filter is the same as Lele's 3-point explicit filter discussed below.

A second set of filters is obtained by approximating the sharp cut-off filter in a least squares sense. For explicit filters of this type, exact values for the coefficients can be found, if one lets the number of grid points $N \rightarrow \infty$. They are given by $a_0 = 1/2$, $a_\ell = 2/(\pi\ell)\sin(\pi\ell/2)$ for $\ell \geq 1$. For the implicit filters, the coefficients are found numerically. The only implicit filter discussed here is a 5-point implicit–7-point explicit filter, for which the coefficients are

$$
\begin{array}{lll}
a_0 = 0.5 & a_1 = 0.85965917 & a_2 = 0.48688422 \\
a_3 = 0.20038708 & \alpha_1 = 0.0 & \alpha_2 = 0.97376845
\end{array}
$$

The least squares filters are compared to the sharp cut-off filter in Figure 1(b). As the number of filter coefficients is increased, better agreement with the sharp cut-off filter is obtained, as expected. However, the filter transfer functions are oscillatory and display large amplitude overshoots, which is undesirable.

A third set of filters is obtained following Lele (1992) in which certain constraints are placed on the filter transfer function. In order to compare these filters to the sharp cut-off filter with $\Delta/\Delta x = 2$, the extra constraint $\widehat{g}(N/4) = 1/2$ was imposed. The filters obtained are shown in Figure 1(c). In general the filters are smooth, monotonically decreasing from 1 to 0. However, the filters are not as sharp as the 5-pt implicit method obtained from the least squares approach.

The filters obtained from Lele's formulation are very smooth but are not very sharp, while the filters found from the least squares approach are much sharper but exhibit oscillations. In an attempt to reach a compromise between the two filters we constructed filters obtained by combining the filter coefficients in the following way

$$
a_n = \psi a_n^{LS} + (1-\psi)a_n^{Lele} \quad , \quad \alpha_n = \psi \alpha_n^{LS} + (1-\psi)\alpha_n^{Lele} \quad . \tag{6}
$$

Figure 1(d) shows the transfer function for the 5-pt implicit–7-pt explicit filter with $\psi = 0.95$. This compromise filter gives a transfer function that is sharper than that from the Lele formulation but does not have as large an overshoot as that from the least squares approach.

3. Results

A number of spatial filters from the previous section were tested in actual LES to determine whether they give results similar to those

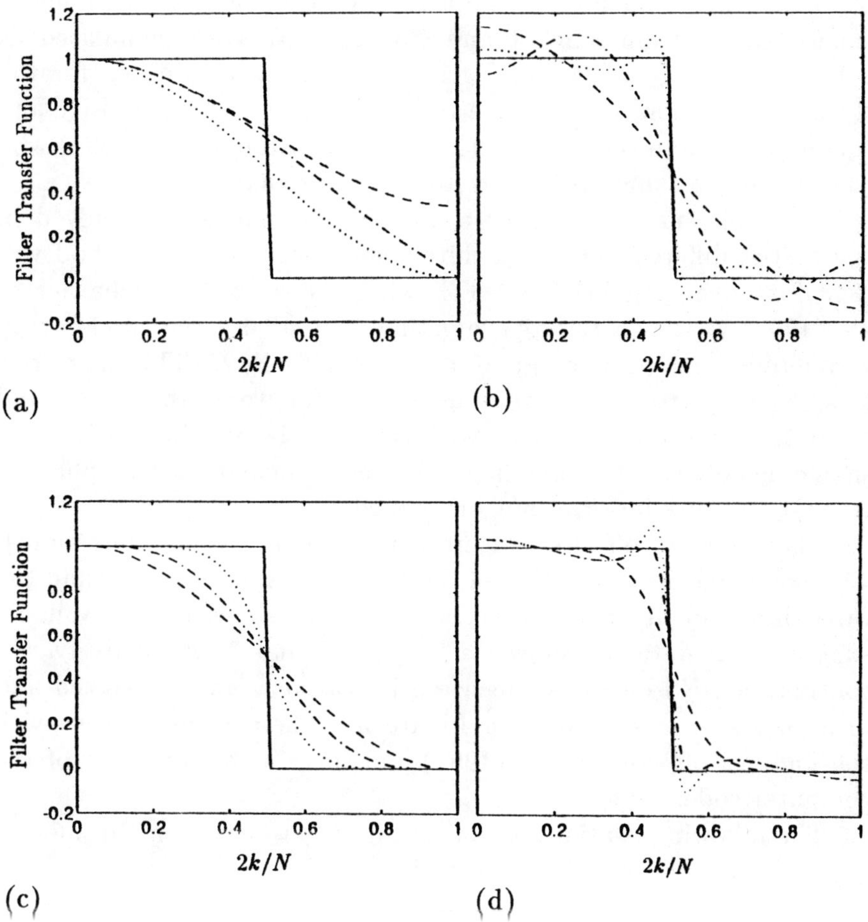

FIG. 1. Comparison of filter transfer functions for spatial filters. ——— , sharp cut–off filter. (a)Top hat filters. ········ , 3-pt explicit trapezoid; – – – – , 3-pt explicit Simpson; —·— , spectral. (b)Sharp cut-off least squares filters. – – – – , 3-pt explicit; —·— , 7-pt explicit; ········ , 5-pt implicit–7-pt explicit. (c)Lele filters. – – – – , 3-pt explicit; ········ , 7-pt explicit; —·— , 5-pt implicit–7-pt explicit. (d) —·— , compromise filter ($\psi = 0.95$); ········ , 5-pt implicit–7-pt explicit least squares filter; – – – – , 5-pt implicit–7-pt explicit Lele filter.

of the sharp cut-off filter. The isotropic grid-generated turbulence experiment of Comte-Bellot and Corrsin (1971) was simulated as a temporal decay on a $(32)^3$ grid. The initial conditions for the state variables and constitutive relations for pressure, viscosity and conductivity were the same as those used by Moin et al. (1991) who considered the same problem in testing the dynamic model.

A pseudo-spectral Fourier collocation scheme was employed in the spatial differencing and a third-order Runge-Kutta method was used for advancing in time. Following the work of Erlebacher et al. (1992) we truncated at every time step all Fourier modes with wavenumber magnitudes greater or equal to N/2. This was done in an attempt to control aliasing errors that were otherwise found to contaminate the flow fields with time and lead to instabilities. It should be noted that in inhomogeneous flow calculations the spherical truncation procedure can not be applied.

The sharp cut-off filter was first employed in the dynamic model. The predicted time evolution of the turbulent kinetic energy and the three-dimensional energy spectra compare well with the available experimental data, as shown in Figures 2 and 3, respectively. In contrast, a DNS on the same size grid was very under-resolved and unstable, as is also shown in Figure 3. These results agree with the findings of Moin et al. (1991) and serve as a validation of the computer code.

The following spatial filters (see Figure 1) were then considered

(1) 3-pt explicit (trapezoid),
(2) 3-pt explicit (Simpson),
(3) 3-pt explicit (least squares),
(4) 7-pt explicit (least squares),
(5) 5-pt implicit–7-pt explicit (least squares),
(6) 5-pt implicit–7-pt explicit (Lele),
(7) 5-pt implicit–7-pt explicit (compromise, $\psi = 0.95$).

In Figure 4, the turbulent kinetic energy evolutions are presented against that from the sharp cut-off filter and the experiment. A comparison of the energy spectra is shown in Figure 5. All implicit filters are found to perform similarly to the sharp cut-off filter with the exception of filter (6) which is more dissipative at higher modes. The 7-pt explicit filter performed poorly by providing even greater amounts of dissipation. The 3-pt explicit filters were found to be unstable.

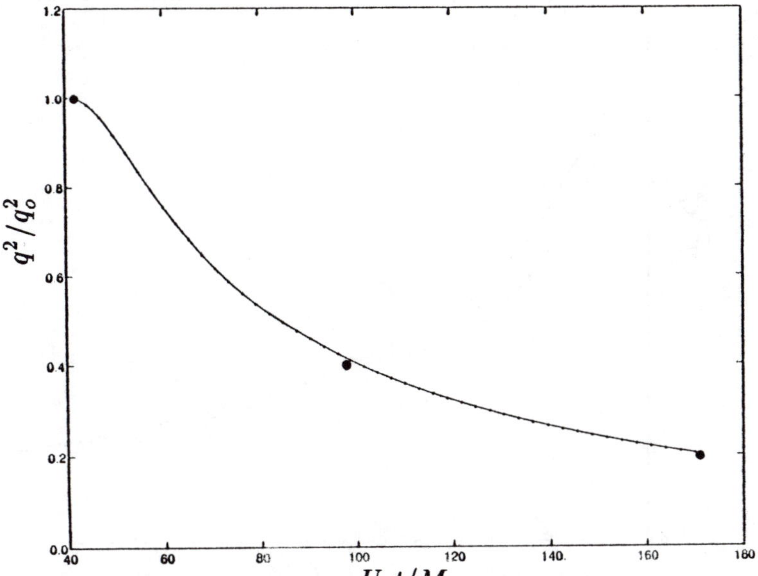

FIG. 2. Comparison of time evolution of resolved turbulent kinetic energy to filtered data of Comte-Bellot and Corrsin (1971). • experiment; ——, LES(sharp cut–off filter).

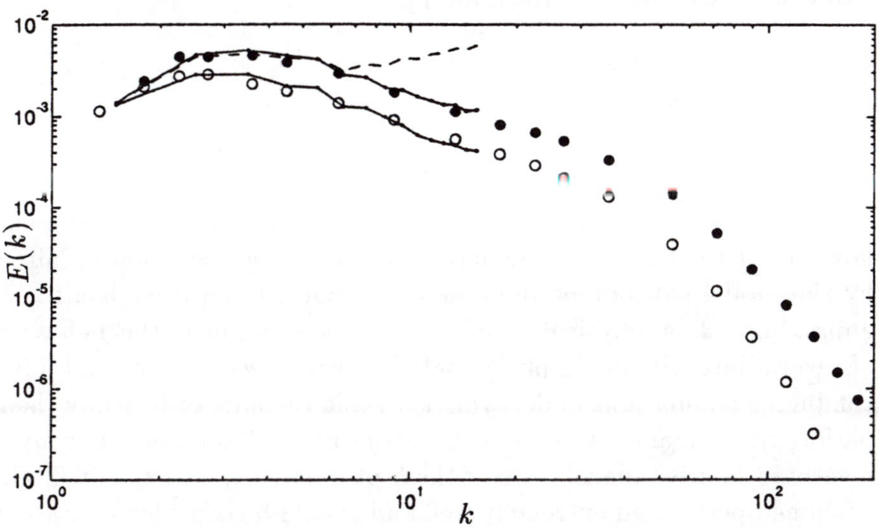

FIG. 3. Comparison of three-dimensional energy spectrum to data of Comte-Bellot and Corrsin (1971). • experiment at $U_0t/M = 98$; o experiment at $U_0t/M = 171$; ——, LES(sharp cut–off filter) at $U_0t/M = 98$ and 171; ----, DNS at $U_0t/M = 98$.

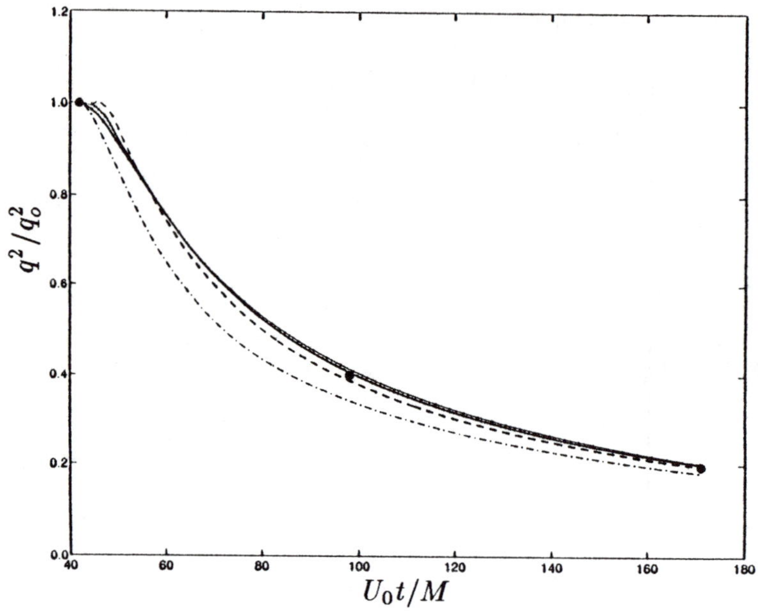

FIG. 4. Time evolution of resolved turbulent kinetic energy. • filtered data of Comte-Bellot and Corrsin (1971); —•— , LES(sharp cut–off filter); —·— , LES(filter 4); ········ , LES(filter 5); ---- , LES(filter 6); —— , LES(filter 7).

4. Conclusions and Future Plans

In performing LES of inhomogeneous compressible turbulent flows using the dynamic SGS model, the filtering operation required by the model can not be done in wave space, but rather has to be approximated in physical space. We have examined the behavior of several implicit and explicit spatial filters in wave space and have conducted simulations of decaying isotropic turbulence to study their performance against that of the sharp cut-off filter defined conveniently in Fourier space. Two of the 5-pt implicit–7-pt explicit filters examined performed extremely well and should be considered as good candidates for future use. These filters have transfer functions that are much sharper than the other filters near the cut-off mode, but exhibit oscillations which, as suggested by the simulations, are not of serious concern. The 7-pt explicit filter was found in the simulations

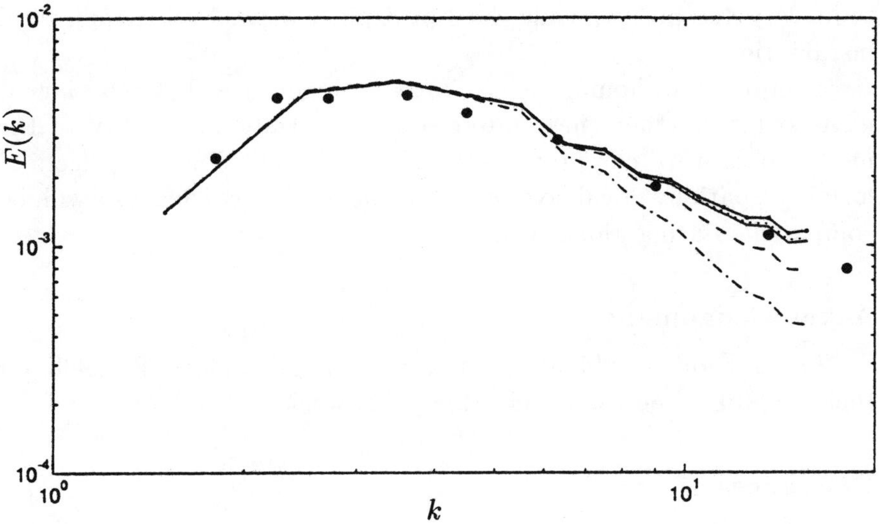

(a)

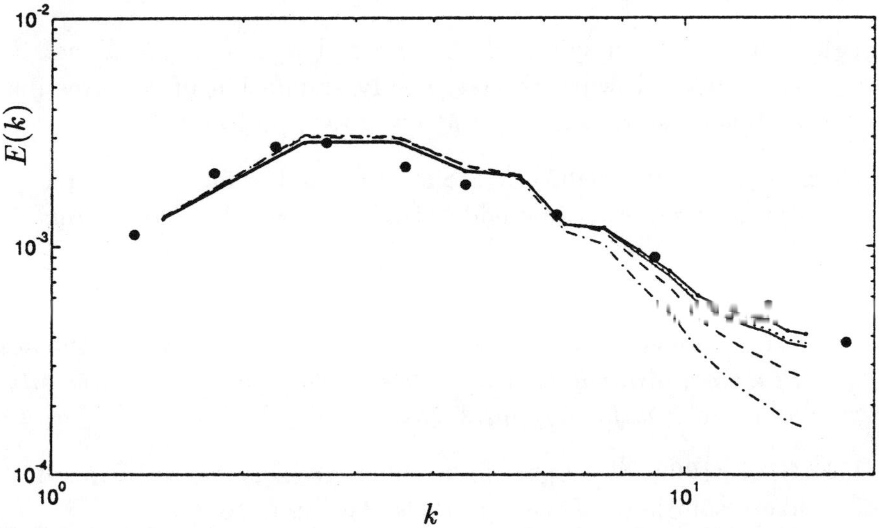

(b)

FIG. 5. Three-dimensional energy spectrum. • data of Comte-Bellot and Corrsin (1971); ——•—— , LES(sharp cut–off filter); —·— , LES(filter 4); ········ , LES(filter 5); ---- , LES(filter 6); —— , LES(filter 7). (a) $U_0t/M = 98$, (b) $U_0t/M = 171$.

to be very dissipative, while the 3-pt explicit filters lead to numerical instabilities.

Compressible homogeneous shear turbulence will be considered next to test further these filters and also study the ability of the model to capture compressibility effects. In addition, the effects of forming spatially local averages for the model coefficients will be compared to using global values.

Acknowledgements

The authors would like to thank T. Lund and U. Piomelli for their helpful suggestions concerning this work.

References

Comte-Bellot, G. and Corrsin, S., 1971. "Simple Eulerian time correlation of full and narrow band velocity signals in grid generated, isotropic turbulence," *J. Fluid Mech.* **48**, pp. 273–337.

Erlebacher, G., Hussaini, M. Y., Speziale, C. G., and Zang, T. A., 1992. "Toward the large-eddy simulation of compressible turbulent flows," *J. Fluid Mech.* **238**, pp. 155–185.

Germano, M., Piomelli, U., Moin, P., and Cabot, W., 1991. "A dynamic subgrid-scale eddy-viscosity model," *Phys. Fluids A.* **3**, pp. 1760–1765.

Ghosal, S., Lund, T. S., and Moin, P., 1992. "A local dynamic model for large eddy simulation," *Center for Turbulence Research, Annual Research Briefs-1992, Stanford University, Stanford, California*, pp. 3–25.

Lele, S. K., 1992. "Compact finite difference schemes with spectral-like resolution," *J. Comp. Phys.* **103**, pp. 16–42.

Lilly, D. K., 1992. "A proposed modification of the Germano subgrid-scale closure method," *Phys. Fluids A.* **4**, pp. 633–635.

Moin, P., Squires, K., Cabot, W., and Lee, S., 1991. "A dynamic subgrid-scale model for compressible turbulence and scalar transport," *Phys. Fluids A.* **3**, pp. 2746–2757.

ON ELLIPTIC RELAXATION NEAR WALL MODELS

A. O. Demuren[1] *and R. V. Wilson*

Old Dominion University
Norfolk, Virginia 23529

ABSTRACT

Near-wall models based on elliptic relaxation equations are incorporated into high-Reynolds number second-moment closures so that they can be integrated all the way down to the wall. Resulting models enable reasonably good predictions of the mean flow and Reynolds stress fields in developed turbulent channel flows. In particular, the high degree of anisotropy in the near-wall region is predicted. However, the models fail to reproduce certain subtle features of the Reynolds stress anisotropy field in the immediate proximity of the wall. Moreover, significant effects of the elliptic relaxation model are seen near the center of the channel.

1. Introduction

Turbulent flows in the vicinity of walls are characterized by very strong anisotropy of the Reynolds stresses. There is also a low Reynolds number region in the immediate proximity of walls. These distinctive effects are usually modeled together in the form of low-Reynolds number modifications to establish high-Reynolds number models. The modifications usually involve damping functions of the van Driest type which are based on the wall-normal distance or on the turbulent Reynolds number. A variety of such near-wall models are reviewed for two-equation, eddy-viscosity based models by Patel et al. (1985) and for second-moment closure models by So et al. (1991). Durbin (1991, 1993) has pointed out that the strong anisotropy in the Reynolds stresses would result mainly from the blocking effect of the wall and as such could be modeled by applying appropriate boundary conditions on the Reynolds stresses at the walls which are then made to relax elliptically towards nearly homogeneous flow far

[1]Research was supported by the National Aeronautics and Space Administration under NASA Contract No. NAS1-19480 while the authors were in residence at the Institute for Computer Applications in Science and Engineering (ICASE), NASA Langley Research Center, Hampton, VA 23681-0001.

M. Y. Hussaini et al. (eds.), Transition, Turbulence and Combustion, Vol. II, 61–71.

from the walls. In a sense, effects of two-point velocity gradient correlations in the 'rapid' pressure-strain correlations are simulated via elliptic relaxation equations within the framework of single-point closures. Proposed models were calibrated with the *DNS* channel flow results of Kim (unpublished), and applied successfully to a range of wall-bounded flows. However, it is somewhat disturbing that some 'well established' coefficients of the underlying high-Reynolds number model had to be changed to obtain good agreement with experimental data.

A distinction should be made between near-wall and wall-proximity models. The former enables integration of the governing equations through the buffer layer and the viscous sublayer all the way down to the wall, whereas the latter is applicable only down to the log-layer and must use wall functions to bridge the near-wall region containing the viscous sublayer and the buffer layer. Most second-moment closures require the use of so-called wall-reflection terms in order to predict correctly the mild level of Reynolds stress anisotropy in the log-layer. That is a distinct problem from the subject of the present study although, it is sometimes misinterpreted as such. Demuren and Sarkar (1993) have demonstrated that the need for special wall-proximity modifications such as the addition of wall-reflection terms, which is widely accepted to be undesirable, is a shortcoming of the model. They showed that the pressure-strain model of Speziale, Sarkar, and Gatski (1991) (denoted *SSG*), developed for homogeneous shear flows, could predict the observed mild level of Reynolds stress anisotropy in the log-layer of plane channel flows without any wall-proxity modifications. Clearly, the flow in the log-layer is nearly homogeneous, and a method based on a well-calibrated quasi-homogeneous second-moment closure should be able to predict it. In the present study, the elliptic relaxation modeling approach of Durbin (1993) is evaluated to determine how it can be used to extend the *SSG* pressure-strain model to the near-wall region. We shall however apply the constraint that 'standard' coefficients employed by Demuren and Sarkar (1993) should not be modified.

2. Mathematical Formulation

2.1. Governing equations

The equations governing the mean flow and turbulence fields are described in cartesian tensor notation for high-Reynolds number flow by Demuren and Sarkar (1993). We adopt the same notation in this paper, and discuss them only where they are changed by the near-wall model.

For the high-Reynolds number flow, the transport equations for the Reynolds stress components are written as:

$$\frac{\partial}{\partial x_l}(U_l\overline{u_iu_j}) = \nu\frac{\partial^2}{\partial x_l^2}(\overline{u_iu_j}) + D_{ij} + P_{ij} + \pi_{ij} - \frac{2}{3}\delta_{ij}\varepsilon \tag{1}$$

where D_{ij} is the turbulent diffusion, P_{ij} is the production, π_{ij} is the pressure-strain correlation, ε is the scalar dissipation rate of the turbulent kinetic energy k, and δ_{ij} is the Kronecker delta. To accommodate near-wall effects, Durbin (1993) proposed to replace $\pi_{ij}-2/3\delta_{ij}\varepsilon$ by $\mathcal{P}_{ij} - \overline{u_iu_j}\varepsilon/k$, so that the transport equation for $\overline{u_iu_j}$ can now be written as:

$$\frac{\partial}{\partial x_l}(U_l\overline{u_iu_j}) = \nu\frac{\partial^2}{\partial x_l^2}(\overline{u_iu_j}) + D_{ij} + P_{ij} + \mathcal{P}_{ij} - \frac{\overline{u_iu_j}}{k}\varepsilon \tag{2}$$

where

$$\mathcal{P}_{ij} = kf_{ij} \tag{3}$$

and f_{ij} is obtained from the solution of the elliptic differential equation:

$$L^2\frac{\partial^2}{\partial x_l^2}(f_{ij}) - f_{ij} = -\frac{\pi_{ij}}{k} - \frac{1}{T}\left(\frac{\overline{u_iu_j}}{k} - \frac{2}{3}\delta_{ij}\right). \tag{4}$$

In equation (4), L is a turbulent length scale given by:

$$L = C_L \max\left\{\frac{k^{3/2}}{\varepsilon}, C_\eta\left(\frac{\nu^3}{\varepsilon}\right)^{1/4}\right\} \tag{5}$$

and T is a turbulent time scale given by:

$$T = \max\left\{\frac{k}{\varepsilon}, C_T\left(\frac{\nu}{\varepsilon}\right)^{1/2}\right\}. \tag{6}$$

Durbin argued that far from the wall, the elliptic part of equation (4) should be zero and $T = k/\varepsilon$, so that equations (1) and (2) would be equivalent. π_{ij} was obtained from the simpler pressure-strain model Launder, Reece, and Rodi (1975) (hereafter denoted LRR).

D_{ij} was obtained from the diffusion model by Daly and Harlow (1970) (hereafter denoted DH).

Following Demuren and Sarkar (1993), pressure-strain models of the linear or quasi-linear type such as the LRR and SSG models can be written in terms of the anisotropy tensor $b_{ij} \equiv (\overline{u_i u_j}/2k - 1/3\delta_{ij})$, the rate of strain tensor $S_{ij} = 1/2(\partial U_i/\partial x_j + \partial U_j/\partial x_i)$, the rotational tensor $W_{ij} \equiv 1/2(\partial U_i/\partial x_j - \partial U_j/\partial x_i)$, and the production of k, P_k in the general form:

$$\begin{aligned}\pi_{ij} &= \alpha_0 \varepsilon b_{ij} + \alpha_1 \varepsilon \left(b_{ik} b_{jk} - \frac{1}{3} II \delta_{ij} \right) + \alpha_2 k S_{ij} + \alpha_3 P_k b_{ij} \\ &+ k \left\{ \alpha_4 \left(b_{ik} S_{jk} + b_{jk} S_{ik} - \frac{2}{3} \delta_{ij} b_{kl} S_{kl} \right) \right. \\ &\left. + \alpha_5 \left(b_{ik} W_{jk} + b_{jk} W_{ik} \right) \right\} .\end{aligned} \tag{7}$$

The standard coefficients for the LRR model are: $\alpha_0 = -3.6$, $\alpha_1 = \alpha_2 = \alpha_3 = 0$, $\alpha_4 = \alpha_5 = 1.20$. However, Durbin (1993) used a much smaller value $\alpha_0 = -2.44$, for the "return-to-isotropy" coefficient, with the justification that this value should decrease in the near-wall region.

The standard coefficients for the SSG model are: $\alpha_0 = -3.4$, $\alpha_1 = 4.2$, $\alpha_2 = 0.8 - 1.3II^{1/2}$, $\alpha_3 = -1.8$, $\alpha_4 = 1.25$, $\alpha_5 = 0.40$. These values were used in the previous study of Demuren and Sarkar (1993), and are also used in the present one.

The DH diffusion model can be written as:

$$D_{ij} = \left\{ C_{s_1} \frac{k}{\varepsilon} \left[\overline{u_k u_l} (\overline{u_i u_j})_{,l} \right] \right\}_{,k} \tag{8}$$

where $(\ )_{,l}$ represents the first derivative with respect to x_l. In their study of high-Reynolds number models, Demuren and Sarkar (1993) obtained better distribution of Reynolds stress anisotropy by using a diffusion model proposed by Mellor and Herring (1973) (denoted MH). This can be written as:

$$D_{ij} = \left\{ C_{s2} \frac{k^2}{\varepsilon} \left[(\overline{u_i u_j})_{,k} + (\overline{u_i u_k})_{,j} + (\overline{u_j u_k})_{,i} \right] \right\}_{,k} . \tag{9}$$

The coefficients C_{s_1} and C_{s_2} have values of 0.22 and 0.072, respectively. Again, Durbin (1993) used a lower value of $C_{s_1} = 0.19$.

The scalar dissipation ε is obtained from:

$$\frac{\partial}{\partial x_l}(U_l\varepsilon) = \nu\frac{\partial^2}{\partial x_l^2}(\varepsilon) + D_\varepsilon + \frac{C^*_{\varepsilon_1}P_k - C_{\varepsilon_2}\varepsilon}{T} \tag{10}$$

where $C^*_{\varepsilon_1} = C_{\varepsilon_1}\left(1 + a_1\frac{P_k}{\varepsilon}\right)$ and

$$D_\varepsilon = \left\{C_{s_\varepsilon}\frac{k}{\varepsilon}[\overline{u_k u_l}(\varepsilon)_{,l}]\right\}_{,k}. \tag{11}$$

The coefficients proposed by Durbin (1993) are:

$$C_{\varepsilon_1} = 1.44,\ C_{\varepsilon_2} = 1.90,\ a_1 = 0.1,\ C_{s_\varepsilon} = 0.14.$$

The value of $C_{s_\varepsilon} = 0.15$, proposed by LRR is used in the present study.

2.2. Boundary conditions

The computational domain is bounded by a wall and a symmetry plane along which boundary conditions are required. For the mean flow, the no slip condition is specified at the wall and zero normal derivative is specified at the symmetry plane. For the turbulence field boundary conditions are required for ε, $\overline{u_i u_j}$, and f_{ij}. The latter would produce the all-important wall-blocking effect. At the wall all components of $\overline{u_i u_j} = 0$ and at the symmetry plane, $(\overline{u_i u_j}) = 0$, if $i \neq j =$ normal, otherwise $(\overline{u_i u_j})_{,\text{normal}} = 0$.

Boundary conditions for f_{ij} are more difficult to specify. Durbin (1993) used order of magnitude analysis arguments to show that at the wall $(\overline{u_i u_j})_{,\text{normal}} \approx 0$ if i or j is in the normal direction (i.e., i, or $j = 2$, in the present notation); otherwise $f_{ij} = 0$. In the former case f_{ij} can be calculated from the finite-difference approximation

$$f_{ij} = \frac{-2\nu^2\overline{u_i u_j}_{(2)}}{\varepsilon_{(1)}\Delta y^4_{(2)}} \tag{12}$$

where $ij = 12$ or 22.

The computed value of f_{22} produces the wall-blocking effect. But since $f_{11} = f_{33} = 0$, we have the inconsistent result that $f_{ii} \neq 0$, which would in turn lead to $\overline{u_i u_i} \neq 2k$. Durbin avoided this problem by computing only $\overline{u_1^2}$, $\overline{u_2^2}$, and k, and obtaining $\overline{u_3^2}$ from $2k - (\overline{u_1^2} +$

$\overline{u_2^2}$). This inconsistency is removed in the present study by simply setting $f_{11} = f_{33} = -0.5 f_{22}$. The justification for this is that the blocking effect of the wall leads to a transfer of component turbulent energy from the 22 component to both 11 and 33 components. At the symmetry plane, $f_{ij} = 0$, if $i \neq j =$ normal; otherwise $(f_{ij})_{,\text{normal}} = 0$.

At the wall, ε is calculated from the finite-difference approximation of $\nu(k_{,n})_{,n}$, and at the symmetry plane $(\varepsilon)_{,n} = 0$.

3. Results and Discussion

Computations of the fully developed plane channel flow were performed using the elliptic relaxation model and combinations of both the *LRR* and *SSG* pressure-strain models and the *DH* and *MH* diffusion models. A Reynolds number, based on center line velocity and channel half width, of 7,890 was chosen so that computed results could be compared to *DNS* data of Kim (unpublished). Although several model combinations were considered, results would be presented here only for two cases. One uses the *LRR* pressure-strain model and the *DH* diffusion model with adjusted coefficients as proposed by Durbin (1993). The other uses the *SSG* pressure-strain model and the *MH* diffusion model with the standard coefficients as utilized in the high-Reynolds number channel flow computations of Demuren and Sarkar (1993).

Computed streamwise velocity profiles are compared to *DNS* data in Figure 1. Both models predict correctly the sublayer region, but the *SSG* model produces better agreement with data in the log-layer. The actual, unnormalized velocity profiles predicted by both models are similar in the log-layer, but the *LRR* model produces a velocity gradient which is too large in the near wall region. Hence, the friction velocity is overpredicted.

Computations of the 4 non-zero components of the Reynolds stress tensor are compared to *DNS* data in Figure 2. The values are normalized with the friction velocity. Both models appear to predict the distributions quite well. However, it is difficult to examine how well the structure of the turbulence field is predicted with the components plotted in this way. Comparison of the Reynolds stress anisotropy field would be much more informative. This is shown in Figure 3. Although the predictions by both models agree reasonably well with *DNS* data, there are significant discrepancies. Firstly, *DNS* data show that the turbulence goes from a 2-component state in the immediate vicinity of the wall to nearly single-component state in the buffer layer around $y^+ \sim 10$, and then to 3-component state by the log-layer. The predictions fail to reproduce the transition from the 2-component to the single-component state. Rather, they show nearly single-component state adjacent to the wall, transitioning smoothly to the 3-component state through the buffer layer into the log-layer. Clearly, turbulent kinetic energy is generated in the 11 component and then transferred to other components via the pressure-strain. This effect is captured by the models, but there is a subtle process which takes place in the highly viscous sublayer region not reproduced by either model. It is doubtful that near-wall models based solely on elliptic relaxation of the wall-blocking effect can reproduce such a process. Secondly, *DNS* data show that the anisotropy levels are virtually constant in the log-layer ($0.2 < y/\delta < 0.8$ or $80 < y^+ < 320$), with a steady return to isotropy throughout the log-layer, and too rapid return-to-isotropy near the center of the channel. The latter is in fact more rapid than found in the high-Reynolds number computations of Demuren and Sarkar (1993). The reason for this is that the elliptic relaxation model produced an effect on the Reynolds stress levels throughout the channel cross-section, rather than only in the near-wall region as one would expect. Basically, the model accentuates changes in the anisotropy level, which signify deviation from homogeneous flow. The strongest changes occur near the wall and there the model has the most effect. But there are significant changes near the center of the channel as well, and there the model produces an undesirable effect. Clearly, the model needs to be reformulated or at least recalibrated to alleviate this problem. On the average, though, the *SSG* model seems to yield better agreement with *DNS* data.

4. Concluding Remarks

A near-wall model based on the elliptic relaxation concept has been implemented for both the *LRR* and *SSG* pressure-strain models and the *DH* and *MH* diffusion models. Computations of the developed channel flow were then made with combinations of the *LRR* and *DH* models and the *SSG* and *MH* models and the results compared to *DNS* data. Both model combinations give reasonable predictions of the mean flow and the Reynolds stress anisotropy. In particular, the predictions fail to reproduce the change from 2-component state to nearly one-component state, observed in the simulation in the immediate vicinity of the wall. Further, the rate of return-to-isotropy near the center of the channel is overpredicted by both model combinations. On the average, the *SSG-MH* model combination performs somewhat better than the *LRR-DH* combination. It is notable that model coefficients were not changed from the standard high-Reynolds number form in the former, whereas they had been adjusted in the latter to yield better agreement with *DNS* data.

Since the elliptic relaxation model is designed to simulate the effect of wall blocking, it appears undesirable that it should affect the anisotropy level throughout the cross-section. Hence, the model needs reformulation or recalibration to ensure that its effect is confined to the near-wall region. For the same reason, it should not be necessary to modify the coefficients of the underlying high-Reynolds number model.

References

Daly, B. J. and Harlow, F. H., 1970. "Transport equations of turbulence," *The Physics of Fluids* **13**, pp. 2634-2649.

Demuren, A. O. and Sarkar, S., 1993. "Perspective: Systematic study of Reynolds stress closure models in the computations of plane channel flows," *Journal of Fluids Engineering* **115**, pp. 5-12.

Durbin, P. A., 1991. "Near-wall turbulence modeling without damping functions," *Theoretical Computational Fluid Dynamics* **3**, pp. 1-13.

Durbin, P. A., 1993. "A Reynolds stress model for near-wall turbulence," *Journal of Fluid Mechanics* **249**, pp. 465-498.

Launder, B. E., Reece, G. J., and Rodi, W., 1975. "Progress in the development of a Reynolds stress turbulence closure," *Journal of Fluid Mechanics* **68**, pp. 537-566.

Mellor, G. L. and Herring, H. J. "A survey of mean turbulent field closure," *AIAA Journal* **11**, pp. 590-599.

Patel, V. C., Rodi, W., and Scheurer, G., 1985. "Turbulence models for near-wall and low Reynolds number flows: A review," *AIAA Journal* **23**, pp. 1308-1319.

So, R. M. C., Lai, Y. G., Zhang, H. S., and Hwang, B. C., 1991. "Second-order near-wall turbulence closures: A Review," *AIAA Journal* **29**, pp. 1819-1835.

Speziale, C. G., Sarkar, S., and Gatski, T. B., 1991. "Modelling the pressure-strain correlation of turbulence: An invariant dynamical systems approach," *Journal of Fluid Mechanics* **227**, pp. 245-272.

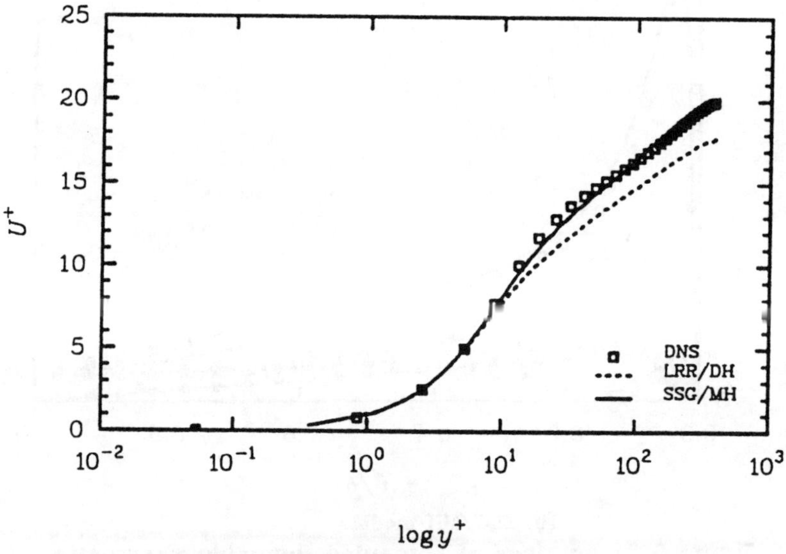

Figure 1 Comparison of computed velocity profiles to DNS data of Kim (unpublished) at Re=7890.

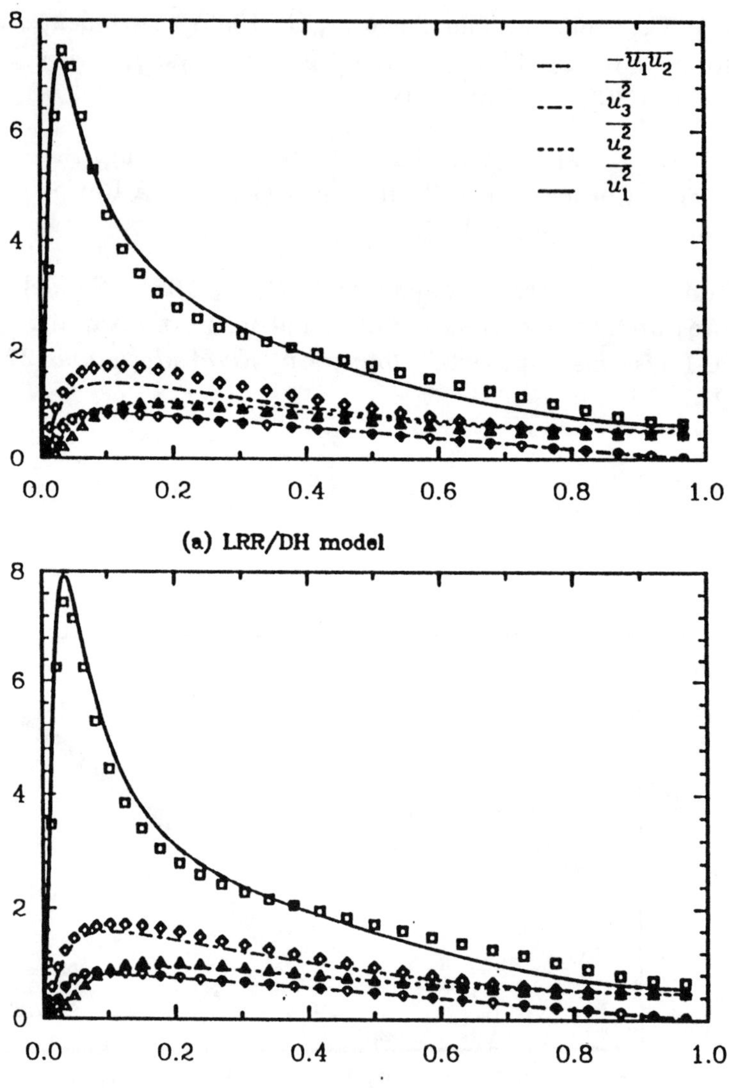

(b) SSG/MH model

Figure 2 Comparison of computed Reynolds stresses to DNS data of Kim(unpublished) at Re=7890; (symbols – DNS, lines – computations).

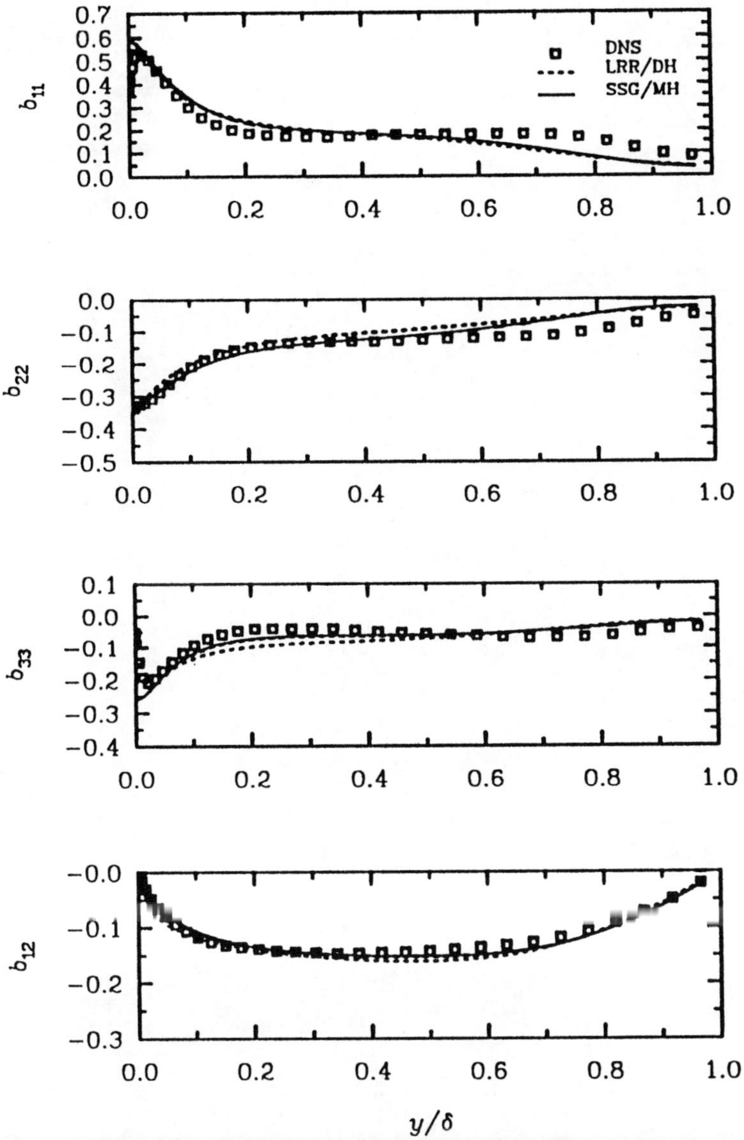

Figure 3 Comparison of computed Reynolds stress anisotropies to DNS data of Kim(unpublished) at Re=7890.

SOME TURBULENCE STATISTICS RELEVANT TO COMPRESSIBLE TURBULENCE MODELING

S.P.G. Dinavahi

Analytical Services & Materials, Inc.
NASA Langley Research Center, Hampton, VA 23681-0001

J. R. Ristorcelli & G.E. Erlebacher

Institute for Computer Applications in Science and Engineering
NASA Langley Research Center, Hampton, VA 23681-0001

ABSTRACT

Some statistics relevant to modeling are presented from the temporal direct numerical simulations (DNS) of a Mach 4.5 boundary-layer flow. Correlations, due to the effects of compressibility, appearing in the second moment equation are investigated. These include the pressure-dilatation, the dilatational dissipation, and other related correlations. The turbulent Mach number is about $M_t \simeq 0.15$ over a large fraction of boundary layer and $\rho_{rms}/\bar{\rho} \simeq 0.2$.

1. Introduction

A phenomenological study of the wall bounded compressible turbulence problem is conducted using the data from direct numerical simulations (DNS). The goal of this study is to explore the importance of new terms related to compressibility in the second moment evolution equations. This includes assessing some of the usual assumptions made in compressible turbulence modeling as well as computing various correlations amongst these quantities to see if the data suggests any suitable way of parameterizing the unclosed terms. Different scalings of these new quantities are explored with the intention of providing insight with which to guide the closure of the unknown correlations appearing in the moment equations.

The DNS conducted is that of an external boundary-layer flow along the axis of a cylinder at Mach 4.5. This geometry represents the axisymmetric analog of a flat-plate boundary-layer flow with a sharp leading edge. All variables are normalized by the free-stream values and the initial displacement thickness. Pressure is normalized

M. Y. Hussaini et al. (eds.), Transition, Turbulence and Combustion, Vol. II, 73–82.

by the pressure head $\rho_\infty U_\infty^2$. The "curvature" of the cylinder has a moderate value of 0.1. The data is generated using the temporal DNS. The calculations were started from a disturbed laminar state formed by the superposition of primary and secondary instability waves. The reader is referred to Dinavahi, Pruett & Zang (1993) for a more detailed discussion of the numerical procedure and the creation of the database.

A word about averaging in the context of temporal DNS is in order. In temporal DNS the roles of x and t are interchanged through a Galilean transformation. As a result, the flow evolves in time rather than in x and periodic boundary conditions are applied in all 3 spatial directions. Averaging is carried out in the two homogeneous directions– the streamwise, x, and the azimuthal, θ. As a result, any averaged quantity is function of only the wall normal distance, y. According to our notation, the Reynolds-average of a function $u^*(x,y,\theta)$ is denoted as $< u^*(x,y,\theta) > = U(y)$ and the Favre-average as $\{u^*(x,\theta,y)\} = V(y)$.

2. Preliminary exposition

In this article, upper case letters are used to denote mean quantities except in the case of the mean density, $< \rho >$, as it has no convenient upper case form. The averaging operation is indicated using the angle brackets for the time average, $< uu >$, and the curly brackets for the density-weighted or Favre average $\{uu\} =< \rho^* uu > < \rho >^{-1}$ where the asterisk denotes the full field, $\rho^* =< \rho > +\rho'$. The dependent variables are decomposed according to

$$u_i^* = U_i + u_i \quad \text{where } < u_i >= 0 \tag{1}$$

$$u_i^* = V_i + v_i \quad \text{where } \{v_i\} = 0 \tag{2}$$

$$\rho^* = < \rho > + \rho' \quad \text{where } < \rho' >= 0 \tag{3}$$

$$p^* = P + p \quad \text{where } < p >= 0 \tag{4}$$

$$T^* = T + \theta \quad \text{where } \{\theta\} = 0 \tag{5}$$

where (U,u) and (V,v) are used to indicate the Reynolds and Favre decompositions of the velocity field. As both the Reynolds and Favre velocities appear in the evolution equation for a compressible turbulence it is necessary to carry both the Favre and Reynolds decompositions. The time-average of the fluctuating Favre velocity, $< v_i >$, characterizes the difference between the Favre and Reynolds velocities, V_i and U_i, as well as the difference between the instantaneous fluctuating portions of these two fields. Here and henceforth the prime on the fluctuating density is dropped. From the definition given in Eqs. (1)-(5), $< \rho v_i >= - < \rho >< v_i >$, *i.e.* the time average of the fluctuating Favre velocity and the mass flux are equivalent quantities (apart from a scaling by the local mean density). For the compressible turbulence problem in the Favre setting the anisotropy tensor is based on Favre averages

$$b_{ij} = \{v_i v_j\}/\{v_p v_p\} - 1/3\delta_{ij} \,. \tag{6}$$

The time average of the Navier-Stokes equation can be written in the following form:

$$D/Dt\,(< \rho > V_i >) = -P_{,i} + \Sigma_{ij,j}\,(U) - (< \rho > \{v_i v_p\})_{,p} \tag{7}$$

where D/Dt denotes the operator defined by $D/Dt(\psi) = (\psi_{,t} + (\psi\, V_p)_{,p}$ where ψ refers to any variable. To a very good approximation the viscosity is only a function of the mean temperature. The viscous terms, $\Sigma_{ij}(U) = < \mu > [U_{i,j} + U_{j,i} - 2/3 U_{q,q}\, \delta_{ij}]$, are given in terms of the time-averaged mean velocity while the problem is posed in terms of the Favre mean velocity. The time and Favre velocities are related by $U_i = V_i + < v_i >$.

The Reynolds stress equations are

$$\begin{aligned}
D/Dt(< \rho > \{v_i v_j\}) = & \\
- & < \rho > \{v_i v_p\} V_{j,p} - < \rho > \{v_j v_p\} V_{i,p} \\
+ & \Pi_{ij} + 2/3 < p v_{k,k} > \delta_{ij} \\
- & [< p v_i > \delta_{pj} + < p v_j > \delta_{ip}]_{,p} \\
- & [< \rho > \{v_i v_j v_p\} - < v_j \sigma_{ip} > - < v_i \sigma_{jp} >]_{,p} \\
+ & < v_j > [-P_{,i} + \Sigma_{ik,k}\,(U)] \\
+ & < v_i > [-P_{,j} + \Sigma_{jk,k}\,(U)] \\
- & < \rho > \varepsilon^d_{ij} - 2/3 < \rho > \varepsilon \delta_{ij}
\end{aligned} \tag{8}$$

where it is understood that the mean viscous term is $\Sigma_{ij}(U) = \Sigma_{ij}(V) + \Sigma_{ij}(< v >)$ and

$$\sigma_{ij} = \Sigma_{ij}(u) =< \mu > [u_{i,j} + u_{j,i} - 2/3 u_{q,q}\, \delta_{ij}]. \tag{9}$$

The following definitions have been used to write the equations in the form given above.

1. The deviatoric part of the pressure-strain correlation is defined as
$$\Pi_{ij} =< p\,(v_{i},_{j} + v_{j},_{i}\,) > -2/3 < pv_{k},_{k} > \delta_{ij}\ . \qquad (10)$$

2. The mean acceleration, $D/Dt\ V_i$, is rewritten in terms of the right hand side of the mean momentum equations.

3. The dissipation of the Favre kinetic energy, $1/2\{v_p v_p\}$, is
$$< u_{j},_{p}\,\sigma_{jp} > = < \rho > \varepsilon =< \rho > (\varepsilon_s + \varepsilon_c + \varepsilon_t) \qquad (11)$$
where $< \rho > \varepsilon_s =< \mu >< \omega_j\omega_j >$ is the solenoidal dissipation, and $< \rho > \varepsilon_c = 4/3 < \mu > < u_{p},_{p}\, u_{q},_{q} >$ is the compressible or dilatational dissipation. The transport term is $< \rho > \varepsilon_t = 2 < \mu > (< u_i u_j >,_{jl} - < u_{p,p} u_l >,_{l}\,)$ which is zero in a homogeneous turbulence. The two identities $< u_{i},_{p}\, u_{p},_{i} > = < u_i u_p >,_{ip} -2 < u_{i},_{i}\, u_p >,_{p} + < u_{i},_{i}\, u_{p},_{p} >$ and $u_{i},_{p} = u_{p},_{i} - \epsilon_{ijp}\omega_j$ where the vorticity is $\omega_j = \epsilon_{ijk} u_{j},_{k}$ have been used.

4. The deviatoric part of the dissipation tensor is
$$< \rho > \varepsilon^d_{ij} = < u_{i},_{p}\,\sigma_{jp} > + < u_{j},_{p}\,\sigma_{ip} > - 2/3\varepsilon\delta_{ij}\ . \qquad (12)$$

Inspection of the Reynolds stress equations indicates several terms which will be influenced by the compressibility of the flow. The relevance of the mass flux to the mean flow equations has already been discussed. The present investigation will, in addition, focus on terms appearing in the Reynolds stress equations that result from a non-trivial equation of state (*i.e.* $< pv_k >$ and $< \rho v_k >$), or from the fact that the velocity field is no longer divergence free (*i.e.* $< pv_{k},_{k} >$ and $< u_{p},_{p}\, u_{q},_{q} >$).

In addition, the possibility of a general polytropic gas law for the fluctuating fields will also be tested. A polytropic coefficient can be defined to assess the entropy associated with the fluctuating pressure field $p/P = n\rho/ < \rho > = (n/(n-1))\vartheta/T$. If $n = 0$ the fluctuating field is isobaric (Strong Reynolds Analogy); if $n = 1$ isothermal; if $n = \gamma$ adiabatic. Squaring the expression for p and averaging produces
$$< pp > /P^2 = n^2 < \rho\rho > / < \rho >^2\ . \qquad (13)$$

3. Mean statistics and turbulence parameters

It is important to establish whether the flow under consideration is fully turbulent. If not, it is then important to establish the state of development of the turbulence relative to a fully turbulent flow.

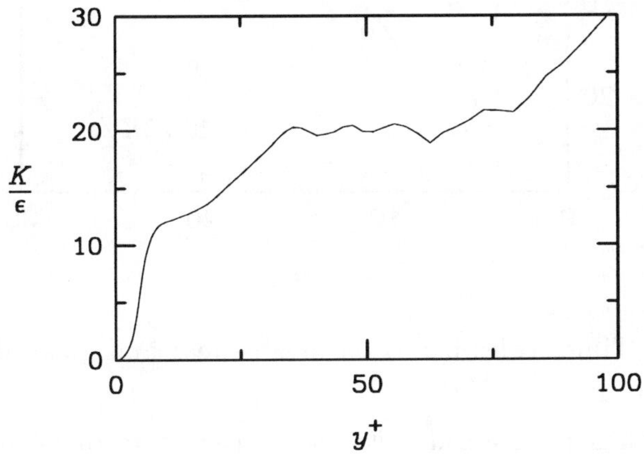

Figure 1: Ratio of kinetic energy to dissipation rate as a function of y^+.

The ratio k/ε characterizes the time scale of the turbulence and is presumed to scale with the integral time scale. This quantity is plotted as a function of y^+ in Figure 1. The ratio rises very sharply near the wall and is close 20 in the neighborhood of $y^+ = 50$. The non-dimensional parameter Sk/ε where $S = V_{1,2}$, relates the turbulence time scale to the mean-flow time scale associated with the shear. A similar timescale appropriate for compressible turbulence, is the ratio of the mean dilatation to the turbulence time scale, i.e. Dk/ε where $D = V_{k,k}$. Of course, it is not yet clear how important a role it will play, Both the above ratios are shown in Figure 2. As expected, the time scale associated with the mean dilatation is much shorter than that associated with the mean shear. If the turbulence were isotropic, the mean dilatation time scale goes to zero. In general, we do not expect this term to play much of a role in this flow since the mean dilatation is small.

A plot of skin friction versus time (not shown) shows that transition is initiated at $t = 150$ and completed by $t = 175$. The flow then relaxes towards a turbulent state. All results correspond to $t = 230$. This gives an indication of how far past the transition region the

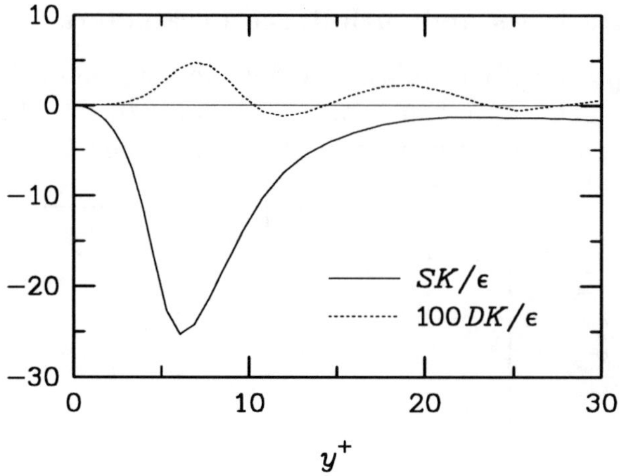

Figure 2: Time scales based on mean shear and mean dilatation.

simulation has progressed. To assess the state of development of turbulence the mean velocities are compared to the known log-law scaling of $u^+(y^+)$ in fully developed turbulent boundary layers (not shown). We follow the definitions proposed by Fernholz and Finley (1980) for u^+ and y^+. However the velocity profile still has inflection points associated with the process of transition. The turbulence is still developing and the mean velocity profile reflects this fact. Nonetheless as the flow is approximately three eddy turnovers past transition, the results from the simulation will be quite relevant to the compressible turbulent regime.

The variation in the mean pressure is very small through the boundary layer except very near the wall where it is governed by the wall temperature and the density. The mean density increases more or less monotonically away from the wall, where it has a value of approximately 0.25 and attains a value of 1 in the free stream. This strong variation in the mean density is one characteristic found in supersonic flows. A strong variation in the mean density is often accompanied by strong density fluctuations (upwards of 20%, as observed in transitional flows (Dinavahi, Ristorceilli & Erlebacher, 1993).

An important quantity in the parameterization of turbulence models, which gives an idea of the equilibrium nature of the turbulent flow, is the ratio $\mathcal{P}/(<\rho>\varepsilon)$ where $\mathcal{P}$ is the rate of production of kinetic energy $<\rho>k = 1/2 <\rho> \{v_j v_j\}$:

$$\mathcal{P} = (<\rho>k)_{,t} = (1/2 <\rho> \{v_j v_j\})_{,t} = -\{v_i v_p\} V_{i,p} \,. \tag{14}$$

The inflection profile in the mean velocity V_1 produces oscillations in $V_{1,2}$ which in turn transmit themselves to $\mathcal{P}$. This confirms the lack of a turbulent equilibrium.

Of primary interest is the Favre mean anisotropy tensor which naturally arises in compressible turbulence closures. Figure 3 shows the values of the anisotropy tensor, b_{ij}, based on the Favre average. We also computed b_{ij} based on u_i and Reynolds averaging, and obtained virtually the same normal variation as that calcuated from Eq.(6).

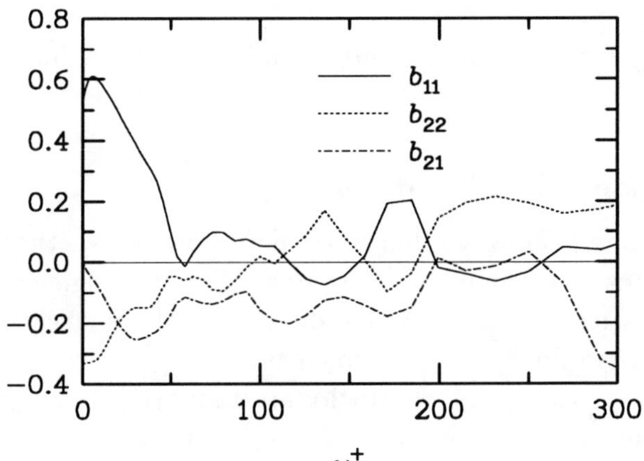

Figure 3: Anisotropy tensor.

The compressibility of the turbulence is indicated by the turbulent Mach number $M_t = (q^2/3)^{1/2}/c$ where $q^2 = \{v_j v_j\}$ and c is the local mean speed of sound. A plot of M_t *vs* y^+ is shown in Figure 4. Also shown in this plot is the normlized density variance. Its behavior is unexpectedly similar to M_t.

The polytropic gas coefficient n has a value close to 1.4 for $y^+ > 200$. When $10 < y^+ < 200$, $n \approx 0.2$. However, there is a thin layer at the wall in which n increases to 1.2 before dropping down close to 1 at the wall. This is to be contrasted to the results of Lee (1992) on the passage of the turbulence through a shock and the homogeneous shear of Blaisdell (1991) where the fluctuations approximately followed an adiabatic gas law, i.e., $n \simeq 1.4 \simeq \gamma$.

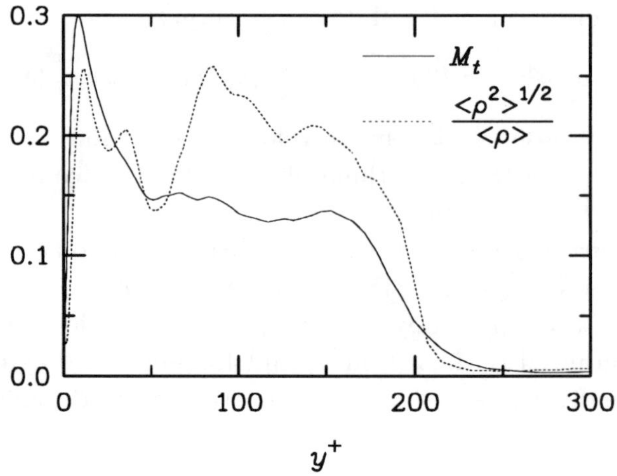

Figure 4: Turbulent Mach number and normalized fluctuating density.

4. Dilatation related statistics

Inspection of the Reynolds stress equations shows that in the simplest compressible turbulent flow possible – homogeneous isotropic turbulence with constant mean density – there are contributions due to compressibility. These contributions, the pressure-dilatation $< pu_{k,k} > = < pd >$, and dilatational dissipation, ϵ_c, are due to the non-zero fluctuating divergence, $d = u_{k,k}$ in a compressible flow. As these quantities appear in the Reynolds stress equation as scalars, their main effect is on the kinetic energy of the turbulence. Figure 5 shows these quantities normalized by the solenoidal dissipation in order to assess their relative importance in the energy budget.

It is not generally appreciated the degree to which the two statistics, $< pd >$ and $< \rho > \varepsilon_c$, typically treated separately, are related. This becomes apparent when a correlation coefficient, $\mathrm{Cor}_{<pd>} = < pd > / < pp >^{1/2} < dd >)^{1/2}$. Additional information about the nature of the pressure dilatation can be found by investigating the correlation $\mathrm{Cor}_{<ppd>} =< ppd > / < pp > < dd >^{1/2}$. This quantity, combined with $\mathrm{Cor}_{<pd>}$, gives information about the relative skewness of the distribution. Both correlations are shown in Figure 6.

A derivation of the complete set of second-order moment equations for the heat fluxes, mass fluxes, pressure fluxes, temperature,

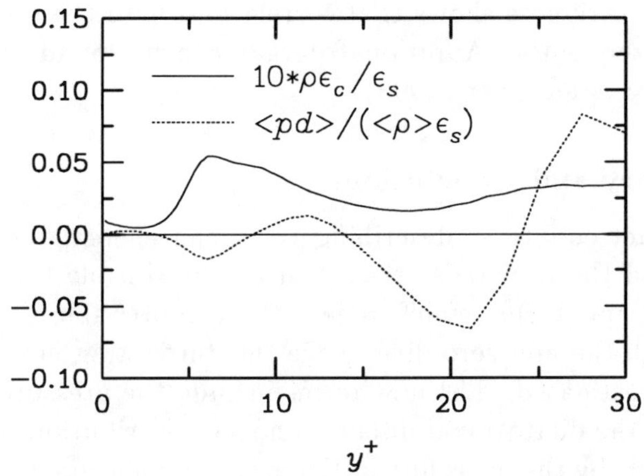

Figure 5: Pressure-dilatation and compressible dissipation rate ϵ_c normalized by ϵ_s.

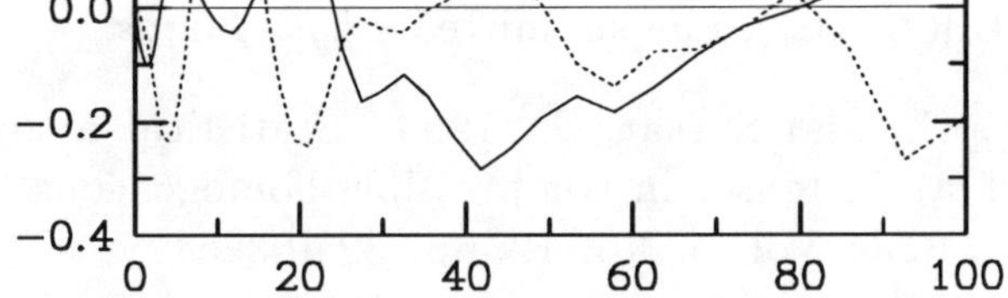

Figure 6: Pressure-dilatation correlations.

and density variances shows that correlations with the dilatation appear very frequently. Additional details can be found in Dinavahi, Ristorcelli & Erlebacher (1993).

5. Summary and Conclusions

The exact equations describing the mean energy, the mean momentum and the Reynolds stresses in a compressible turbulence are given. The new terms which reflect the compressible nature of the flow through the non-zero divergence the fluctuating field are identified and investigated. The new terms include the pressure-dilatation covariance, the dilatational dissipation, the contribution to the mean viscous terms by the mass flux in the mean momentum equations and the various components of the pressure and viscous work terms in the mean energy equation.

References

Blaisdell, G.A. (1991). Numerical simulations of compressible turbulence. PhD thesis, MAE, Stanford University, Stanford CA.

Dinavahi, S. P. G., Pruett, C.D., Zang, T. A., 1993. Direct numerical simulation and data analysis of a Mach 4.5 transitional boundary layer flow. Submitted to Phys. Fluids.

Dinavahi, S. P. G., Ristorcelli, J. R., and Erlebacher,G., 1993. Turbulence statistics relevant to compressible turbulence modeling. ICASE Report. To be submitted *Phys. Fluids.*

Erlebacher, G., and Sarkar, S., 1993. Statistical analysis of the rate of strain tensor in compressible homogeneous turbulence. *Phys. Fluids*, Vol. 5, No. 12, pp. 3240-3254.

Fernholz, H. H., and Finley, P. J., 1980. A critical commentary on mean flow data for two-dimensional compressible turbulent boundary layers. AGARD-AG-253.

Lee, S., 1992. Interaction of isotropic turbulence with a shock wave. PhD thesis, MAE, Stanford University, Stanford CA.

TESTING A MODEL FOR THE TURBULENT MASS FLUX

J. R. Ristorcelli

Institute for Computer Applications in Science and Engineering
NASA Langley Research Center
Hampton, Virginia 23681-0001

S. P. G. Dinavahi

Analytical Services & Materials, Inc.
NASA Langley Research Center
Hampton, Virginia 23681-0001

ABSTRACT

The turbulent mass flux, or equivalently the time mean of the Favre velocity fluctuation, appears in the evolution equations of compressible $k - \varepsilon$ and Reynolds stress closures. Mathematically it represents the difference between unweighted and density weighted time averages of the velocity field and is therefore a measure of the effects of compressibility. An algebraic truncation of its evolution equation, in which there appears only one term requiring closure, produces an algebraic stress type model for the mass flux. The mass flux is found to be proportional to the mean density gradients with a tensor eddy viscosity that depends on the mean velocity gradients and the Reynolds stresses. The representation of the mass flux is tested against DNS data of a wall bounded flow at $Ma = 4.5$.

1. Introduction

This article presents and tests an algebraic expression for the *non-zero first-moment* of the fluctuating velocity field, the Favre fluctuation mean, $< v_i >$. The need to consider this quantity is motivated by its frequent contributions in the first and second-order moment equations in flows with mean density gradients. As such the present discussion is pertinent to two-equation $k - \varepsilon$ type closures as well as the Reynolds stress closures.

In the compressible mean flow equations the mass flux makes a contribution to the viscous terms. This contribution, usually neglected as a result of assuming that the Reynolds-mean and Favre-mean velocities are approximately equal, is not appropriate as will

M. Y. Hussaini et al. (eds.), Transition, Turbulence and Combustion, Vol. II, 83–93.

be seen from DNS data to be presented. The ramifications of this assumption have been explored in Ristorcelli (1993a).

The algebraic model for the specific mass flux, or equivalently, the Favre fluctuation mean was derived in Ristorcelli (1993b) which also includes many observations on the mathematical nature of $< v_i >$.

2. Preliminary Exposition

The notation is now introduced. In general, upper case letters will be used to denote mean quantities except in the case of the mean density, $< \rho >$, since ρ which has no convenient upper case form. The averaging operation is indicated using the angle brackets for time means, $< uu >$, and the curly brackets for the density-weighted or Favre mean, $\{uu\}$, where $\{uu\} =< \rho^* uu >< \rho >^{-1}$ and the asterisk denotes the full field, $\rho^* =< \rho > +\rho'$. The dependent variables are decomposed according to

$$\begin{aligned} u_i^* &= U_i + u_i && \text{where } < u_i >= 0 \\ u_i^* &= V_i + v_i && \text{where } \{v_i\} = 0 \\ \rho^* &= < \rho > + \rho' && \text{where } < \rho' >= 0 \\ p^* &= P + p && \text{where } < p >= 0 \end{aligned}$$

where (U, u) and (V, v) are used to indicate the Reynolds and Favre decompositions of the velocity field. Both the Reynolds mean and the Favre mean velocities appear in the evolution equation for a compressible turbulence: it is, therefore, necessary to carry both the Favre and the Reynolds decompositions of the velocity field before any modeling approximations are made. The Favre variables appear naturally in terms involving body forces while the Reynolds variables appear naturally in surface force terms such as the pressure and the viscous stresses.

From the definitions above it is easily shown that the time-mean of the fluctuating Favre velocity, $< v_i >$, characterizes the difference between the Favre-expectation and time-expectation of the mean velocities, V_i and U_i, as well as the difference between the instantaneous fluctuating velocities:

$$\begin{aligned} U_i &= V_i + < v_i > \\ u_i &= v_i - < v_i > . \\ < \rho v_i > &= - < \rho >< v_i > . \end{aligned}$$

The last equality follow from the definition of the Favre-average of the Favre-deviation, $\{v_i\} = 0$. Thus the time-average of the fluctuating

Favre velocity and the mass flux are, apart from a scaling by the local mean density, equivalent quantities. The two phrases mass flux and Favre fluctuation mean will be used interchangeably. The primes on the fluctuating density have been dropped.

Mathematically $< v_i >$ represents the difference between unweighted and density weighted time-averages of the velocity field and is therefore a measure of the effects of compressibility through variations in density. It plays an important role in parameterizing the anisotropic effects of compressibility associated with the mean dilatation and gradients in the mean density. Experimentally it is an important and essential quantity that allows Favre-averaged numerical results to be related to time-averaged experimental results. These facts, discussed in more detail below, motivate the present investigation.

The mass fluxes appear in several places in the compressible turbulence equations. They contribute to 1). the viscous terms in the mean momentum equation and in the mean energy equation it contributes to 2). the viscous terms 3). the pressure work terms 4). the viscous work terms and the 5). pressure flux. Thus assuming that $U_i \simeq V_i$ is equivalent to neglecting the mass flux in five different locations in the mean equations. In the Reynolds stress equations the mass flux 6). determines the relative importance of a new production mechanism due the acceleration of the mean flow. 7). In the same term it also contributes to the mean viscous term. Also appearing in the mean energy equation and the Reynolds stress equations are the pressure fluxes. 8). The pressure fluxes are, in part, determined by the mass flux (in full if the fluctuations are assumed to be adiabatic). A more comprehensive discussion of these issues is given in Ristorcelli (1993a).

3. An Algebraic Expression for the Favre Fluctuation Mean

The details of a derivation of an evolution equation for the specific mass flux have been given in Ristorcelli (1993b). An algebraic truncation of the evolution equation, exact in the limit of homogeneous turbulence or in the limit of dominating production terms:

$$\begin{aligned} < v_i > (\mathcal{P} - \varepsilon)/k &= - < v_p > V_{i,p} + \{v_i v_p\} < \rho >_{,p} < \rho >^{-1} \\ &\quad + < v_i v_{p,p} > . \end{aligned}$$

Note that both the mean velocity and mean density gradients appear as production terms in the above expression. The equation has been obtained by assuming a structural equilibrium in which $D/Dt\ [< v_i >< v_j > /\{v_k v_k\}] = 0$ allowing convective derivatives, $D/Dt < v_i >$ to be expressed in terms of the right hand side of the evolution equation for the turbulence energy:

$$\frac{D}{Dt} < v_i > = \frac{< v_i >}{\{v_p v_p\}} \frac{D}{Dt} \{v_q v_q\} = < v_i > (\mathcal{P} - \varepsilon)/k.$$

It remains to close the last term on the right hand side. The correlation with the divergence is represented by a linear relaxation model. A linear relaxation with time scale τ_d

$$< v_i v_{p},_{p} > = - < v_i > /\tau_d$$

is chosen. The time scale, τ_d, in the model for $< v_i v_{p},_{p} > = - < v_i > /\tau_d$ may be thought of as a dilatational time scale. Preliminary results indicate that the general form of the model will not change. Computations using the acoustic time scale, $\tau_d = M_t k/\varepsilon$, to represent the dilatational time scale have been successful.

Substituting for the unknown correlation with the fluctuating divergence in the algebraic truncation of the evolution equation for $< v_i >$ produces

$$< v_p > (\delta_{ip} + \tau V_{i},_{p}) = \tau \{v_i v_p\} < \rho >,_p < \rho >^{-1}$$

where $\tau = (M_t k/\varepsilon)/(1 + M_t(\mathcal{P}/\varepsilon - 1))$. The model is now a set of three coupled linear algebraic equations of the form $A_{ip} < v_p > = b_i$: the mass flux in one direction is influenced by the mass flux in another direction. In addition, the contraction of the density gradient on the Reynolds stress allows for countergradient transfer. Performing the inversion the model can be written in symbolic form as

$$< v_i > = \tau \mathcal{T}_{ij} \{v_j v_p\} < \rho >,_p < \rho >^{-1}$$

where $\mathcal{T}_{ij} = (\delta_{ij} + \tau V_{i},_{j})^{-1}$. This is an anisotropic eddy viscosity model in which the eddy viscosity tensor $\nu_{jp} = \mathcal{T}_{ji}\{v_i v_p\}$ is a function of the Reynolds stresses and the mean flow gradients. Though this form suggests the structure of the model it is not the most suitable form for computation. Recourse to the Cayley-Hamilton theorem allows the inverse of a matrix to be written in terms of the first and second powers of the matrix:

$$III_A \mathbf{A}^{-1} = \mathbf{A}^2 - I_A \mathbf{A} + II_A \mathbf{1}.$$

Substituting $\mathbf{A} = \mathbf{1} + \tau\nabla\mathbf{V}$ produces an expression for the inverse

$$III_A(\mathbf{1} + \tau\nabla\mathbf{V})^{-1} = (1 - I_A + II_A)\mathbf{1} + (2 - I_A)\tau\nabla\mathbf{V} + \tau^2\nabla\mathbf{V}\nabla\mathbf{V}$$

and the final model can be written, in ascending powers of modified dilatational to mean flow time scales, as

$$< v_i >= \tau[\nu_0\delta_{ij} + \nu_1\tau V_{i,j} + \nu_2\tau^2 V_{i,k} V_{k,j}]\{v_j v_p\} < \rho >_{,p} < \rho >^{-1}$$

where the viscosity coefficients are functions of the invariants of, $\mathbf{A} = \mathbf{1} + \tau\nabla\mathbf{V}$, and are given by $\nu_0 = (1 - I_A + II_A)/III_A$, $\nu_1 = (2 - I_A)/III_A$, and $\nu_2 = 1/III_A$. The invariants are given by

$$\begin{aligned} I_A &= < A > \\ II_A &= 1/2(< A >^2 - < A^2 >) \\ III_A &= 1/6(< A >^3 - 3 < A >< A^2 > + 2 < A^3 >) \end{aligned}$$

for which $<>$ indicates the trace of the enclosed matrix. The time scale is defined as $\tau = (M_t k/\varepsilon)/(1 + M_t(\mathcal{P}/\varepsilon - 1))$. Note that all the viscosity coefficients are known for a given mean field; they are not phenomenological parameters that require experimental data. Only one phenomenological assumption - to obtain the relaxation model for the correlation with the fluctuating dilatation - has been made.

4. Implementations of the Mass Flux Model in Simple Flows

An expression for the model for the mass flux is now investigated in three limiting cases: in an isotropic turbulence with negligible velocity gradients, in an anisotropic turbulence with small velocity gradients and in an anisotropic turbulence with a unidirectional mean shear.

Case 1: Isotropic turbulence with small velocity gradients

In the case $\tau\nabla\mathbf{V} << 1$ and that the turbulence is isotropic the eddy viscosity tensor assumes the form $\tau\{v_j v_p\} \sim M_t(k/\varepsilon)2k\delta_{jp}$ and the model takes the form

$$< v_i >\sim M_t(k/\varepsilon)k < \rho >_{,i} < \rho >^{-1} .$$

This can be compared to the usual eddy viscosity model: $< v_i >= (\mu_T/ < \rho >^2 \sigma_\rho) < \rho >_{,j}$ in which $\mu_T = C_\mu f_\mu < \rho > k^2/\epsilon$ and thus

$$< v_i >\sim (k/\varepsilon)k < \rho >_{,i} < \rho >^{-1} .$$

The usual eddy viscosity form misses the M_t scaling which is necessary if the $< v_i >$ are to vanish in the absence of compressibility effects. Thus, apart from the M_t scaling, a scalar viscosity assumption will work, in the limit of an isotropic turbulence without mean velocity gradients. Note that this form in a boundary layer flow where the density gradient is cross-stream, cannot predict an axial mass flux. It only allows for a flux down the gradient transfer.

Another major shortcoming of the eddy viscosity assumption is realizability and its impact on computability in compressible closures. Gatski (1993) has used a scalar eddy viscosity representation and found it to be computationally destabilizing. Zeman and Coleman (1991) also point out that inadequate representations of the mass flux can destabilize computations in flows when the acceleration term is important. This occurs, for example, in the passage through a shock or in flows in which the mean strain or mean dilatation is important. This issue is treated in greater detail in Ristorcelli (1993b).

Case 2: Small velocity gradients, $\tau\nabla\mathbf{V} << 1$

In the case $\tau\nabla\mathbf{V} << 1$ and that the turbulence is anisotropic the expression for the mass flux becomes

$$< v_i >= \tau\{v_j v_p\} < \rho >_{,p} < \rho >^{-1} .$$

It is interesting to compare the expression for the mass flux to Zeman's (1993). Recalling that $< \rho v_i >= - < \rho >< v_i >$ Zeman's model is written

$$< \rho v_i >= -M_t(q^2/\epsilon)\{v_i v_p\} < \rho >_{,p} .$$

The present model and Zeman's model, in the limit of small velocity gradients and for an equilibrium turbulence in which $\mathcal{P} = \varepsilon$, are equivalent. Comparisons of this model with the DNS data in a wall bounded compressible flow shows that it does not successfully capture the results known from DNS. The neglected terms involving the velocity gradients, in as much as they represent production terms in the evolution equation for the mass flux, are essential. Computations with the neglected velocity gradient capture the near wall behavior very nicely, as will be seen in the next section.

Case 3: Simple mean shear

For the simple shear a surprisingly simple expression for the mass flux is possible. In the simple shear $V_{i,j} = V_{1,2}\,\delta_{i1}\delta_{j2}$. The invariants of **A** are $I_A = 3$, $II_A = 3$, $III_A = 1$ and the viscosity coefficients take on the following shockingly simple values $\nu_0 = 1, \nu_1 = -1, \nu_2 = 1$ and, as the square of the density gradient is zero, the mass flux model becomes

$$< v_i >= \tau[\delta_{ij} - \tau V_{1,2}\,\delta_{i1}\delta_{j2}]\{v_j v_p\} < \rho >_{,p} < \rho >^{-1} .$$

The streamwise and cross-stream components of the Favre fluctuation mean, in a flow with only a cross-stream density gradient, become

$$\begin{aligned} < v_1 > &= \tau[\{v_1 v_2\} - \tau V_{1,2}\{v_2 v_2\}] < \rho >_{,2} < \rho >^{-1} \\ < v_2 > &= \tau\{v_2 v_2\} < \rho >_{,2} < \rho >^{-1} . \end{aligned}$$

Note that the production terms, proportional to $V_{1,2}$, are included in the expressions and thus the streamwise mass flux will be larger than the cross-stream as is the case for the normal Reynolds stresses in a unidirectional shear.

5. Results and Discussion

Compressible turbulence models typically invoke the approximation $U_i \simeq V_i$ to close the mean momentum and energy equation. In the mean momentum equations this assumption affects the viscous terms; in the mean energy equation this assumption affects the work done by the mean pressure forces and the mean viscous stresses. The approximation $U_i \simeq V_i$ involves assuming that $< v_i ><< V_i$ as $U_i = V_i + < v_i >$. The DNS of a wall boundary layer after transition of Dinavahi *et al.* (1993) was used to test this assumption. In their simulation $Ma = 4.5$ and $Re_\delta = 10^4$. Their DNS code is temporal: the computation is started with a laminar flow that subsequently undergoes transition. At this point in time the simulation is only three eddy turnover past transition: $\varepsilon\delta x/(kU) \simeq 3$. The turbulence field is not what one might call fully developed. However three eddy turnovers past transition will be enough for our purposes.

Figure 1 shows the three velocities that enter the relationship $U_i = V_i + < v_i >$. From the figure the approximate relationship $U_1 \simeq V_1$ used as a modeling assumption is justified. Note that the non-zero streamwise mass flux, $< \rho v_1 > = - < \rho > < v_1 >$ is in the same direction as the mean flow even though the mean density gradient is cross-stream. From the figure it is clear the approximate relationship $U_2 \simeq V_2$ used as a modeling assumption is not justified. Not only are the two mean velocities not similar they are not even in the same direction. This is due to the inhomogeneity of the mean and fluctuating density field and shows that the mean transverse momentum transfer, V_2, is in the direction *opposite* to the mean fluid particle motion, U_2. This means that eddy viscosity type arguments used to close unknown terms must be made much more carefully. This unexpected, but not unsurprising, phenomena will be seen to be the typical situation in complex compressible flows for the mean velocities along the mean density gradients.

Figure 2 shows how the algebraic representation for the mass flux, given in case 3 of the previous section and derived in Ristorcelli (1993b), compares to the DNS data. The model predictions for the streamwise mass flux are surprisingly good. The peak is captured very well in location and magnitude. This is because the production terms, proportional to $V_{1,2}$, are large and captured exactly. Note that a streamwise mass flux is engendered by a cross-stream density gradient. This is a behavior that a simple (isotropic) eddy viscosity model cannot predict.

The cross-stream component, $< v_2 >$, using this form is less successful. This primarily because there are no large production terms in the expression for $< v_2 >$ and the sensitivity to the assumed linear relaxation approximation for the dilatational correlation becomes manifest. The linear relaxation approximation used to close the term $< v_i v_{p,p} >$ is expected to be most effective in an equilibrium turbulence. The present turbulence is only three eddy turnover past transition and it is clear, from data shown in Ristorcelli *et al.* (1993), that the effects of the initial conditions have not yet been fully dispersed by the turbulence.

We had not expected the mass flux to be important in this unidirectional flow: as the wall is approached $< v_2 > /V_2 \simeq 0.25$. Apparently the four fold variation in mean density across the boundary layer is large enough to create a non-negligible mass flux. The mass flux is expected to be important in flows with large density gradients (combustion), separation (inflection points), in cold wall flows, and in flows with mean dilatation, shocks, adverse pressure gradients, or strong streamwise accelerations.

References

Dinavahi, S. P. G. and Pruett, C. D., 1993. "Analysis of direct numerical simulation data of a Mach 4.5 transitional boundary layer flow," ASME Fluids Engineering Conference, June 21-24, Washington DC.

Gatski, T., 1993. Private communication.

Lee, S., 1992. "Interaction of isotropic turbulence with a shock wave," Ph.D. Thesis, MAE, Stanford University, Stanford, CA.

Ristorcelli, J. R., 1993a. "Carrying the mass flux terms in the first and second moment equations of a compressible turbulence," ICASE report, submitted to *Phys. Fluids.*

Ristorcelli, J. R., 1993b. "A representation for the turbulent mass flux contribution to Reynolds-stress and two-equation closures of compressible turbulence."

Ristorcelli, J. R., Dinavahi, S. P. G., and Erlebacher, G. E., 1993. "Turbulence statistics relevant to compressible turbulence modeling," ICASE report, to be submitted to *Phys. Fluids.*

Zeman, O., 1993. "A new model for super/hypersonic turbulent boundary layers," AIAA 93-0897, 31st Aerospace Sciences meeting, Reno, NV.

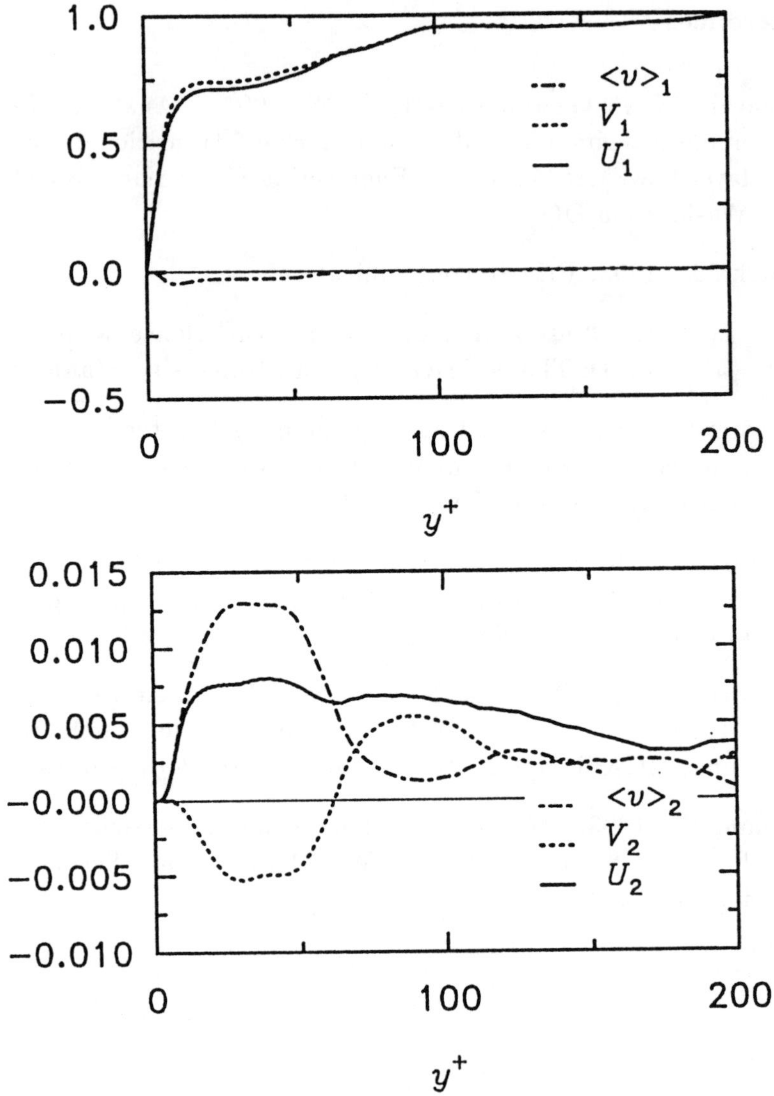

Figure 1. The Reynolds mean, Favre mean and Favre fluctuation mean velocity profiles.

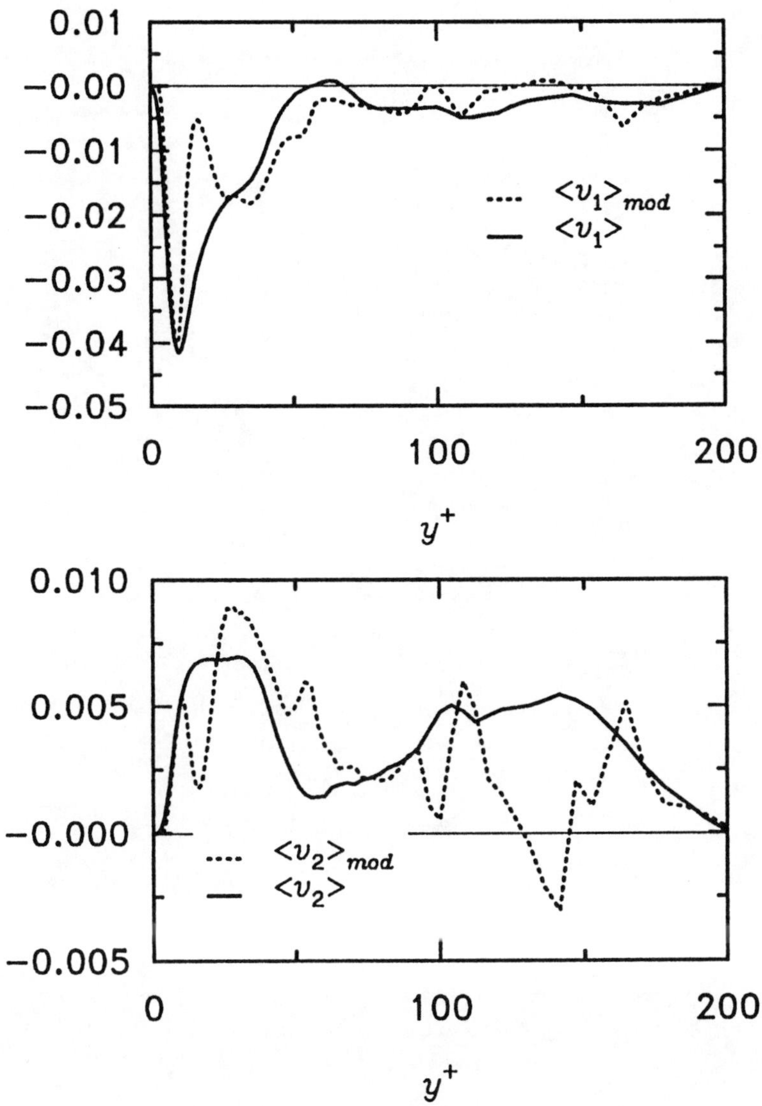

Figure 2. The Favre fluctuation mean as computed using the DNS and as predicted by the model.

ASSESSMENT OF THE SSG PRESSURE-STRAIN MODEL IN COMPLEX TURBULENT SHEAR FLOWS

Bassam A. Younis[1]

Department of Civil Engineering
City University
London EC1V 0HB - United Kingdom

Thomas B. Gatski

NASA Langley Research Center
Hampton, Virginia 23681-0001

Charles G. Speziale[1]

Department of Aerospace and Mechanical Engineering
Boston University
Boston, Massachusetts 02215

ABSTRACT

The pressure-strain model of Speziale, Sarkar, and Gatski (1991) is here employed as part of a complete second-order closure and is used to predict the behavior of a number of complex turbulent shear flows involving rotational body forces and streamline curvature in lateral and longitudinal planes. Comparisons are made with experiments and those that show the model, unmodified in any way and used without a wall-reflection term, succeeds in predicting the correct response of the mean-flow and the turbulence fields to the imposed distortion.

1. Introduction

It is now widely recognized that second-order closure models are essential for the accurate prediction of the majority of turbulent flows encountered in Aeronautics. However, the acceptance and serious use of such models by industry has so far been very limited partly because of the computational effort involved in solving seven differential transport equations but also, and more fundamentally, because

[1]Research was supported by the National Aeronautics and Space Administration under NASA contract No. NAS1-19480 while the authors were in residence at the Institute for Computer Applications in Science and Engineering (ICASE), NASA Langley Research Center, Hampton, VA 23681-0001.

M. Y. Hussaini et al. (eds.), Transition, Turbulence and Combustion, Vol. II, 95–104.

of the use in such models of a "wall reflection" term to account for the modifying effects of a rigid wall on the fluctuating pressure field in its vicinity. This term invariably involves an empirical wall-damping function, the evaluation of which, while straightforward for a uni-directional flow over a single flat wall, becomes difficult in the presence of corners and nearly impossible when complex geometries are involved. The ability to represent the effects of the fluctuating pressure field without the use of the troublesome wall-reflection term must therefore be regarded as prerequisite to the routine use of second-order closure models in general-purpose CFD codes.

Of the numerous proposals for modeling the difficult fluctuating pressure-strain described in the literature, that of Speziale, Sarkar, and Gatski (1991) satisfies certain conditions which are necessary for a model to be equally applicable to both wall-bounded and homogeneous shear flows. Those conditions are set out in Abid and Speziale (1993) and a practical demonstration of the model's performance in a simple wall-bounded flow has been provided by Demuren and Sarkar (1993) who applied it to the fully-developed flow in a straight channel. The present contribution extends the model assessment to turbulent shear flows whose structure is distorted by the imposition of body forces or by rates of strain additional to the usual mean shear. We report below the results for three very different cases: the flow in a rotating channel, a two-dimensional boundary layer developing over a convex-curved surface and a three-dimensional boundary layer developing over an infinite swept wing. Two further test flows were considered in the 1993 ICASE/LaRC Workshop on Transition, Turbulence, and Combustion: the three-dimensional flow over an axially-rotated cylinder and the swirling free jet. The results for those cases are reported separately.

2. Computational Details

The Reynolds-averaged equations for steady, incompressible turbulent flow are given in Cartesian tensor notation as:

$$\frac{\partial U_j}{\partial x_j} = 0 \tag{1}$$

$$\rho\frac{\partial U_i}{\partial t} + \rho U_j\frac{\partial U_i}{\partial x_j} = \frac{\partial}{\partial x_j}(\mu\frac{\partial U_i}{\partial x_j} - \rho\overline{u_i u_j}) + F_i - \frac{\partial P}{\partial x_i}. \tag{2}$$

The transport equation for $\overline{u_i u_j}$ can be written as:

$$U_k \frac{\partial \overline{u_i u_j}}{\partial x_k} = \overbrace{P_{ij}}^{production} + \overbrace{D_{ij}}^{diffusion} + \overbrace{\Phi_{ij}}^{pressure-strain} - \overbrace{\epsilon_{ij}}^{dissipation} . \tag{3}$$

Diffusion is here assumed to be entirely by turbulent fluctuations and is modeled by a gradient-transport hypothesis. Dissipation is assumed to be isotropic at high turbulence Reynolds number and the pressure-strain term is approximated with the standard SSG formulation. The complete modeled $\overline{u_i u_j}$ equation now reads:

$$\begin{aligned} U_k \frac{\partial \overline{u_i u_j}}{\partial x_k} &= C_s \frac{k}{\epsilon} \frac{\partial}{\partial x_k} \left(\overline{u_k u_l} \frac{\partial \overline{u_i u_j}}{\partial x_l} \right) - \left(\overline{u_i u_k} \frac{\partial U_j}{\partial x_k} + \overline{u_j u_k} \frac{\partial U_i}{\partial x_k} \right) \\ &- (C_1 \epsilon + C_1^* P_k) b_{ij} + C_2 \epsilon (b_{ik} b_{kj} - \frac{1}{3} b_{mn} b_{mn} \delta_{ij}) \\ &+ [C_3 - C_3^* (b_{mn} b_{mn})^{\frac{1}{2}}] k S_{ij} \\ &+ C_4 k \left(b_{ik} S_{jk} + b_{jk} S_{ik} - \frac{2}{3} b_{mn} S_{mn} \delta_{ij} \right) \\ &+ C_5 k (b_{ik} W_{jk} + b_{jk} W_{ik}) - \frac{2}{3} \delta_{ij} \epsilon \end{aligned} \tag{4}$$

where b_{ij}, S_{ij} and W_{ij} are, respectively, the turbulence anisotropy, the mean rate of strain and the mean vorticity tensors.

The scalar energy-dissipation rate is obtained from the solution of the standard equation:

$$U_j \frac{\partial \epsilon}{\partial x_j} = \frac{\partial}{\partial x_j} \left(C_\epsilon \frac{k}{\epsilon} \overline{u_j u_l} \frac{\partial \epsilon}{\partial x_l} \right) + C_{\epsilon_1} \frac{\epsilon}{k} P_k - C_{\epsilon_2} \frac{\epsilon^2}{k} . \tag{5}$$

The complete model contains a number of coefficients which are assigned values recommended by the originators, viz

C_1	C_2	C_1^*	C_3	C_3^*	C_4	C_5	C_{ϵ_1}	C_{ϵ_2}	C_s	C_ϵ
3.4	4.2	1.8	0.8	1.3	1.25	0.4	1.44	1.83	0.22	0.183

The mean-flow and the turbulence-model equations are solved simultaneously with a finite-volume procedure for boundary-layer flows. The inlet boundary conditions are taken from the experiments. Near the wall, the numerical solutions of the momentum equations are matched to the logarithmic "law of the wall." Also, the gradients of all the turbulent stresses in the direction normal to the wall are set to zero (consistent with Couette flow assumptions) while the boundary values for ϵ are fixed from local equilibrium considerations.

3. Results and Discussion

3.1. Flow in a rotating channel

Data for model assessment are provided by the experiments of Johnston, Halleen, and Lezius (1972) in a wide aspect-ratio rectangular channel rotated about a spanwise axis. Fully-developed flow conditions are assumed and the rotational terms are therefore present only in the equations for the turbulent stresses. Those terms act as either 'sources' or 'sinks' – destabilizing the flow on the pressure side, where turbulence activity is increased, and stabilizing it on the suction side where turbulence is suppressed or even extinguished altogether for sufficiently high rotation rates. This in turn distorts the distribution of mean velocity in the channel where the position of maximum velocity is shifted towards the suction side. This can be seen very clearly in Figure 1 where results for two Rossby numbers (Ro=$\Omega D/U_{ave}$) are presented. Figure 1 also shows comparisons with non-rotating channel data as well as with the predictions of Launder, Tselepidakis and Younis (1987) obtained with the Gibson-Launder (1978) pressure-strain model (dashed line) which required substantial wall-damping modification. Overall, the two closure models yield virtually identical results which are in very close accord with the measurements except, perhaps, for the high rotation case where roll instabilities may well have been present in the experiments.

3.2. Boundary layer over convex curved surface

In the experiment of So and Mellor (1973), a two-dimensional boundary layer is initially developed over a flat plate and then on a convex-curved wall with longitudinal radii of curvature that vary with streamwise distance. This flow therefore provides a severe test for the pressure-strain model as it highlights its response to the sudden introduction, and the sustained application, of an extra rate of strain. The stabilizing effects of streamline curvature cause the wall skin-friction coefficient to fall well below its flat-plate value as can be seen in Figure 2 where the SSG and GL (dashed line) models are compared with the data. The figure also shows the cross-stream distribution of mean velocity and turbulent shear stress at the last measurement station in the curved section. The complete collapse of the turbulent shear stress in the outer part of the flow, where the mean shear is still finite but the curvature effects are strongest, is particularly well predicted. The predicted and measured profiles of the normal stresses at the same streamwise location are compared in Figure 3. The extensive reduction in those quantities wrought by the curvature is reasonably well captured by the SSG model though some quantitative differences are apparent. The figure also shows comparative results obtained with the GL pressure-strain model (dashed line) in which the use of a standard wall-reflection term produces normal-stress distributions that are in serious error in the near wall region.

3.3. Boundary layer over an infinite swept wing

Bradshaw and Pontikos (1985) obtained detailed measurements in a three-dimensional boundary layer generated under infinite-swept wing conditions with a 35 degree angle of sweep. The development of the flow is simultaneously influenced by significant adverse pressure gradients and by streamline curvature in the lateral planes. The pressure gradient effects are felt more strongly by the slow moving fluid near the wall and hence the resultant velocity profile is very skewed, with an angle of skewness relative to the free stream that varies with height. The predicted and measured variations of this angle with height are compared in Figure 4. The maximum flow skewness occurs near the wall and is predicted by SSG to be at 16.5 degrees which is slightly below the measured value. Another aspect of the present flow which is reproduced here is the misalignment between the directions of the turbulent shear stresses and the corresponding rates of strain. This is an important feature of three-dimensional flows generally and one which invalidates basic assumptions in mean-field closure models. Figure 4 shows that the model captures the skewness of the mean rates of strain fairly accurately across the whole flow but does not succeed in reproducing the correct degree of skewness of the turbulent stresses except, perhaps, in the outer region of the boundary layer. The GL model (dashed line) with wall damping yields similar results to the SSG model which suggests that this feature of the predictions may well be due to some other closure assumption (e.g. turbulent transport or dissipation) or even to the use of the standard log-law in a flow involving strong adverse pressure gradients.

4. Concluding Remarks

This study demonstrates that the SSG model fairly closely predicts the main features of a number of complex turbulent shear flows without modification and without resort to the troublesome wall-reflection terms found necessary in alternative second-order closures. The model is therefore particularly suited for use in general CFD codes though its performance in adverse pressure-gradient flows argues in favor of its extension to incorporate the effects of low Reynolds number near-wall turbulence.

References

Abid, R. and Speziale, C. G., 1993. "Predicting equilibrium states with Reynolds stress closures in channel flow and homogeneous shear flow," *Phys. Fluids A* **5**, p. 1776.

Bradshaw, P. and Pontikos, N. S., 1985. "Measurements in the turbulent boundary layer on an infinite swept wing," *J. Fluid Mech.* **159**, p. 105.

Demuren, A. O. and Sarkar, S., 1993. "Systematic study of Reynolds stress closure models in the computations of plane channel flows," *ASME J. Fluids Eng.* **115**, p. 5.

Gibson, M. M. and Launder, B. E., 1978. "Ground effects on pressure fluctuations in the atmospheric boundary layer," *J. Fluid Mech.* **86**, p. 491.

Johnston, J. P., Halleen, R. M., and Lezius, D. K., 1972. "Effects of spanwise rotation on the structure of two dimensional fully developed turbulent channel flow," *J. Fluid Mech.* **56**, p. 533.

Launder, B. E., Tselepidakis, D. P., and Younis, B. A., 1987. "A second-moment closure study of rotating channel flow," *J. Fluid Mech.* **183**, p. 163.

So, R. M. C. and Mellor, G. L., 1973. "Experiments on convex curvature effects in turbulent boundary layers," *J. Fluid Mech.* **60**, p. 43.

Speziale, C. G., Sarkar, S., and Gatski, T. B., 1991. "Modeling the pressure-strain correlation of turbulence: An invariant dynamical systems approach," *J. Fluid Mech.* **227**, p. 245.

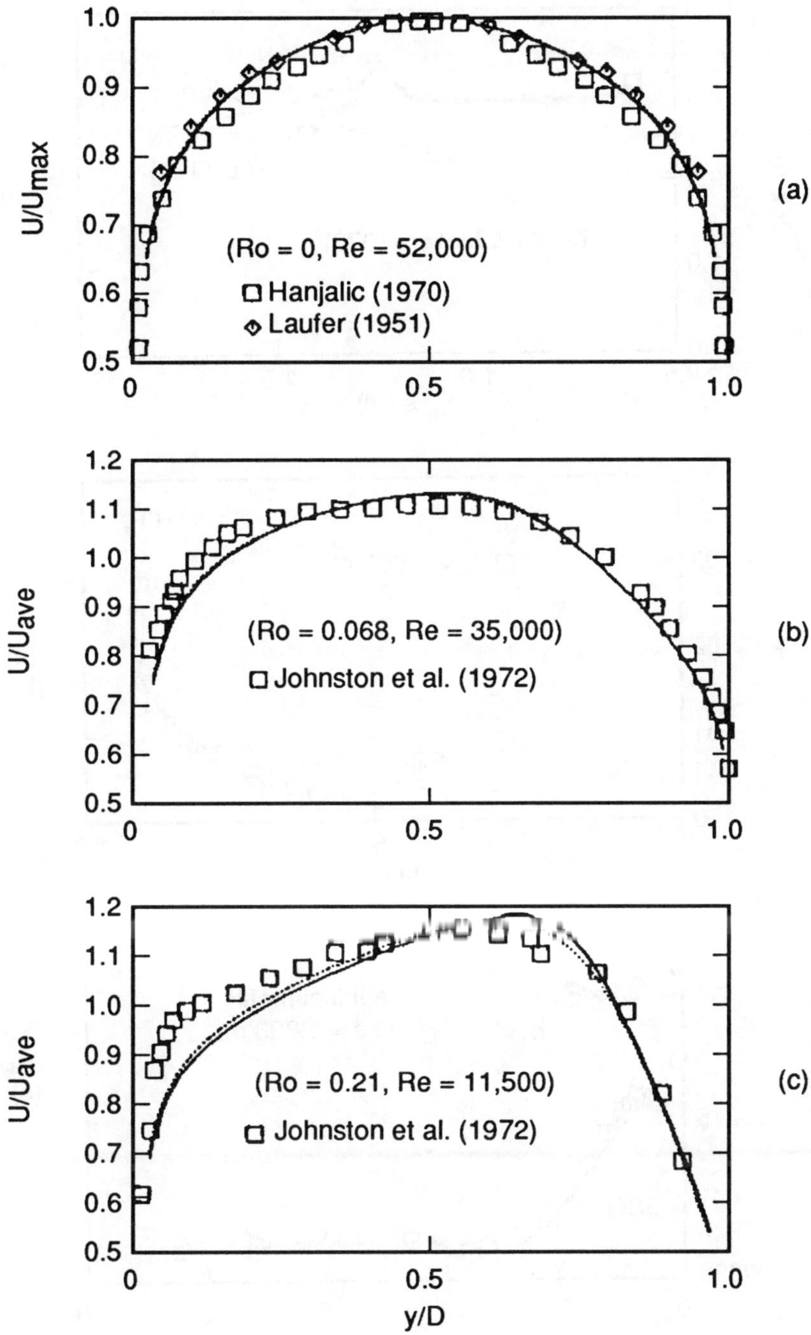

Figure 1. Effect of rotation on mean-velocity distribution.

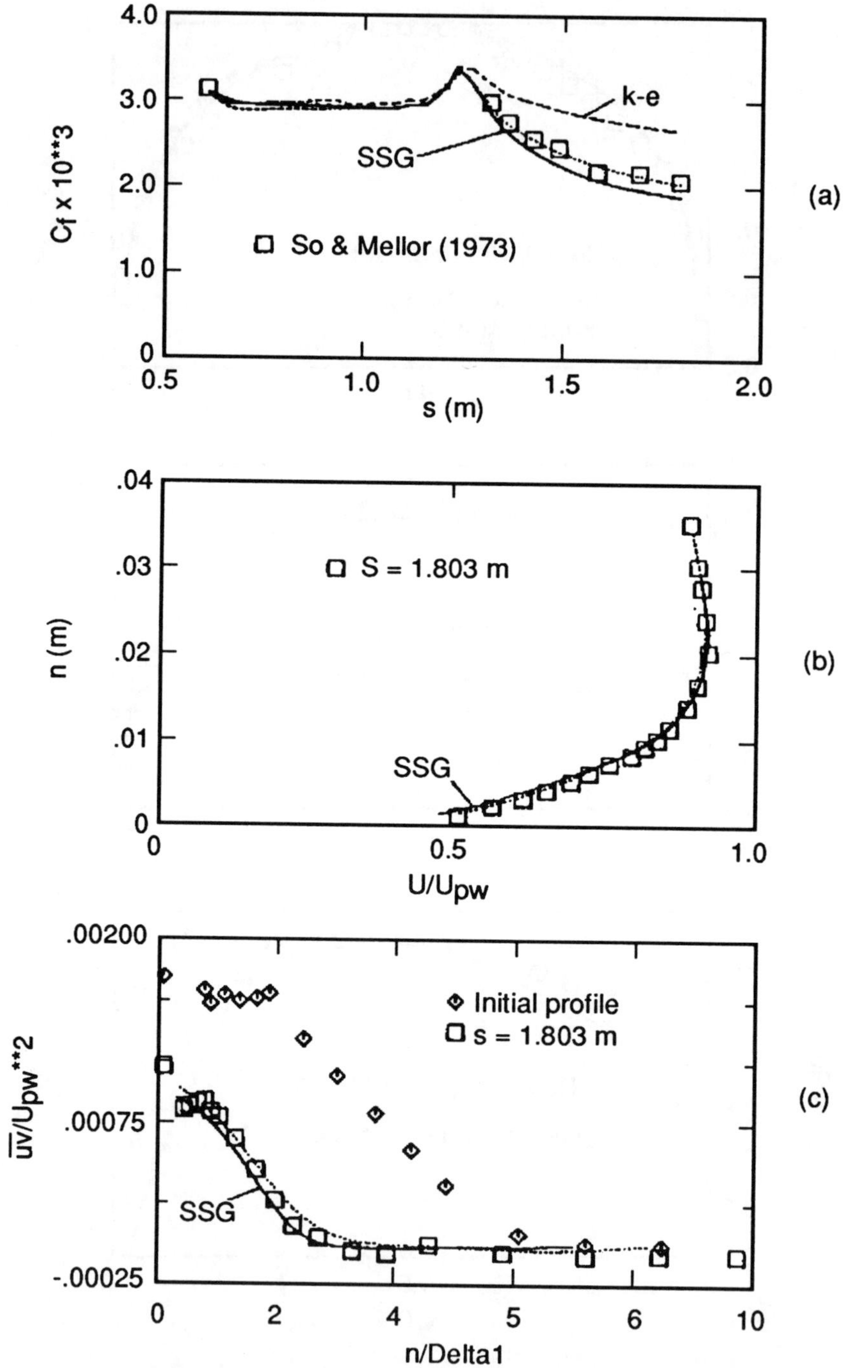

Figure 2. Skin-friction, mean velocity and turbulent shear-stress distribution in curved boundary layer.

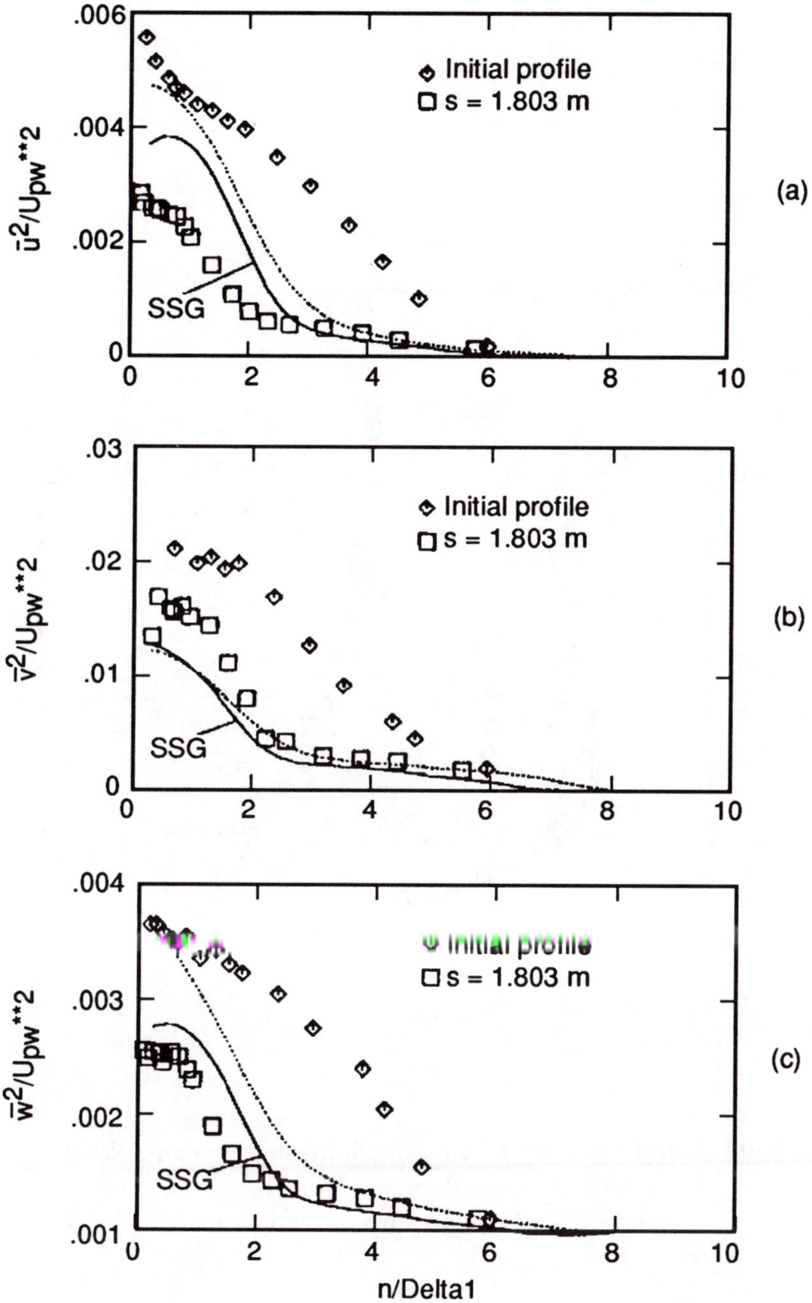

Figure 3. Predicted and measured normal turbulent stresses in curved boundary layer. Data of So and Mellor (1973).

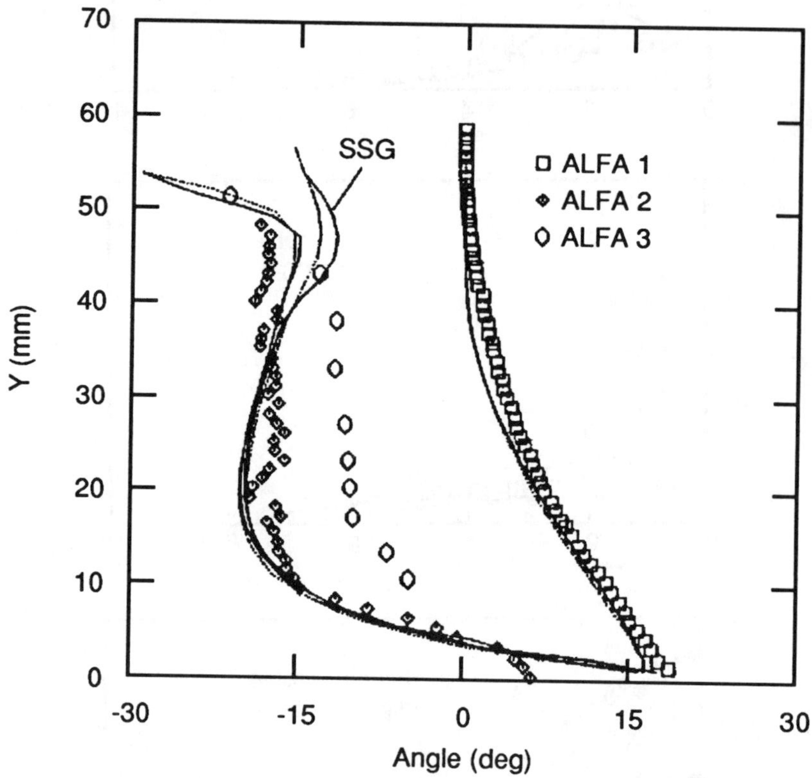

Figure 4. Infinite swept wing. Inclination of mean-velocity, turbulent shear stress and mean rate of strain vectors. Data of Bradshaw and Pontikos (1985).

RENORMALIZATION GROUP THEORY AND TURBULENCE MODELING

GROUP SUMMARY: RENORMALIZATION GROUP THEORY AND TURBULENCE MODELING

Ye Zhou

Institute for Computer Applications in Science and Engineering
NASA Langley Research Center
Hampton, Virginia 23681-0001

The technique of renormalization group (RNG) theory has achieved great success in the field of critical phenomena. Indeed, Professor K. G. Wilson was awarded the 1982 Nobel Prize in Physics. Recently, the methodology of RNG has attracted considerable attention as a systematic approach to subgrid scale modeling in large-eddy simulations and Reynolds stress modeling. The broad scientific objectives of this group were to improve the understanding of the fundamentals of renormalization group approaches and to use this knowledge to construct better models for complex turbulent flows of technological interest. The RNG theory itself is highly mathematical and can be divided into ϵ-RNG, recursive RNG, and conditional average RNG according to the various assumptions and approximations. All of these approaches were considered during the Workshop. The group consisted of G. Berkooz (Cornell University), S. H. Lam (Princeton University), J. Lee (Wright Patterson Air Force Base), W. D. McComb (University of Edinburgh, UK), R. Rubinstein (ICOMP, NASA Lewis Research Center), S. Thangam (Stevens Institute of Technology), G. Vahala (College of William and Mary), and Y. Zhou (ICASE, NASA Langley Research Center). D. N. Riahi (University of Illinois) who was a member of the advanced asymptotics group also contributed a paper for inclusion.

The paper by Zhou, Vahala, McComb, and Lam is a direct consequence of the fruitful and animated discussions that took place at these meetings. This paper gives a brief introduction to renormalization group, as applied successfully to the theory of phase transitions, and sketches the history of the attempts to apply the method to turbulence. Authors also presented a succinct statement, and critical assessment, of current RNG theories of turbulence.

Berkooz, Elezgaray, Vahala, and Zhou made some preliminary investigations into the possibility of applying RNG theory to nonhomogeneous flows and wall bounded flows. By using a basis compatible

M. Y. Hussaini et al. (eds.), Transition, Turbulence and Combustion, Vol. II, 107–108.

with the production of turbulence near a wall and with the boundary conditions, this approach has the potential of creating RNG based models which will be valid at the wall and will capture more of the physics of the flow.

Lam examined the attempts by ϵ-RNG theory to provide quantitative information on the decay law of homogeneous isotropic turbulence. This paper is followed by the Comment of V. Yakhot (Princeton University) and Lam's reply to this comment.

Rubinstein discussed the connection between the ϵ expansion and the distant interaction approximation in the ϵ-RNG.

Thangam, Zhou, and Vahala analyzed a recursive RNG based turbulence model for Reynolds stress. The model used a finite wavenumber truncation scheme to account for the spectral distribution of energy and included both local and nonlocal interactions. Turbulent flow past a backward facing step is chosen as a test case and the results based on detailed computations demonstrate the viability of the proposed model.

D. N. Riahi, applied ϵ-RNG theory to an incompressible linearly stratified uniform flow past an obstacle. Using this RNG procedure, both the eddy viscosity and eddy mass diffusivity are determined.

A special panel discussion, organized by Y. Zhou, was held during the Workshop on 29 June, 1993. The theme of the panel discussion was to provide an open discussion of the fundamentals of RNG approaches and to review the status of RNG methods for turbulence modeling. S. A. Orszag and V. Yakhot (Princeton University) came to ICASE on June 29 specifically to participate in this panel discussion. The panel discussion consisted of presentations by Y. Zhou, V. Yakhot, S. A. Orszag, and G. Vahala along with a general discussion moderated by W. D. McComb. The report by Zhou and Speziale (Boston University) summarizes the activities and provides an overview of the RNG methods in turbulence modeling.

RNG ON WAVELET AND NON FOURIER REPRESENTATIONS - APPLICATIONS TO NONHOMOGENEOUS AND WALL BOUNDED FLOWS[1]

Gal Berkooz

Center for Applied Math.
and Dept. of Mechanical and Aerospace Engineering
Cornell University
Ithaca, New York 14853

Juan Elezgaray

CRPP/CNRS Bordeaux, France

George Vahala

College of William and Mary
Williamsburg, Virginia 23185

Ye Zhou

Institute for Computer Applications in Science and Engineering
NASA Langley Research Center
Hampton, Virginia 23681-0001

ABSTRACT

We make some preliminary observations on the possibility of applying RNG theory to nonhomogeneous flows and wall bounded flows. Non Fourier basis play a prominent role in this program. With a wavelet basis we are able to capture more of the physics of a nonhomogeneous flow. Specifically, the wavelet representation allows for different physical length scales at different locations in the flow field. By using a basis compatible with the production of turbulence near a wall and with the boundary conditions, we have the potential of creating RNG based models which will be valid at the wall and capture

[1]The work of G.B was supported by AFOSR 49620-92-J-0287. The work by G.V. and Y.Z. was supported by the National Aeronautics and Space Administration under NASA Contract No. NAS1-19480 while authors were in residence at the Institute for Computer Applications in Science and Engineering (ICASE), NASA Langley Research Center, Hampton, VA 23681. G.V. was also supported in part by DOE.

M. Y. Hussaini et al. (eds.), Transition, Turbulence and Combustion, Vol. II, 109–118.

more of the physics of the flow. A good candidate basis is the basis of energy-stability eigenfunctions. The eigenfunctions of the energy stability problem have been shown to be close to the POD eigenfunctions. However, the energy stability eigenfunctions are computable from coarse features of the flow.

We observe that since the Fourier basis is the only basis in which the derivative of a mode is multiple of the mode itself and the multiplication of two modes may be expressed as sums or differences of two other modes using a non Fourier basis complicates the RNG procedure. This implies the necessity of a computer in performing the RNG calculations, prior to the simulation of the resultant models.

1. Introduction

The use of a Fourier basis in a RNG theory (Yakhot and Orszag, 1986; Zhou et al., 1988, 1989; Zhou and Vahala, 1992, 1993) depends on homogeneity in two crucial points. The first point is the assumption that the length scales of the flow are uniform throughout physical space. If they are not uniform it is not consistent to use a cutoff based on global spatial averages as done in the elimination procedure of Fourier modes. Furthermore, the definition of "energy containing scales" for the whole flow is not clear, as the scales vary from point to point in physical space. The second point is in the assumption that the two point correlation tensor in a Fourier representation is diagonal, or equivalently that the time dependent coefficients of different Fourier modes are uncorrelated. This implies the flow is homogeneous, see (Lumley, 1971; Berkooz et al., 1993). For wall bounded flows the problem is further complicated by the fact that most algebraic stress-strain relationships do not satisfy the boundary conditions at the wall. The result of this deficiency is the introduction of empirical "fixes" such as wall functions etc.

The use of a non Fourier basis may overcome some of these problems. Specifically, the use of a wavelet basis allows for the introduction of different length scales at different locations in physical space while retaining the spectral (wave space) structure required for a systematic elimination of smaller scales. However, a complication is introduced in that the two point correlation tensor is not diagonal in the wavelet representation. We will show in the sequel that this introduces additional coupling terms and complicates the RNG procedure. From experience with a one dimensional model problem, the Kuramoto Sivashinsky Equation, we believe the new coupling terms will be small, since the two point correlation tensor is "close" to diagonal (i.e. the off diagonal entries are small).

The use of an eigenfunction basis such as the POD basis (Lumley, 1971; Berkooz et al., 1993) has the advantage that the two point correlation tensor will be diagonal in both non homogeneous and wall bounded flows. However, the calculation of the POD basis requires extensive experimental input. Recently (Poje and Lumley, 1993) have shown that the basis obtained by solution of the energy stability problem may give a good approximation to the POD basis, at least in the wall region of the boundary layer. The energy stability eigenfunctions are computable from coarse properties of the flow. Hence, an RNG procedure based on this basis has the potential of capturing the physics of near wall flows and having a consistent stress strain relationship at the wall. In this note we will not pursue eigenfunction based RNG further.

2. RNG in a Wavelet Representation

We start with a decomposition of the velocity field into wavelets:

$$\mathbf{u}(\mathbf{x}, t) = \sum_{l,m} a_{l,m}(t)\psi_{l,m}(\mathbf{x}). \tag{1}$$

The wavelets $\psi_{l,m}(\mathbf{x})$ have the property of localization in scale (denoted by the index l) and space (denoted by the index m). We will not attempt a review of wavelets, for this the reader is referred to the review (Farge, 1992) and references therein.

2.1. The recursive RNG formalism applied to a wavelet representation

The derivative of a wavelet includes contributions from all wavelet scales. Specifically if one considers compactly supported wavelets, or wavelets with an exponential decay in physical space, as would be the case in this work, the Fourier spectrum of the wavelet will decay algebraically. This implies that the *wavelet* spectrum of the derivative of a wavelet will decay algebraically. For simplicity of the exposition we will proceed with a one dimensional example, the Kuramoto-Sivashinsky Equation.

The Kuramoto-Sivashinsky Equation is given by

$$u_t + \frac{1}{\mu} u_{xxxx} + u_{xx} + u u_x = 0,$$

$$u \text{ periodic on } [0, L].$$

(An equivalent form has $\frac{1}{2}(u_x)^2$ as the nonlinearity).

In the wavelet basis Ψ_{lm}, the KS equation takes the form

$$\dot{a}_{lm} = L(lm|l'm')a_{l'm'} + N(lm|l'm'|l''m'')a_{l'm'}a_{l''m''} \tag{2}$$

where the linear term

$$L(lm|l'm') \equiv - < \Psi_{lm}|\Psi_{l'm',xx} + \Psi_{lm}|\Psi_{l'm',xxxx} > \tag{3}$$

and the nonlinear term

$$N(lm|l'm'|l''m'') \equiv < \Psi_{lm}|\Psi_{l'm'}\Psi_{l''m'',x} > . \tag{4}$$

The l-label indicates the particular wavelet generation while the m-label gives specific spatial localization information within that particular wavelet generation. It is convenient to order the wavelet generations in standard RNG notation $l_0 > l_1 > ....$, where l_0 is the next higher generation from l_1. (Note that this notation is just the opposite of the one used in the wavelet literature!)

The two point correlation tensor is not diagonal, i.e. the coefficients of the wavelet modes are correlated. This is a reflection of the fact that for the Kuramoto-Sivashinsky equation, whose statistics are known to be homogeneous (see Pomeau et al. (1984), Zaleski (1989), Kevrekidis et al. (1990)) the energy optimal basis is the Fourier basis. However, using uncorrelatedness of the Fourier basis and the algebraic decay of the power spectrum of the wavelets in Fourier space we can conclude that the correlations of the wavelet modes decay algebraically, with an exponent that grows as the degree of smoothness of the wavelet increases. In Berkooz et al. (1993) the two point correlation tensor for the Kuramoto-Sivashinsky equation represented in spline wavelets with exponential decay in physical space was studied in detail, numerically and analytically. The class of spline wavelets used in Berkooz et al. (1993) is described in Perrier and Basdevant (1989).

Keeping the two properties of the wavelet basis in mind we perform one step of small scale elimination, removing the highest wavelet generation l_0 (i.e the first subgrid shell in RNG terminology) we have

$$\dot{a}^{>}_{l_0 m} = L(l_0 m|l'm')[a^{>}_{l_0 m'}\delta(l_0 - l') + a^{<}_{l'm'}]$$

$$+N(l_0 m|l'm'|l''m'')[a^{>}_{l_0 m'}\delta(l_0 - l') + a^{<}_{l'm'}][a^{>}_{l_0 m''}\delta(l_0 - l'') + a^{<}_{l''m''}], \quad (5)$$

where superscript $>$ on the wavelet expansion coefficient refers to the subgrid generation that is being removed. The superscript $<$ refers to those wavelet generations that are resolvable. Assume rapid time variation of small scale modes:

$$0 = L(l_0 m|l_0 m')a^{<}_{l_0 m'} + L(l_0 m|l'm')a^{<}_{l'm'}$$

$$+N(l_0 m|l_0 m'|l_0 m'')a^{>}_{l_0 m'}a^{>}_{l_0 m''} + N(l_0 m|l_0 m'|l''m'')a^{>}_{l_0 m'}a^{<}_{l''m''}$$

$$+N(l_0 m|l'm'|l_0 m'')a^{<}_{l'm'}a^{>}_{l_0 m''} + N(l_0 m|l'm'|l''m'')a^{<}_{l'm'}a^{<}_{l''m''}. \quad (6)$$

Due to the spectral gap of one octave between the different wavelet generations the assumption of rapid time variation of the small scales is more justifiable than for Fourier modes.

The operator L for the removal of the first generation is nothing more than the wavelet spectrum of the wavelets themselves. We know that the off diagonal terms decay algebraically (in the case of the spline wavelets this exponent may be high). In order to keep the analysis in as simple a form as possible we will assume L is diagonal, otherwise the algebra becomes quite cumbersome. The small scale equation becomes

$$-a^{>}_{l_0 m} = L^{-1}(l_0 m|l_0 m)N(l_0 m|l_0 m'|l''m'')a^{>}_{l_0 m'}a^{<}_{l''m''}$$

$$+L^{-1}(l_0 m|l_0 m)N(l_0 m|l'm'|l_0 m'')a^{<}_{l'm'}a^{>}_{l_0 m''}$$

$$+L^{-1}(l_0 m|l_0 m)N(l_0 m|l'm'|l''m'')a^{<}_{l'm'}a^{<}_{l''m''} + \ldots. \quad (7)$$

The resolvable scale equation is, for $l < l_0$

$$\dot{a}^{<}_{lm} = L(lm|lm)a^{<}_{lm} + N(lm|l'm'|l''m'')a^{<}_{l'm'}a^{<}_{l''m''}$$

$$+N(lm|l_0 m'|l''m'')a^{>}_{l_0 m'}a^{<}_{l''m''} + N(lm|l'm'|l_0 m'')a^{<}_{l'm'}a^{>}_{l_0 m''}$$

$$+N(lm|l_0 m'|l_0 m'')a^{>}_{l_0 m'}a^{>}_{l_0 m''}. \quad (8)$$

For simplicity we will assume the wavelet coefficients are uncorrelated. $< a^{>}_{l_0 m'}a^{>}_{l_0 m''} >= Q_{l_0 m'}\delta(m' - m'')$. As Berkooz et al. (1993)

have shown numerically the decay of correlations as a function of distance in spectral space is fairly rapid.

Hence, after performing averages, the effect of removing the l_0-wavelet generation on the resolvable generations l with $l < l_0$, is

$$
\begin{aligned}
&\dot{a}^{<}_{lm} - L(lm|lm)a^{<}_{lm} \\
&+N(lm|l_0m'|l_0m'')L^{-1}(l_0m'|l_0m')N(l_0m'|l_0m''|\tilde{l}''\tilde{m}'')Q_{l_0,m''}a^{<}_{\tilde{l}''\tilde{m}''} \\
&+N(lm|l_0m'|l_0m'')L^{-1}(l_0m''|l_0m'')N(l_0m''|l_0m'|\tilde{l}''\tilde{m}'')Q_{l_0,m'}a^{<}_{\tilde{l}''\tilde{m}''} \\
&+N(lm|l_0m'|l_0m'')L^{-1}(l_0m'|l_0m')N(l_0m'|l_0\tilde{m}'|l''m'')Q_{l_0,m''}a^{<}_{\tilde{l}'\tilde{m}'} \\
&+N(lm|l_0m'|l_0m'')L^{-1}(l_0m''|l_0m'')N(l_0m''|\tilde{l}'\tilde{m}'|l_0m')Q_{l_0,m'}a^{<}_{\tilde{l}'\tilde{m}'} \\
&= N(lm|l'm'|l''m'')a^{<}_{l'm'}a^{<}_{l''m''} + \alpha_{lm} + \beta_{lm}
\end{aligned} \tag{9}
$$

where α_{lm} and β_{lm} are terms cubic in $a^{<}a^{<}a^{<}$, and are a natural outcome of the subgrid wavelet elimination:

$$
\begin{aligned}
\alpha_{lm} &= -N(lm|l_0m'|l''m'')a^{<}_{l''m''}L^{-1}(l_0m'|l_0m') \\
&\quad N(l_0m'|l'\tilde{m}'|\tilde{l}''\tilde{m}'')a^{<}_{\tilde{l}'\tilde{m}'}a^{<}_{\tilde{l}''\tilde{m}''} \\
\beta_{lm} &= -N(lm|l'm'|l_0m'')L^{-1}(l_0m''|l_0m'')a^{<}_{l'm'} \\
&\quad N(l_0\tilde{m}''|\tilde{l}'\tilde{m}'|\tilde{l}''\tilde{m}'')a^{<}_{\tilde{l}'\tilde{m}'}a^{<}_{\tilde{l}''\tilde{m}''}.
\end{aligned} \tag{10}
$$

Other terms generated by RNG are zero after the average procedure.

We note that cubic terms are a natural result of scale elimination. Myers et al. (1993) have recently derived low dimensional models for the dynamics of the Kuramoto-Sivashinsky equation using invariant manifolds techniques. They find that cubic terms are critical to the success of the dimension reduction procedure. Moreover, it is critical to have enough high scales removed in order to generate the appropriate cubic terms. For example, a four dimensional invariant manifold required four generations of wavelets (fifteen modes) to generate the appropriate cubic terms.

2.2. The energy budget

It is convenient to consider the energy budget in the Fourier spectral space. Fourier transform the KS equation, one has

$$\frac{\partial v(k)}{\partial t} + (\nu_0 k^4 - k^2)v(k) = -i\sum_j jv(j)v(k-j) \tag{11}$$

where the dissipation coefficient $\nu_0 = 1$. In the RNG procedure, the effects of removing subgrid shells are two fold: (a) a renormalized dissipation coefficient, and (b) the introduction of higher order nonlinearities.

For the subgrid scale, $k_1 < k < k_0$, we have

$$(\nu_0 k^4 - k^2)v^>(k) = -i\sum_j j[\underbrace{v^>(j)v^<(k-j)}_{I} + \underbrace{v^<(j)v^>(k-j)}_{II} + \underbrace{v^<(j)v^<(k-j)}_{III} + ...]. \tag{12}$$

The large scale $k < k_1$

$$(\nu_0 k^4 - k^2)v^<(k) = -i\sum_j j[v^<(j)v^<(k-j) + \underbrace{v^>(j)v^<(k-j)}_{A} + \underbrace{v^<(j)v^>(k-j)}_{B} + \underbrace{v^>(j)v^>(k-j)}_{C}]. \tag{13}$$

Substituting the solution of (12) into (13) and using

$$< v^>(j)v^>(j') >= Q(j)\delta(j-j') \tag{14}$$

we have

$$\frac{\partial v^<(k)}{\partial t} + (\nu_1(k)k^4 - k^2)v^<(k) = -i\sum_j jv^<(j)v^<(k-j) - \sum_{jj'}\Big[\frac{jj'}{j^2(\nu_0 j^2 - 1)}v^<(k-j)v^<(j)v^<(j-j') + \frac{jj'}{(k-j)^2[\nu_0(k-j)^2 - 1]}v^<(j)v^<(j')v^<(k-j-j')\Big] \tag{15}$$

where the effective dissipation coefficient is

$$\nu_1(k) = \nu_0 + \delta\nu_0(k) \tag{16}$$

with

$$\delta\nu_0(k)k^4 = \sum_j \frac{j^2 - jk}{j^4 - j^2} Q(k-j) + \sum_j \frac{j^2 - jk}{(k-j)^2[(k-j)^2 - 1]} Q(j). \tag{17}$$

Note the appearance of cubic nonlinearities in Eq. (15). Although a direct analysis of energy transfer caused by terms (A) and (B) (which lead to the cubic nonlinearities) has not been performed for KS equation, its counterpart in Navier-Stokes equation has been examined by (Zhou and Vahala, 1993). They found that these terms play a dominant role in removing energy from the resolvable scale near the cutoff wavenumber.

From the experience with Navier-Stokes equation, we know that term (C) removes the energy in a manner that is consistent with the concept of eddy viscosity. This is why the contribution from term (C) contributes to the eddy dissipation, not to the growth term.

In wavelet representation the energy budget can be obtained from Eq. (8) by multiplying through by $a^<_{lm}$ and performing subgrid averaging. The wavelet energy transfer equation for the resolvable scale will then take the form

$$[\frac{\partial}{\partial t} - 2L(lm|lm)]E_{lm} = T^{<<}_{lm} + T^{><}_{lm} + T^{>>}_{lm} \tag{18}$$

where

$$T^{<<}_{lm} = N(lm|l'm'|l'')a^<_{l'm'}a^<_{l''m''}a^<_{lm} \tag{19}$$

is the rate of energy transfer to wavelet generation l and spatial location m from the resolvable wavelet generations and spatial locations. $T^{><}_{lm}$ is the energy transfer rate to the wavelet generation l and spatial location m from the interaction of resolvable-subgrid wavelet generations $a^<a^>$, while $T^{>>}_{lm}$ is the transfer rate to (l,m) wavelet from the subgrid-subgrid wavelet generations $a^>a^>$.

It should be noted the information from Eq. (18) on energy transfer will give a detailed space-time nature of the energy transfer and should yield information on which are the important subgrid interaction wavelets–both in generation and spatial location. This feature is most relevant when dealing with inhomogeneous flows.

3. Summary

The goal of this work was to examine the possible contributions of a wavelet, or other non-Fourier basis, based RNG procedure. At the onset we realized the potential of such a procedure: Capturing more of the physics of nonhomogeneous and wall bounded flows. Specifically we do not assume uniformity of scales throughout the domain and we can satisfy wall boundary conditions.

We went on to examine the details of an RNG procedure for a one dimensional equation, the Kuramoto-Sivashinsy equation. Here the main difficulty lies in the fact that the two point correlation tensor is not diagonal in the wavelet representation, and the derivative operator is not diagonal either. The first difficulty is fundamental and comes from the physics of the equation. Here we have to be wise in the choice of the wavelet so as to minimize the correlation. The second difficulty is technical, to overcome it we would have to use a computer in the process of deriving the RNG equations.

In summary, we believe the approach indicated in this paper has the potential of overcoming some of the main problems standing in the way of applying a consistent RNG approach to wall bounded and nonhomogeneous flows.

References

Berkooz, G., Elezgaray, J., and Holmes, P., 1992. "Coherent structures in random media and wavelets," *Physica D* **61**, p. 47.

Berkooz, G., Holmes, P., and Lumley, J. L., 1993. "The proper orthogonal decomposition in turbulent flows," *Annu. Rev. Fluid Mech.* **25**, p. 507.

Farge, M., 1992. "Wavelet transforms and their applications to turbulence," *Annu. Rev. Fluid Mech.* **24**, p. 395.

Kevrekidis, I. G., Nicolaenko, B., and Scovel, C., 1990. "Back in the saddle again: a computer assisted study of the Kuramoto-Sivashinsky equation," *SIAM Jour. Appl. Math.* **50**, p. 760.

Lumley, J. L., 1971. *Stochastic tools in turbulence.* Academic Press.

Myers, M., Holmes, P., Elezgaray, J., and Berkooz, G., 1993. "Wavelet projections of the kuramoto-sivashinsky equation i: Heteroclinic cycles and modulated traveling waves for short systems," in preparation.

Perrier, V. and Basdevant, C., 1989. "Periodical wavelet analysis: A tool for inhomogeneous field investigation. theory and algorithms," *Recherche Aerospatiale* **3**, p. 54.

Poje, A. and Lumley, J. L., 1993. "A model for large scale structures in turbulent shear flows," Technical Report FDA93-03, Sibley School of Mechanical and Aerospace Engineering, Cornell University.

Pomeau, Y., Pumir, A., and Pelce, J., 1984. "Intrinsic stochasticity with many degrees of freedom," *J. Stat. Phys.* **37**, p. 39.

Yakhot, V. and Orszag, S. A., 1986. "Renormalization group analysis of turbulence theory. I. Basic theory," *J. Sci. Comput.* **1**, p. 3.

Zaleski, S., 1989. "A stochastic model for the large scale dynamics of some fluctuating interfaces," *Physica D* **34**, p. 427.

Zhou, Y. and Vahala, G., 1992. "Local interactions in renormalization methods for Navier-Stokes turbulence," *Phys. Rev. A* **46**, p. 1136.

Zhou, Y. and Vahala, G., 1993. "Reformulation of recursive renormalization group based subgrid modeling of turbulence," *Phys. Rev. E* **47**, p. 2503.

Zhou, Y., Vahala, G., and Hossain, M., 1988. "Renormalization group theory for the eddy viscosity in subgrid modeling," *Phys. Rev. A* **37**, p. 2590.

Zhou, Y., Vahala, G., and Hossain, M., 1989. "Renormalized eddy viscosity and Komogorov's constant in forced Navier-Stokes turbulence," *Phys. Rev. A* **40**, p. 5865.

ON RNG THEORY AND THE DECAY LAW OF HOMOGENEOUS ISOTROPIC TURBULENCE

S. H. Lam

Department of Mechanical and Aerospace Engineering
Princeton University
Princeton, New Jersey 08544

ABSTRACT

This paper reviews the attempts by RNG theory to provide quantitative information on the decay law of homogeneous isotropic turbulence.

1. Introduction

The paper by Yakhot and Orszag (1986) [YO] extended the so-called "renormalization group" (RNG) method—used by Forster, Nelson, and Stephen (1976) [FNS] to analyze the response of a randomly stirred fluid—to turbulent flows at large Reynolds numbers. By postulating a "correspondence principle" and using an unconventional "ϵ-expansion" procedure, YO obtained a number of quantitative results. A critique of this paper was published by Lam (1992).

A standard model problem in the theory of turbulence is the decay law of homogeneous isotropic turbulence. This author believes that RNG methodology is inherently incapable of providing substantive theoretical information on this problem. However, the YO theory was able to derive the decay law directly from their RNG theory,

$$K = (t - t_o)^{\frac{1}{1-C_{\xi 2}}},$$

(where K is kinetic energy of all eddies, t is time and t_o is an arbitrary constant) and obtained $C_{\xi 2} = 1.7215$—which is in reasonable agreement with experimental data. In 1992, Smith and Reynolds (1992) [SR] found an algebraic error in the YO derivation, and claimed that when the error is corrected the theory would yield $C_{\xi 2} = 5.65$, an unacceptable value. In early 1993, Yakhot and Smith (1992) [YS] developed a new RNG-based theory (which needed additional assumptions) and obtained $C_{\xi 2} = 1.68$. The value of $C_{\xi 2}$ is important in turbulence theory because it is frequently used in the equations

M. Y. Hussaini et al. (eds.), Transition, Turbulence and Combustion, Vol. II, 119–131.

of the so-called K-epsilon turbulent flow models (Yakhot, Orszag et al., 1992).

It will be shown here that the new YS theory used certain approximations which are both inaccurate and unnecessary. When these approximations are removed, the self-consistent $C_{\xi 2}$ of this new theory is again unacceptable.

2. The Two-Equation Formulation

At sufficiently high Reynolds numbers, the energy spectrum $E(k)$ is theoretically expected to follow the Kolmogorov scaling law for $k \geq k^*(t)$—where $k^*(t)$ denotes the lower edge of the inertial range. The $E(k)$ profile shape for $k \leq k^*(t)$ is, in general, problem dependent. The function $k^*(t)$ is a property of the solution. In YO, k^* is denoted by Λ_f, while in YS it is denoted by Λ_L.

Following Reynolds (1976), the "two-equation" formulation of the homogeneous isotropic turbulence problem is:

$$\frac{dK(0)}{dt} = -\mathcal{E}, \tag{1a}$$

$$\frac{d\mathcal{E}}{dt} = -C_{\xi 2}\frac{\mathcal{E}^2}{K(0)}, \tag{1b}$$

where $\mathcal{E}$ is the energy dissipation rate, $K(0)$ is the kinetic energy including all wave numbers, and $C_{\xi 2}$ is a dimensionless constant. At this level, the value of $C_{\xi 2}$ must be supplied by additional closure considerations. As we shall show presently, k_*(t) and $C_{\xi 2}$ are related (see (13d) below).

Equation (1a) is essentially the definition of $\mathcal{E}$. Equation (1b) is a closure hypothesis. We shall see later that the YS theory does not need to postulate (1b); it claims that RNG methodology can *derive* a relation (see (16a,b)) which, as will be shown later (see (17a)), is in the same form as (1b).

In the following subsections, we shall supplement (1a) and (1b) by detailed information on the profile of the turbulent kinetic energy spectrum. We shall see that $C_{\xi 2}$ can be determined by some innocent-looking assumptions on the profile—without the assistance of any RNG-based methodology.

2.1. The two-equation solutions

Without loss of generality, the turbulent kinetic energy spectrum $E(k)$ can be described as follows:

$$E(k) = e(\xi;t)E_*^+, \quad 0 \le \xi \le 1, \tag{2a}$$
$$E(k) = \xi^{-5/3}E_*^+, \quad 1 \le \xi \le \infty, \tag{2b}$$

where

$$\xi \equiv \frac{k}{k_*} \tag{2c}$$

and

$$E_*^+ \equiv \pi_2 \mathcal{E}^{2/3} k_*{}^{-5/3} \tag{2d}$$

is the value of E just to the right of k_*, and $e(\xi;t)$ is some unspecified integrable dimensionless function of ξ. The dimensionless π_2 is usually denoted as C_K in the turbulence literature. Note that for $k \ge k_*$ the Kolmogorov spectrum has been assumed:

$$E(k) = \pi_2 \mathcal{E}^{2/3} k^{-5/3}, \quad (k > k_*), \tag{3a}$$
$$K(k) \equiv \int_k^\infty E(k)dk = \frac{3}{2}kE(k), \quad (k > k_*). \tag{3b}$$

To prepare for a possible discontinuity across k_*, the value of E just to the left of k_* is denoted by E_*^- to distinguish it from its value just to the right. Hence

$$e(1;t) = \frac{E_*^-}{E_*^+}.$$

The q-th moments of $e(\xi;t)$ are defined as:

$$c_q = c_q(t) \equiv \int_0^1 \frac{e(\xi;t)}{e(1;t)}\xi^q d\xi, \quad q = 0, 1, 2, \ldots.$$

The shape of the low k spectrum can thus be fully described by the parameters E_*^-/E_*^+, k_* and c_q, $q = 0, 1, 2, \ldots$, all may be function of time.

The total kinetic energy $K(0)$ can be computed in terms of k_* and c_o:

$$K(0) \equiv \int_0^\infty E(k)dk = \left(1 + \frac{2E_*^-}{3E_*^+}c_o\right) K_*^+ = c_{oo}K_*^+, \tag{4}$$

where

$$K_*^+ \equiv K(k_*) = \frac{3}{2} k_* E_*^+, \tag{5a}$$

$$c_{oo} \equiv 1 + \frac{2E_*^-}{3E_*^+} c_o. \tag{5b}$$

Eliminating $E(k)$ between (3a) and (3b) and evaluating at k_*, one obtains:

$$\mathcal{E} = \left(\frac{2K_*^+}{3\pi_2} \right)^{3/2} k_*. \tag{6}$$

Taking the logarithmic derivative of (4) with respect to time and using (1b), one obtains:

$$\frac{1}{c_{oo}} \frac{dc_{oo}}{dt} + \frac{1}{K_*^+} \frac{dK_*^+}{dt} = -\frac{\mathcal{E}}{K(0)}. \tag{7}$$

Taking the logarithmic derivative of (6) with respect to time and using (7), one obtains:

$$\frac{d\mathcal{E}}{dt} = -\frac{3}{2} \frac{\mathcal{E}^2}{K(0)} + \left(\frac{1}{k_*} \frac{dk_*}{dt} - \frac{3}{2} \frac{1}{c_{oo}} \frac{dc_{oo}}{dt} \right) \mathcal{E}. \tag{8}$$

Equating the right hand sides of (8) and (1b), one obtains:

$$\left(\frac{1}{k_*} \frac{dk_*}{dt} - \frac{3}{2} \frac{1}{c_{oo}} \frac{dc_{oo}}{dt} \right) = \left(\frac{3}{2} - C_{\xi 2} \right) \frac{\mathcal{E}}{K(0)}. \tag{9}$$

Using (7) to eliminate $\mathcal{E}/K(0)$, and (5a) to eliminate K_*^+ in favor of E_*^+, we can solve for $C_{\xi 2}$ to obtain:

$$C_{\xi 2} = \frac{5 + 3m_1}{2(1 + m_1 + m_2)}, \tag{10}$$

where m_1 and m_2 are defined by:

$$m_1 \equiv \frac{k_*}{E_*^+} \frac{dE_*^+}{dk_*}, \tag{11a}$$

$$m_2 \equiv \frac{k_*}{c_{oo}} \frac{dc_{oo}}{dk_*}. \tag{11b}$$

Assuming m_1 and m_2 are constants, one obtains from (11a) and (11b):

$$E_*^+ \propto k_*^{m_1}, \tag{12a}$$

$$c_{oo} \propto k_*^{m_2}. \tag{12b}$$

Under the assumption that $C_{\xi 2}$ is a constant, the solutions of (1a) and (1b) are easily obtained:

$$K(0) \propto t^{x_1}, \qquad x_1 = \frac{1}{1 - C_{\xi 2}}, \tag{13a}$$

$$\mathcal{E} \propto t^{x_2}, \qquad x_2 = \frac{C_{\xi 2}}{1 - C_{\xi 2}}, \tag{13b}$$

$$E_*^+ \propto t^{x_3}, \qquad x_3 = \frac{5/2 - C_{\xi 2}}{1 - C_{\xi 2}}, \tag{13c}$$

$$k_* \propto t^{x_4}, \qquad x_4 = \frac{C_{\xi 2} - 3/2}{1 - C_{\xi 2}}, \tag{13d}$$

$$\nu_T \propto t^{x_5}, \qquad x_5 = \frac{2 - C_{\xi 2}}{1 - C_{\xi 2}}, \tag{13e}$$

The values of m_1 and m_2 must be provided by additional closure assumptions. Once they are available, $C_{\xi 2}$ can be computed from (10). Equation (13d) is the relation between $k_*(\mathrm{t})$ and $C_{\xi 2}$ mentioned earlier.

2.2. The power law energy spectrum for $k < k_*$

We consider now a power law energy spectrum $E(k)$ for the low wave number region:

$$E(k) = A(t)k^n, \qquad k \leq k_*, \tag{14a}$$

where n is a positive constant and $A(t)$ is some unspecified but differentiable function of t. One obtains straightforwardly:

$$A(t) = \frac{E_*^-}{k_*^n} = \frac{E_*^-}{E_*^+}\frac{E_*^+}{k_*^n}, \tag{14b}$$

$$e(\xi;t) = \frac{E_*^-}{E_*^+}\xi^n, \tag{14c}$$

$$c_q = \frac{1}{1+n+q}, \qquad q = 0, 1, \ldots. \tag{14d}$$

Substituting (13c,d) into (14b), one obtains:

$$A(t) \propto \frac{E_*^-}{E_*^+}t^{x_6}, \quad x_6 = x_3 - nx_4 = \frac{2(1+n)C_{\xi 2} - (3n+5)}{2(C_{\xi 2} - 1)}. \tag{15}$$

Reynolds (1976) gave a thorough discussion of this problem, and recommended $n = 2$ based on the idea (Bradshaw, 1972) that the large scale turbulence should be "equi-partitioned," $E_*^- = E_*^+$ to ensure continuity of $E(k)$, and A be "permanent" (*i.e.* $x_6 = 0$) in recognition of the asymptotically long time scales for low k dynamics. These closure assumptions yielded $m_1 = 2$, $m_2 = 0$, $C_{\xi 2} = 11/6 = 1.833$, $c_o = 1/3$, and $c_{oo} = 11/9$.

Note that if $m_1 = 2$ and $m_2 = 0$ were chosen without any additional detailed information on $e(x)$ (or n), the same $C_{\xi 2} = 11/6$ would be obtained. In other words, it is the shape of $E_*^+(k_*)$ and $c_{oo}(k_*)$ that matter; the shape of $e(\xi)$ (or the value of n) does not matter. For fixed m_1 and m_2, the main results (except for $A(t)$ or x_6 which alone depends on n) are not affected if n is changed from 2 to any other constant. Up to this point, no RNG concept had been used.

3. The RNG-Based Discontinuous YS Model

In the new YS theory for this problem, it was found necessary to place an infra-red cutoff at $k = \Lambda_L$ for the YO random force. In addition, a power law profile (14a) with $n = 2$ was chosen for the energy spectrum $E(k)$ for $k \leq \Lambda_L$. Most importantly, YS approximated $K(0)$ by $K(\Lambda_L)$ in (1a). This turns out to be a poor approximation which, as will be shown later, is exacerbated by the presence of a substantial discontinuity in E(k) at $k = \Lambda_L$. In the following presentation, $K(0)$ and $K(\Lambda_L)$ shall maintain separate identities.

The new YS analysis essentially retained (1a). Instead of (1b), they derived, using their RNG methodology, the following closure relations (*e.g.* by imposing an "infrared cut-off" for the random force, etc.):

$$\frac{d\mathcal{E}}{dt} = -T_2, \tag{16a}$$

where

$$T_2 \approx (2\nu_T(\Lambda_L))^2 \int_0^{\Lambda_L} k^4 E(k) dk, \tag{16b}$$

$$\mathcal{E} \approx 2\nu_T(\Lambda_L) \int_0^{\Lambda_L} k^2 E(k) dk, \tag{16c}$$

and Λ_L is the so-called infrared cutoff wavenumber of the random force. In addition to the above, the YS theory chose the cutoff precisely at k_*:

$$\Lambda_L = k_* \tag{16d}$$

and used the turbulent viscosity $\nu_T(k_*)$ scaling law:

$$\nu_T(k_*) = \pi_1 \mathcal{E}^{1/3} k_*^{-4/3}. \tag{16e}$$

The numerical value for π_1 used was taken from the original YO theory. Note that the concept of infra-red cutoff and the relation (16c) are not needed or used in RNG theory anywhere else, but they are crucial to the YS derivation of (16b,c,d). We shall not deal with how these results are derived and justified, but shall focus only on their consequences.

Using (10) and (14) in (16b,c), we obtain by straightforward calculations:

$$T_2 \approx 6c_4(\pi_1\pi_2)^2 \frac{E_*^-}{E_*^+} \frac{\mathcal{E}^2}{K(\Lambda_L)}, \quad (c_4 = \frac{1}{n+5}, K(\Lambda_L) = K_*^+) \tag{17a}$$

$$\mathcal{E} \approx 2c_2\pi_1\pi_2 \frac{E_*^-}{E_*^+}\mathcal{E}, \quad (c_2 = \frac{1}{n+3}). \tag{17b}$$

Note that only the product $\pi_1\pi_2$ appears in (17a) and (17b). Equation (17b) yields a time-independent discontinuity of E(k) at $k = k_*$:

$$\frac{E_*^-}{E_*^+} = \frac{1}{2c_2\pi_1\pi_2}, \tag{18}$$

and its value is 3.173, computed using the first column of (52f) in Lam, $\pi_1\pi_2 = 1/\pi_6 = 1/1.269$. In other words, *the energy spectrum used in the YS analysis is discontinuous at k=k_*.*

Substituting (10) into (17a) and comparing the result with (1b), one obtains (for $n = 2$):

$$C_{\xi 2} = \frac{3c_4\pi_1\pi_2}{c_2} \frac{K(0)}{K(\Lambda_L)} \approx 1.68 \frac{K(0)}{K(\Lambda_L)} \tag{19a}$$

where

$$\frac{K(0)}{K(\Lambda_L)} = 1 + \frac{c_o}{3c_2\pi_1\pi_2} \approx 1.705. \tag{19b}$$

[YS] ignored (19b) and used $K(0)/K(\Lambda_L) \approx 1$ in (19a) to obtain $C_{\xi 2} \approx 1.68$. Clearly, *the approximation $K(0)/K(\Lambda_L) \approx 1$ is inaccurate and unnecessary*, and the factor $K(0)/K(\Lambda_L)$ when properly included in (19a) is responsible for the unacceptably high value of $C_{\xi 2} = 2.880$.

4. The RNG-Based Continuous YS Model

The $E(k)$ discontinuity found above is merely a manifestation of having too many closure relations. Multiplying (2d) by (16e), we obtain:

$$\mathcal{E} = \frac{\nu_T(k_*)E_*^+ {k_*}^3}{\pi_1\pi_2} \tag{20}$$

which is in conflict with (16c) and forced the discontinuity of $E(k)$ at k_*.

A continuous YS model (i.e. $E_*^- = E_*^+$ required) can be constructed by using the YS equations (16a,b,c) but relaxing the requirement that $\Lambda_L = k_*$, (16d), treating k_*/Λ_L as an additional unknown. Using straightforward calculations, one obtains from (16b):

$$C_{\xi 2} = 6(\pi_1\pi_2)^2 \left[c_4 \left(\frac{k_*}{\Lambda_L}\right)^{10/3} + \frac{3}{10}\left(1 - \left(\frac{k_*}{\Lambda_L}\right)^{10/3}\right)\right] \frac{K(0)}{K(\Lambda_L)}, \tag{21a}$$

where

$$\frac{K(0)}{K(\Lambda_L)} = \left(1 + \frac{2c_o}{3}\right)\left(\frac{\Lambda_L}{k_*}\right)^{2/3}. \tag{21b}$$

Eliminating $\mathcal{E}$ between (16c) and (20), we obtain:

$$2\left[c_2 \left(\frac{k_*}{\Lambda_L}\right)^{4/3} + \frac{3}{4}\left(1 - \left(\frac{k_*}{\Lambda_L}\right)^{4/3}\right)\right] = \frac{1}{\pi_1\pi_2} \tag{21c}$$

which determines k_*/Λ uniquely. Using any columns from (52f) of Lam in (21) for $\pi_1\pi_2$, one can solve for k_*/Λ_L and use it to compute for the corresponding values for $C_{\xi 2}$ from (21a):

$1/(\pi_1\pi_2)$	k_*/Λ_L	$C_{\xi 2}$	$C_{\xi 2}K(\Lambda_L)/K(0)$
1.269	0.310	3.022	1.113
1.500	0.000	∞	0.800

Table 1. Numerical Results of Continuous Model

But these values for $C_{\xi 2}$ are unacceptably high. Also shown are the values of $C_{\xi 2}K(\Lambda_L)/K(0)$ which are now too low.

5. Conclusions

As pointed out by SR, the original YO theory—which did not require additional closure assumptions—failed to produce a reasonable value for $C_{\xi 2}$. The new YS theory, which called for an infra-red cutoff Λ_L on the random force and an assumed form for $E(k)$ for $k \leq \Lambda_L = k_*$, produced two closure relations, (16b) and (16c). The use of (16c) simultaneously with (20) meant that *two* independent definitions for $\mathcal{E}$ were being enforced at the same time. The conflict between them manifested itself through (17b) which forced the large $E(k)$ discontinuity at $k = k_*$. The value $C_{\xi 2} = 1.68$ was obtained by approximating $K(0)$ by $K(k_*)$ which is not justifiable in view of (19b). The value $C_{\xi 2} = 2.880$ obtained by not making this approximation is unacceptably high.

If (16d) is abandoned and k_*/Λ_L is determined by (21c), a continuous E(k) model can be obtained, but all the resulting values for $C_{\xi 2}$ as shown in Table 1 are also unacceptably high. If (16c) and (16d) are both abandoned so that (21c) is ignored and $\pi_1 \pi_2$ is taken from any column of (52f) of Lam (1992), then the theory becomes undetermined: (21a) and (21b) together can generate any reasonable desired value for $C_{\xi 2}$ by adjusting the value of k_*/Λ_L.

As indicated by Reynolds (1976) and supported by additional details here, a two-equation model for this problem can be determined by relatively innocent looking closure assumptions. From the physical point of view, the RNG methodology, which focussed on finding the impacts of eliminating slices of small eddies on long wave length modes, could not possibly play the key role here since the decay rate is mainly controlled by how the peak of the E(k) profile shifts with k_* (*i.e.* m_1).

A crucial test of RNG theory is the reasonableness of the value of $C_{\xi 2}$ obtained. The original YO theory, which needed no additional closure assumption at all, failed the test. The analysis here shows that the new YS theory, which needs additional closure assumptions in the RNG context in addition to a large discontinuity in $E(k)$ at k_*, also fails the test if the *ad hoc* and unnecessary approximation $K(0)/K(\Lambda_L) \approx 1$ is disallowed.

References

Bradshaw, P., 1972. "The understanding and prediction of turbulent flow," *Aeronaut. J.* **76**, p. 403.

Forster, D., Nelson D., and Stephen, M., 1977. "Large-distance and long-time properties of a randomly stirred fluid," *Phys. Rev. A* **16**, p. 732; see also *Phys. Rev. Lett.* **36**, p. 867.

Lam, S. H., 1992. "On the RNG theory of turbulence," *Phys. Fluids A* **4**, p. 1007.

Reynolds, W. C., 1976. "Computation of turbulent flows," *Annu. Rev. Fluid. Mech.* **8**, p. 183.

Smith, L. M. and Reynolds, W. C., 1992. "On the Yakhot-Orszag renormalization group method for deriving turbulence statistics and models," *Phys. Fluids A* **4**, p. 364.

Yakhot, V. and Orszag, S. A., 1986. "Renormalization group analysis of turbulence. I. Basic theory," *J. Sci. Comput.* **1**, p. 3.

Yakhot, V., Orszag, S. A., Thangam, S., Gatski, T. B., and Speziale, C. G., 1992. "Development of turbulence models for shear flows by a double expansion technique," *Phys. Fluids A* **4**, p. 1510.

Yakhot, V. and Smith, L. M., 1992. "The renormalization group, the ϵ-expansion and derivation of turbulent models," *J. Sci. Comput.* **7**, p. 35.

V. Yakhot's Comments on H. Lam's Paper Entitled "On RNG theory and the decay law of homogeneous isotropic turbulence"

The Renormalization Group for turbulence is based on an assumption that the inertial range dynamics of turbulent flows are described by the Navier-Stokes equations driven by the Gaussian white-in-time random force, defined by the correlation function:

$$< f_i(k,t) f_j(k',t' > \propto 2\overline{\epsilon} \frac{\delta(k^2 - k_0^2)}{k^{d-2}} \delta(k+k')\delta(t-t') = D(k)\delta(t-t')$$

where the wave-number $k = k_0$ corresponds to the beginning of the inertial range where the energy is pumped into the system at the rate $\overline{\epsilon}$. It is convenient to introduce Gel'fand's δ-function representation:

$$\delta(k) = \frac{(4-\epsilon)}{4\pi} k^{1-\epsilon}$$

in the limit $\epsilon \to 4$. We can show that in the interval $k > k_0$ the model consists of the Navier-Stokes equations the random force (2) but with $O(1)$ amplitude. The derived energy spectrum is $E(k) \approx k^{-\frac{5}{3}}$ for $k > k_0$ and $E(k) \approx k^2$ in the interval $k < k_0$. The second expression is a symptotically exact, since the system obeys the fluctuation-dissipation theorem in the interval $k < k_0$. The theory shows that the spectrum has a discontinuity at $k = k_0$.

Professor Lam, claiming that the energy spectrum cannot be a discontinuous function of the wave-number, disputes our results. To judge the merits of Professor Lam's argument, let us consider the exact expression for the energy spectrum following from the Navier-Stokes equations:

$$E(k) = D(k) + \int d\Omega \int_0^t d\lambda \overline{N(k,\lambda) \cdot v(-k,t)}$$

where $N(k,t)$ is the Fourier transform of the non-linear term in the Navier-Stokes equations. This expression is integrated over directions of the wave-vector k. It is clear that $E(k)$ is a continuous function in the intervals $k > k_0$ and $k < k_0$. However, at $k = k_0$ the energy spectrum has a discontinuity due to the energy source $D(k) \neq 0$ at only $k = k_0$. The general result that continuity is violated at the points in physical or Fourier space where sources are located is a direct consequence of the conservation laws. The well-known examples can be found in the theory of heat transfer, electromagnetism and even chemical kinetics, the field of Professor Lam's expertise. In statistical hydrodynamics, all correlation functions and fluxes have discontinuities at $k = k_0$.

One can attempt to construct a theory of turbulence without introducing neither boundary conditions nor external energy sources, as Professor Lam did in his well known paper. In this case the energy spectrum cannot have any discontinuities since the only statistically steady solution of the unforced Navier-Stokes equation is $v = 0$.

The argument, given above, does not depend on space dimensionality. To demonstrate the discontinuity in the energy spectrum $E(k)$, we have conducted direct numerical simulation (2048^2 resolved Fourier modes) of a two-dimensional flow stirred by a random noise with $D(k) \neq 0$ in the interval $500 < k < 525$. The initial condition in the numerical experiment was $v = 0$. The time evolution of the energy spectrum is presented on the figure. The discontinuity of $E(k)$ is readily seen in accord with equations of motion.

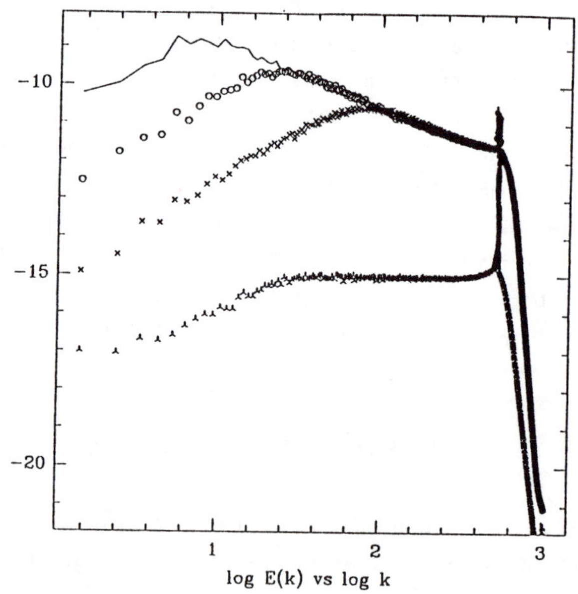

Figure 1: Time evolution of the energy spectrum in a two-dimensional flow, stirred by white–in–time random force with $D(k) \neq 0$ at $500 < k < 525$. The discontinuity of $E(k)$ is seen at all times.

Closing Comments by Author

The main point in Dr. Yakhot's replies appears to be that the $E(k)$ of the YS RNG theory (for homogeneous isotropic turbulence decay) should be discontinuous, and that the discontinuity is caused by a delta function in the postulated correlation function of the random force at k_0. I must confess that I am unable to comprehend this (new) line of reasoning.

RENORMALIZATION GROUP THEORY MODELING FOR A STRATIFIED TURBULENCE

D. N. Riahi

Department of Theoretical & Applied Mechanics
University of Illinois
Urbana, Illinois 61801

ABSTRACT

Renormalization group theory is applied to an incompressible and non-diffusive linearly stratified uniform flow past an obstacle. Turbulent fluxes for the wake flow are accounted for by repeatedly recasting the governing equations with the smallest scales represented by effective larger scales. Expressions for eddy viscosity and for eddy mass diffusivity are determined.

1. Introduction

The present study considers the problem of an incompressible and non-diffusive linearly stratified flow past a body with uniform upstream velocity. It is assumed that the Reynolds number for the flow in the wake behind the body is sufficiently large that the wake flow is in the turbulent regime. Stratified turbulent wake flow problems are important in applications. An example is air polluted turbulence behind a large mountain range leading to the so-called rotors (Lin and Pao, 1979), which are a major factor with regard to aviation safety.

The governing Navier-Stokes equation, under the Boussinesq approximation, the density equation for the incompressibility condition and the non-divergence condition of the velocity vector are given by Hanazaki (1988), who investigated steady stratified laminar flow past a sphere. Hence these equations will not be repeated here. These equations are coupled through the buoyancy term in the momentum equation.

The goal of the present investigation was to apply the renormalization group (RNG) theory (Yakhot and Orszag, 1986) to determine the expressions for the eddy viscosity and eddy diffusivity for mass density which could be valuable – particularly for future large eddy simulation (LES) studies of the above described stratified turbulent

M. Y. Hussaini et al. (eds.), Transition, Turbulence and Combustion, Vol. II, 133–136.

wake flow problem. To our knowledge, this is the first study to determine theoretically in a rational manner the expressions for such eddy quantities in a stratified coupled flow problem. Previous RNG and LES successes (Piomelli et al., 1990; Yakhot and Smith, 1992) encouraged us to undertake the present investigation.

2. Analysis and Results

Following Yakhot and Orszag (1986) and their correspondence principle, we add a Gaussian random force $\boldsymbol{f}$ to the momentum equation so that the resulting system becomes statistically equivalent to the original system in the inertial range (Hinze, 1975). We then define the Fourier decomposition of the velocity $\boldsymbol{u}$, density ρ and pressure p fields in the range $k \leq \Lambda$, where k is the magnitude of the wave number vector $\boldsymbol{k}$ and Λ, is an ultraviolet cutoff. The space-time Fourier-transformed system of the governing equations is a system of equations for the transformed quantities $\tilde{\boldsymbol{f}}(\hat{k}), \rho(\hat{k})$ and $\tilde{\boldsymbol{u}}(\hat{k}), \tilde{p}(\hat{k})$ for $\boldsymbol{f}, \boldsymbol{u}, \rho$, and p, respectively, where $\hat{k} = (\boldsymbol{k}, w)$ and w is the frequency. Next, we assume that the modes vanish for $k > \Lambda$ (Yakhot and Orszag, 1986) and write

$$(\tilde{\boldsymbol{f}}, \tilde{\boldsymbol{u}}, \tilde{\rho}, \tilde{p}) = (\boldsymbol{f}^{<}, \boldsymbol{u}^{<}, \rho^{<}, p^{<}) + (\boldsymbol{f}^{>}, \boldsymbol{u}^{>}, \rho^{>}, p^{>}), \tag{1}$$

where the quantities with the superscript $<$ (long-wavelength modes) belong to the interval $0 < k < \Lambda \exp(-r)$, while the quantities with the superscript $>$ (short-wavelength modes) belong to the interval Λ. $\exp(-r) < k < \Lambda$. Here r is a positive constant. The expression (1) is used in the governing system for the transformed quantities and a formal parameter $\lambda(=1)$ is introduced in front of the nonlinear terms to facilitate the perturbation solution of the system.

The RNG procedure consists of two steps. First, all short-wavelength modes terms in the system are removed, up to order λ^3 and third order or higher of these terms, by repeated substitution of the system for the short-wavelength modes into the system for the long-wavelength modes. Second, averages are taken over the part $\boldsymbol{f}^{>}$ of the random force $\boldsymbol{f}$. The resulting system of equations for $\boldsymbol{u}^{<}$ and $\rho^{<}$ is an approximation which is valid in the limit $\hat{k} \to 0$. After some lengthy algebra and simplification in the limit $\hat{k} \to 0$, we find that equation for $\boldsymbol{u}^{<}$ contains a term like $R\lambda^2 k^2 \boldsymbol{u}^{<}$ and equation for $\rho^{<}$ contains a term like $S\lambda^2 k^2 \rho^{<}$, where the coefficients R and S are given by

$$R = V_o + \frac{D_o}{V_o^2(2\pi)^d}\left(\frac{S_d}{\in \Lambda^\varepsilon}\right)\frac{(d^2 - d - \in)}{[2(d+2)d]}[\exp(\in r) - 1], \qquad (2a)$$

$$S = \frac{(2\pi)^{-d} D_o}{V_o^2}\left(\frac{S_d}{\in \Lambda^\varepsilon}\right)[\exp(\in r) - 1]\left(\frac{-1+d}{d}\right). \qquad (2b)$$

Here $D_o = T_o V_o K_B/\rho_o$, V_o is the kinematic viscosity, T_o is a reference temperature of the fluid, ρ_o is a reference fluid density, K_B is Boltzmann's constant, d is the dimension of space, $\in = 4 + y - d$, $y > -2$ is a constant, ($y = -2$ describes fluid in thermal equilibrium. $y > -2$ corresponds to strongly nonequilibrium flow cases) and $S_d = 2\pi^{d/2}/\Gamma(d/2)$ is the area of a d-dimensional unit sphere (Yakhot and Orszag, 1986). Hence, the expressions for eddy viscosity and eddy diffusivity for mass density are given by $R - V_o$ and S, respectively. The expression for $R - V_o$ is the same as in the homogeneous fluid case with no buoyancy (Yakhot and Orszag, 1986). It is seen from (2) that the ratio $(R - V_o)/S$ depends on y and d. This latter result may indicate both the non-diffusive nature of the incompressibility condition (density equation) and the fact that a classical eddy diffusivity formula for buoyancy driven flows is not supported by the experimental data (Adrian, 1972).

Acknowledgement

The author would like to thank Ron Adrian for valuable discussion on the subject of eddy diffusivity.

References

Adrian, R. J., 1972. "Turbulent thermal convection: Statistical model equation and experiments on convection under a stable layer," Ph.D. Thesis, University of Cambridge, U.K.

Hanazaki, H., 1988. "A numerical study of three-dimensional stratified flow past a sphere," *J. Fluid Mech.* **192**, pp. 393-419.

Hinze, J. O., 1975. *Turbulence*, 2nd Edition, McGraw-Hill Co.

Lin, J. T. and Pao, Y. H., 1979. "Wakes in stratified fluids," *Ann. Rev. Fluid Mech.* **11**, pp. 317-338.

Piomelli, U., Zang, T. A., Speziale, C. G., and Lund, T. S., 1990. "Application of renormalization group theory to the large-eddy simulation of transitional boundary layers," *Instability and Transition*, Vol. II, edited by M. Y. Hussaini and R. G. Voigt, Springer-Verlag, pp. 480-496.

Yakhot, V. and Orszag, S. A., 1986. "Renormalization group analysis of turbulence. I. Basic theory," *J. Sci. Comput.* **1**, pp. 3-51.

Yakhot, V. and Smith, L. M., 1992. "The renormalization group, the ϵ-expansion and derivation of turbulence models," *J. Sci. Comput.* **7**, pp. 35-61.

RNG FOR CFD? AN ASSESSMENT OF CURRENT RNG THEORIES OF TURBULENCE[1]

Ye Zhou

Institute for Computer Applications in Science and Engineering
NASA Langley Research Center
Hampton, Virginia 23681-0001

George Vahala

Department of Physics, College of William and Mary
Williamsburg, Virginia 23185

David McComb

Department of Physics, University of Edinburgh
United Kingdom

Sau-hai Lam

Department of Mechanical and Aerospace Engineering
Princeton University
Princeton, New Jersey 08544

ABSTRACT

This paper is based on discussions carried out at the ICASE/LaRC Workshop. It gives a brief introduction to renormalization group, as applied successfully to the theory of phase transitions, and sketches the history of attempts to apply the method to turbulence. This is followed by a succinct statement, and critical assessment, of current RG theories of turbulence.

[1]The work of Y.Z. and G.V. was supported by the National Aeronautics and Space Administration under NASA Contract No. NAS1-19480 while the authors were in residence at the Institute for Computer Applications in Science and Engineering (ICASE), NASA Langley Research Center, Hampton, VA 23681. G.V. was also supported in part by DOE. Authors are listed in reverse alphabetical order.

M. Y. Hussaini et al. (eds.), Transition, Turbulence and Combustion, Vol. II, 137–161.

1. Purpose and Scope of This Paper

Many of those who take an interest in the computation of turbulent flows have found the development of renormalization group (RG) methods to be a rich source of hope, puzzlement and, increasingly, frustration. This is hardly surprising, in view of the present state of the subject. Some advocates of RG make extravagant claims for the method, as applied by them to turbulence. Whereas, other workers adopt radically different approaches and make much more modest claims. Certainly, in contrast to the situation in statistical field theory, there appears to be a complete absence of consensus in the literature.

This was seen to be a major problem by our discussion group, and it was remarked on several occasions, and by several participants, that there was little cross-reference between our various papers. It was also remarked that the basic theory is often seen as 'difficult' or inaccessible because of its origins in rather esoteric areas of theoretical physics. Accordingly, we felt that for the results of our discussions to be helpful to the computational fluid dynamics community this paper should first, give a clear, simple account of the development of renormalization group; and second a critical comparison of current approaches, in which relative strengths and weaknesses are clearly stated.

However, before proceeding to the main body of this paper, we should say a few words about the nature and constitution of our discussion group, and also make a comment on terminology. In addition to the present authors, who have all published recent, distinct versions of renormalization group for turbulence, our group consisted of Jon Lee, Robert Rubinstein and Siva Thangam. Also, although they did not participate in the ongoing discussions, Victor Yakhot and Steven Orszag attended the workshop for one day in order to give a presentation on their work, and subsequently took part in a lively panel discussion. This contribution was of considerable assistance to the authors in their assessment of the Yakhot-Orszag proposal: see Section 4.

Lastly, we should mention that renormalization group is often referred to as 'RNG' - hence our title! However, as renormalization is normally regarded as just one word, 'RG' is the more correct abbreviation and will be used throughout this paper.

2. What Is Renormalization Group?

In this section we shall explain something of the origins and meaning of RG. In the process we shall define terms and introduce concepts which will be needed when we consider the question of how RG is to be applied to fluid turbulence.

We should begin by noting that RG is often introduced as having originally been a method of removing divergences in quantum field theory. While this is true, it is not very illuminating for those who have little or no knowledge of that subject. We can bridge this gap by considering, as an example, a problem which may be discussed entirely in terms of classical physics. This is the electron gas, which consists of a gas of electrons in a uniform background of positive charge, so that the system is, overall, electrically neutral. Such a gas can be a model for the conduction electrons in a metal, for a plasma or for an electrolyte. This allows us to introduce the general idea of renormalization, and is the subject of Section 2.1. In order to introduce the more specific topic of renormalization group, we consider as an example the phase transition from para- to ferromagnetism. This is dealt with in Section 2.2.

2.1. Renormalization

It is a cardinal principle of physics that the macroscopic properties of a material, such as its dielectric constant or elastic modulus, will not in general depend on the size of the sample of material under consideration. So, for a microscopic theory, this implies that we must be able to take limits, corresponding to the distance between constituent microscopic particles shrinking to zero, or to the overall size of the system going to infinity, without upsetting our theory. However, if we take our example of an electron gas, it is easily seen that problems may arise in both these limits.

Let us suppose, for instance, that we wished to work out the total electrostatic potential energy of all the electrons in the system and that we did that naively by adding up all the contributions from pairs of electrons, each pair contributing the usual Coulomb '$1/r$' law. It is obvious that a continuum limit would require $r \to 0$, with each pairwise potential diverging, and becoming infinite in that limit. Less obviously, the limit of infinite system size also leads to a divergence in the potential energy of a test charge at a point. Suppose we take the test charge to be at the origin and assume that the surrounding charges have a uniform density ρ, say. Then, as the number of electrons in a spherical shell between r and $r + dr$ goes as $\rho r^2 dr$, it is easily seen that this overpowers the Coulomb potential to give a total potential which varies as $2\pi \rho e R^2$, where e is the electronic charge and R is the system size.

The first of these divergences is usually referred to as being 'ultraviolet', because (as we know from Fourier analysis) short distances correspond to high spatial frequencies. Similarly, the long-range divergence in the second case, is referred to as being an 'infra-red divergence'. From our point of view, the second case is the more interesting. Physically the energy at a point does not diverge with system size because a cloud of electrons around each individual electron acts to screen the long-range Coulomb potential and turns it into an exponentially decaying potential. Theoretically, this effect shows up when we take the interactions between individual electrons into account. In perturbation theory it corresponds to summing a certain class of terms to all orders. In fact, the potential (at lowest nontrivial order) becomes $[e \exp -r/l_D]/r$, where l_D is the Debye length and depends on the density of electrons.

This is known as a screened potential. But, an alternative interpretation is to say that we replace the electron charge e by the renormalized electron charge $e \exp -r/l_D$. This replacement of a 'bare' particle by a 'dressed' particle, in order to take account of the effect of interactions, is an important technique in many-body physics and nowadays is widely referred to as *renormalization*. It should be noted that a characteristic feature of the process is that the renormalized properties (as in our example) are scale-dependent. For turbulence, the obvious analogy would be the replacement of the bare molecular viscosity ν_0 by an effective turbulent viscosity ν_T, say. It is well known that turbulent viscosities do depend on (at least) the physical scale of the problem.

This discussion serves to introduce the idea of renormalization. Our next step is to consider what is meant by renormalization *group*, and it turns out that this is just one way of achieving renormalization.

2.2. Renormalization group

As we have already said, RG was originally introduced in field theory as an *ad hoc* method of eliminating divergences. It was based on the obvious equivalence between taking the continuum limit and making a scaling transformation of the basic variables such as position or momentum. However, this theory was regarded as rather opaque in terms of its underlying physics, and from our point of view, it is better to consider its more modern incarnation as a theory of critical phenomena, in which it is seen as a way of dealing with problems involving a large range of length or time scales. Its possible relevance to turbulence should now be beginning to emerge; but we shall first consider another simple physical system: the lattice-spin model of a ferromagnet.

In physical terms, magnetism arises because spins at lattice sites become aligned with each other. This tendency for spins to align is opposed by thermal effects, which tend to promote disorder. Thus alignments occur as random fluctuations on length scales ranging from the lattice spacing (L_0, say) up to some correlation length (ξ, say). The correlation length depends on the temperature, and becomes infinite as the Curie temperature (or critical point) is approached. At the critical point, fluctuations on all scales from L_0 up to the size of the material specimen can occur, and therefore a net overall magnetization can appear.

The theoretical problem is to calculate the Hamiltonian (and hence the thermodynamic properties of the material), but this is not easily achieved due the presence of an interaction term, which is the sum over configurations of all spin interactions. The fact that all lengthscales are (in principle) equally important makes it difficult to know where to begin. Certainly, if any scales are eliminated, then in some way their effect must be retained.

The problem is eased to some extent by considering models, such as the Ising model, in which spin vectors are Boolean in character and are either 'up' or 'down', with no intermediate states permitted. Another helpful feature of this model is that the spins are supposed to have only nearest-neighbour interactions. Then the application of RG to this problem can be understood as a form of coarse-graining, in the following sense. We begin with an interaction Hamiltonian H_0, associated with two spins separated by a distance L_0 (i.e. the lattice spacing). Then we calculate an effective Hamiltonian H_1, associated with regions of size $2L_0$, by averaging out the effects of scales L_0. Next we calculate H_2, associated with regions of size $4L_0$, with the effects of scales less than or equal to $2L_0$ averaged out.

The above process can be expressed in terms of a transformation T, which is applied repeatedly: $TH_0 \rightarrow H_1, TH_1 \rightarrow H_2, TH_2 \rightarrow$

$H_3 \ldots$. At each stage, the length scales are changed: $L_0 \to 2L_0, 2L_0 \to 4L_0 \ldots$, and in order to compensate, the spin variables are also scaled in an appropriate fashion such that the Hamiltonian always looks the same in scaled coordinates. It is this rescaling which leads to renormalization, and the set of transformations $\{T\}$ defines a simple group: hence *renormalization group*. If iterating this transformation leads to the result

$$TH_n = H_{n+1} = H_n,$$

then $H_n = H_N$ (say) is referred to as a *fixed point*. In the case of critical phenomena, this corresponds to a critical point. We shall discuss its significance for turbulence presently; but, before doing so, we conclude this brief discussion of the magnetic case by indicating how its methodology may be made to come closer to a continuum problem (as in fluid motion), which is what concerns us here.

Suppose we make the Fourier transformation from **x**-space to wavevector **k**-space. Further suppose that we have eliminated (in some way) the smallest length scales, less than or equal to Λ^{-1}, say, so that we are left with a Hamiltonian which is a function of continuous variables in **k**-space, $H(\Lambda)$. Then we may formally eliminate further modes in a band $\Lambda/b \leq k \leq \Lambda$, as follows. Denote the three-dimensional spin field by $S_\alpha(\mathbf{k})$, where $\alpha = 1$, 2, or 3, and take the following steps:

1. Integrate out all $S_\alpha(\mathbf{k})$ for which $\Lambda/b \leq k \leq \Lambda$.

2. Rescale the remaining modes of the spin field by enlarging wavevectors by a factor b.

3. Multiply each $S_\alpha(\mathbf{k})$ by a constant factor ζ_b.

The parameter b must satisfy $1 < b < \infty$ and is known as the *spatial rescaling factor*; while ζ_b is the spin rescaling factor, and involves technicalities of the magnetic case, which we shall not pursue here. If we write the above three-step process as

$$H' = \mathcal{T}_b H,$$

then the set $\mathcal{T}_b$, $1 < b < \infty$ is called the renormalization group. It is often remarked that, as we have not defined the inverse of $\mathcal{T}_b$, this is strictly a semi-group.

3. History of the Application of RG to Turbulence

The macroscopic motion of a fluid may be thought of as exhibiting two 'phase transitions'. That is, as we increase the Reynolds number of a given flow, we may expect first a transition to turbulence and second a transition to scaling behaviour of the turbulence. In the latter case, this means the onset of an intermediate range of wavenumbers in which the energy spectrum takes the form of a power-law. It is this latter transition which concerns us here.

It is of course difficult to identify the onset of scaling behaviour as such and a convenient way to think about this problem is to consider the case where the Reynolds number (based on the turbulent microscale, say) is large enough for a power-law spectrum to exist over an appreciable range of wavenumbers. Now define a *local* Reynolds number, using the inverse of wavenumber as the length scale. Then as we scan through the wavenumbers, from the maximum towards smaller values, we are, in effect, increasing the local Reynolds number to the point where power-law behaviour begins (In other words, the boundary between the inertial and viscous ranges). Conceptually, therefore, the simplest application of RG to turbulence involves the progressive elimination of high-k modes, with the fixed point corresponding to a renormalized viscosity in the inertial range (The equivalent, in terms of CFD, would be a subgrid effective viscosity, in large-eddy simulations.). As turbulence is not a Hamiltonian system, we shall carry out these operations on the Navier-Stokes equations.

3.1. RG applied to the Navier-Stokes equations

We may introduce the Fourier components of the velocity field $U_\alpha(\mathbf{k}, t)$ in the usual way, thus:

$$U_\alpha(\mathbf{x}, t) = \sum_{\mathbf{k}} U_\alpha(\mathbf{k}, t) e^{i\mathbf{k}.\mathbf{x}}. \tag{1}$$

Then the solenoidal Navier-Stokes equation for incompressible fluid motion takes the form (e.g. see McComb 1990)

$$(\frac{\partial}{\partial t} + \nu_0 k^2) U_\alpha(\mathbf{k}, t) = \lambda M_{\alpha\beta\gamma}(\mathbf{k}) \int d^3 j\, U_\beta(\mathbf{j}, t) U_\gamma(\mathbf{k} - \mathbf{j}, t), \tag{2}$$

where

$$M_{\alpha\beta\gamma}(\mathbf{k}) = (2i)^{-1}\{k_\beta D_{\alpha\gamma}(\mathbf{k}) + k_\gamma D_{\alpha\beta}(\mathbf{k})\}, \tag{3}$$

and

$$D_{\alpha\beta}(\mathbf{k}) = \delta_{\alpha\beta} - \frac{k_\alpha k_\beta}{|\mathbf{k}|^2}, \tag{4}$$

and λ is a book-keeping parameter, which can be used to facilitate the iterative solution of equation (2), by expanding in powers of λ, and setting $\lambda = 1$ at the end of the calculation. This procedure is well known in statistical and many-body physics and is usually referred to as the λ-expansion.

Turbulence almost invariably occurs because there is a mean rate of shear in a fluid. If we wish to study stationary, isotropic turbulence, then the artificial nature of this problem requires us to add hypothetical stirring forces to the right-hand side of equation (2), in order to maintain the turbulence against viscous dissipation. Denoting the random force by $f_\alpha(\mathbf{k}, t)$, it is usual to specify its distribution as multivariate normal and to choose the force-force covariance to be

$$\langle f_\alpha(\mathbf{k}, t) f_\beta(-\mathbf{k}, t')\rangle = D_{\alpha\beta}(\mathbf{k}) W(k) \delta(t - t'), \tag{5}$$

where $W(k)$ has dimensions of *velocity*^{2}/*time*, and remains to be specified. Then one can show that stationarity requires that the rate at which the stirring forces do work is the same as the rate of viscous dissipation ε, or

$$\int_0^\infty 4\pi k^2 W(k) dk = \varepsilon, \tag{6}$$

where

$$\varepsilon = \int_0^\infty 2\nu_0 k^2 E(k) dk. \tag{7}$$

It has always been normal practice in turbulence theory, to choose $W(k)$ to be peaked near the origin, so that its arbitrary nature is only of importance at low wavenumbers, and a universal energy spectrum can develop at large wavenumbers. More recently, the application of RG to stirred fluid motion (as we shall see in Section 3.2) has introduced theories which depend strongly on the choice of $W(k)$. Accordingly, questions of how $W(k)$ is chosen and justified must then be considered. In particular, in this context, the question of whether or not $W(k)$ should depend on some characteristic length scale (such as the ultra-violet cutoff Λ) will surface later as a major issue.

In order to try to carry out the RG algorithm given above in Section 2.2, we consider the Fourier components to be defined on the interval $0 \leq k \leq k_0$, where k_0 is the largest wavenumber present and is of the order of the Kolmogorov dissipation wavenumber (e.g. see McComb (1990)). Then, for some k_1, such that $k_1 < k_0$, we filter the velocity field at $k = k_1$. This may be expressed in terms of the unit step functions,

$$\theta^-(k) = \begin{cases} 1 & \text{if } 0 \le k \le k_1 \\ 0 & \text{if } k_1 \le k \le k_0; \end{cases} \tag{8}$$

$$\theta^+(k) = \begin{cases} 0 & \text{if } 0 \le k \le k_1 \\ 1 & \text{if } k_1 \le k \le k_0, \end{cases} \tag{9}$$

allowing us to define the following useful filtered forms:

$$U_\alpha^<(\mathbf{k},t) = \theta^-(k)U_\alpha(\mathbf{k},t),$$

$$U_\alpha^>(\mathbf{k},t) = \theta^+(k)U_\alpha(\mathbf{k},t),$$

$$M_{\alpha\beta\gamma}^<(\mathbf{k}) = \theta^-(k)M_{\alpha\beta\gamma}(\mathbf{k}),$$

$$M_{\alpha\beta\gamma}^>(\mathbf{k}) = \theta^+(k)M_{\alpha\beta\gamma}(\mathbf{k}).$$

Then substitution of these forms into equation (2) allows us to decompose the Navier-Stokes equation into separate low-k and high-k forms, viz.,

$$\begin{aligned}(\frac{\partial}{\partial t} + \nu_0 k^2)U_\alpha^<(\mathbf{k},t) = {} & \lambda M_{\alpha\beta\gamma}^<(\mathbf{k}) \int d^3j\, \{U_\beta^<(\mathbf{j},t)U_\gamma^<(\mathbf{k}-\mathbf{j},t) \\ & +2U_\beta^<(\mathbf{j},t)U_\gamma^>(\mathbf{k}-\mathbf{j},t) \\ & +U_\beta^>(\mathbf{j},t)U_\gamma^>(\mathbf{k}-\mathbf{j},t)\},\end{aligned} \tag{10}$$

$$\begin{aligned}(\frac{\partial}{\partial t} + \nu_0 k^2)U_\alpha^>(\mathbf{k},t) = {} & \lambda M_{\alpha\beta\gamma}^>(\mathbf{k}) \int d^3j\, \{U_\beta^<(\mathbf{j},t)U_\gamma^<(\mathbf{k}-\mathbf{j},t) \\ & +2U_\beta^<(\mathbf{j},t)U_\gamma^>(\mathbf{k}-\mathbf{j},t) \\ & +U_\beta^>(\mathbf{j},t)U_\gamma^>(\mathbf{k}-\mathbf{j},t)\}.\end{aligned} \tag{11}$$

If we now try to carry out the first of the three steps outlined at the end of Section 2, for the lattice-spin Hamiltonian, we can adapt that procedure to the present (Navier-Stokes) case, as follows:

- Solve equation (11) for $U^>$.
- Substitute this solution into equation (10) for the $U^<$ and do a partial average over the $U^>$.

- Terms resulting from this procedure which are linear in $U^<$ can be interpreted as contributing an increment to the turbulent viscosity.

The main problems which arise due to mode coupling, are now obvious by inspection. First, the solution of (11) for $U^>$ contains terms in $U^<$. When these are substituted on the right-hand side of (10), the result is a triple nonlinearity in the $U^<$, which can then generate higher-order nonlinearities in subsequent iterations. Second, averaging out the high-wavenumber modes requires the property

$$\langle U^< U^> U^> \rangle = U^< \langle U^> U^> \rangle,$$

which cannot be strictly true, as $U^<$ and $U^>$ are just parts of the same velocity field and are not statistically independent.

It cannot be emphasized too strongly that these two problems are the inherent problems in applying RG to the Navier-Stokes equation. It is the way in which they are tackled which distinguishes one RG theory of turbulence from another. We shall now discuss two main lines of development, the first due to Forster, Nelson and Stephen (1976, 1977: referred to as FNS) and the second originated by Rose (1977); although, strictly speaking, in the latter case, for the linear problem of passive scalar convection.

3.2. Stirred fluids in the long-wavelength limit

The approach of FNS was influenced by an analogy with dynamical critical phenomena, and could be regarded as a way of modifying the turbulence problem to make it resemble a form of Ising model. They took the turbulent modes to be restricted to the range $0 \leq k \leq \Lambda$, where the cutoff wavenumber Λ was chosen to be small enough to exclude inertial cascade effects. The specification of their class of models was completed by the choice of a power law for the force covariance, thus

$$W(k) = W_0(y) k^{-y}, \tag{12}$$

for $k \leq \Lambda$. Here y is dimensionless, but it is important to note that the dimension of W_0 depends on the value chosen for y.

The choice of a value for y amounts to a choice of model, and $y = 2$ gives Model A of FNS, corresponding to thermal equilibrium;

whereas $y = 0$ gives us their Model B, which corresponds to macroscopic stirring of the fluid. We shall confine our attention to Model B.

Essentially FNS evaded the two difficulties inherent in the turbulence problem by considering a problem so artificial that these difficulties are simply not present. They used a modified form of λ-expansion, replacing λ by λ_0, which is then treated as renormalizable. The corresponding zero-order velocity field is simply related to the stirring force by a linear propagator (or Green's function). By choosing the stirring forces to be statistically independent of each other for distinct values of k, they ensure that the averaging problem becomes trivial, so that they can average over the $U^>$ without affecting the $U^<$. In fact, this is an even greater simplification than the Ising model provides for the magnetic case.

On this basis FNS renormalize the viscosity, the stirring forces and the coupling strength. An important feature of their analysis is that the triple nonlinearity in the $U^<$ is found to become negligible as the iteration proceeds to the fixed point. Thus a characteristic feature of their method is the form invariance of the Navier-Stokes equation under renormalization, as the fixed point is approached. There are various restrictions on this result, but the principle one is that it is only valid asymptotically, as k tends to zero.

The important thing to realize is that the FNS theory has nothing to do with turbulence and cannot be applied to it. Their choice of Λ to be small enough to exclude inertial cascade effects would, if applied to real fluid turbulence, imply that Λ must therefore be orders of magnitude smaller than the Kolmogorov dissipation wavenumber for that flow. Its authors only claimed to have obtained some long-wavelength properties of a stirred fluid. In the context of microscopic theories, this is the appropriate limit for universal behaviour; e.g. the onset of spontaneous magnetism, or the occurrence of hydrodynamic behaviour in a gas. But, as we know, this limit is of little interest in macroscopic turbulence, as here the significant universal behaviour is associated with the cascade at large wavenumbers (or short wavelengths). As FNS pointed out, the cascade is inaccessible to their theory; because at large wavenumbers their procedure would soon generate an intractably large coupling parameter.

3.3. Isotropic turbulence

A more realistic approach was initiated by Rose (1977), who considered the subgrid modelling of the turbulent convection of a passive scalar contaminant (ϕ, say). His approach was based upon '...an iterative procedure which models away first the dissipation-range eddies and then the eddies of a slightly larger size, etc.' This meant a genuine recursive mode elimination, involving finite-sized bands of wavenumbers, in the inertial and dissipation ranges of k-space. The averaging out of modes in a band $k_1 \leq k \leq k_0$, was followed by the same procedure for a band $k_2 \leq k \leq k_1$, and so on, with the nth-wavenumber being given by

$$k_n = h^n k_0, \qquad 0 \leq h \leq 1, \tag{13}$$

for arbitrarily chosen h.

Rose made two assumptions, both of which seem quite reasonable. He took $U^>$ to be small compared to $U^<$, and he assumed that the $U^>$ are rapidly varying on the timescales of the $U^<$. A straightforward iterative perturbation method then led to a fixed point, corresponding to a renormalized (or subgrid) scalar diffusivity. However, when we ask how Rose's theory tackles the inherent problems of mode coupling, as outlined at the end of Section 3.1, we find that the first of these does not occur in the form that we discussed.

This is because Rose studied the explicitly linear problem of scalar convection, in which the triple nonlinearity in the $U^<$ does not arise. Instead, he was faced with the analogous term $\langle \phi^< U^< U^< \rangle$, which he treated as part of a new scalar diffusion equation. It is this new equation which exhibits form invariance under RG transformation, but this involves an additional truncation in each cycle of iteration, in order to maintain the form invariance. The second problem, that of performing partial averages over the $U^>$, was simply evaded by treating $U^<$ and $U^>$ as if they were statistically independent of each other. Then the partial average is trivially just a combination of a filter and ensemble average.

A different approach called *iterative averaging* was followed by McComb (1982), who introduced a variant of Reynolds averaging in which the equation of motion is first conditionally averaged term by term and then subtracted from the unaveraged equation. The result is an equation for the high-k modes that does not contain a term nonlinear in the low-k modes, and hence the problem of the triple nonlinearity does not arise (McComb and Shanmugasundaram, 1983). Apart from this, the theory employed the same assumptions for the velocity-field case as Rose had for the scalar case, and the recursion relation led to a fixed-point value of renormalized viscosity independent of the arbitrarily chosen molecular viscosity ν_0.

Later, a renormalized dissipation relationship was derived (McComb, 1986) and took the form

$$\varepsilon = \int_0^{k_n} 2\nu_n(k)k^2E(k)dk. \tag{14}$$

This permitted a calculation of the Kolmogorov constant; but, although the values obtained were in good agreement with experiment, they showed an unphysical dependence (as did Rose's results) on the scale parameter h, as defined by equation (13). This behaviour was attributed to an inadequate treatment of the relationship between the conditional average and the usual ensemble average.

4. Current Approaches to Renormalization Group in Turbulence

In recent years, Zhou and Vahala have been investigating the application of Rose's method directly to the Navier-Stokes equations. Working in conjunction with studies based upon numerical simulations of turbulence, they take the view that the triple nonlinearity is of crucial importance near the wavenumber which divides the $U^<$ from the $U^>$. In contrast, McComb has persisted in eliminating this nonlinearity and has concentrated on developing a systematic way of handling the relation between the conditional average and the ensemble average. This has resulted in an improved theory insofar as there is now a range of h for which the Kolmogorov constant does not depend on the shell thickness. More recently still, Lam has developed a phenomenological RG for turbulence, in which there is a single adjustable parameter. The remainder of this section is taken up with short accounts of the essential features of these three theories.

4.1. Recursive renormalization group

If we return to the equilibrium Ising model, in Sec. 2.2, in which the Hamiltonian H_0 involves only nearest-neighbor interactions, it was shown by Wilson (1975) that the effective Hamiltonians H_n ($n \geq 1$) now involve both nearest-neighbor and diagonal nearest-neighbor interactions. Form invariance of H_n is then obtained. Similarly, in our form of recursive RG, as applied to the Navier-Stokes equations, form invariance can be imposed, when both the quadratic and the RG-induced cubic nonlinearities are retained, by truncating the λ-expansion.

The recursive RG theory (Zhou et al., 1988, 1989; Zhou and Vahala, 1992, 1993a,b) extends Rose's (1977) treatment of the linear problem of passive scalar convection to Navier-Stokes turbulence. The resulting renormalized Navier-Stokes contains a triple velocity product and an eddy viscosity term that exhibits a mild cusp behavior (Zhou et al., 1988, 1989). The triple velocity products are absent only when there is a spectral gap between the subgrid and resolvable scales (Zhou et al., 1988). The recursive RG has the potential for capturing a variety of dynamical features that depends on the interplay between local and nonlocal triad interactions (Zhou, 1993a,b).

The RG iterative procedure eliminates the small scales recursively. It was shown that the modifications to the Navier-Stokes equation include (1) a renormalized eddy viscosity and (2) a triple product in the fluid velocity. The eddy viscosity is calculated by means of a difference equation or, in the limit $h \to 1$, by means of a differential equation (Zhou and Vahala, 1993b). Beyond the second iteration, the RG-induced triple products also contribute to the eddy viscosity in the renormalized Navier-Stokes equation. The subgrid velocity autocorrelation function is related to the energy spectrum of the Kolmogorov inertial range. It is of particular interest to determine the effect of the new triple velocity product on the resolvable scale energy transfer and the difference, if any, from that of the usual Navier-Stokes quadratic velocity product, and we shall now do this by considering their individual contributions to the eddy viscosity.

The traditional concept of eddy viscosity (Kraichnan, 1976) is that the subgrid turbulence simply enhances the viscous term in the momentum equation, and that the value of the enhanced eddy viscosity can be determined by equating the resulting increase in the energy dissipation rate to the rate at which energy is transferred from resolved to subgrid scales. Within recursive RG theory (Zhou and Vahala, 1993a), it has been shown that the triple velocity product in the renormalized momentum equation, which produces a fourth order velocity product in the energy equation, removes energy from the resolved scales. Within the recursive RG analysis the averaged fourth-order product can be decomposed into a product of averaged second-order products and formally takes the form of a gradient diffusion process with eddy viscosity ν_T. The triple term contribution to the eddy viscosity is zero as $k \rightarrow 0$ and increases rapidly as $k \rightarrow k_c$. Here k_c is the cutoff wavenumber which separates the resolvable and subgrid scales. The spectral eddy viscosity is simply the sum of the contributions from the momentum equation and the triple nonlinear term. The resulting eddy viscosity appears to be in qualitative agreement with the closure theory and direct numerical measurements. In particular, it predicts the correct asymptotic behaviors of the eddy viscosity as $k \rightarrow 0$ and $k \rightarrow k_c$. The methodology has been applied to the passive scalar being advected by incompressible turbulence (Zhou and Vahala, 1993b).

It is of particular interest to determine the effect of the triple velocity product (or the second term on the RHS of Eq. 10) on the resolvable-scale energy transfer and the difference, if any, from that of the usual Navier-Stokes quadratic velocity product (the third term on the RHS of Eq. (10)), and we do this by considering their individual contributions to the eddy viscosity.

Energy transfer and spectral eddy viscosity can be analyzed using results from direct numerical simulations by introducing an artificial cut at a wavenumber k_c that is smaller than the maximum resolved wavenumber k_m of the simulation. With this fictitious separation between the subgrid and resolvable scales, it is possible to evaluate the effect of the subgrid $k_c < k < k_m$ on the resolved scales $k < k_c$. We form an energy equation from the momentum equation and introduce the following notation: $T^{><}(k)$ and $T^{>>}(k)$ represent the spectrum of energy transfer to mode $\mathbf{k}$ resulting from interactions with *one* and *both* modes above the cutoff k_c respectively. Measurements of numerical simulation databases indicate the following:

• $T^{>>}(k)$ removes energy throughout the resolvable scales in a manner consistent with the notion of eddy viscosity.

• $T^{><}(k)$ removes energy from the last resolved octave that was transferred there by the resolved scale transfer; that is, it allows the local flow of energy through k_c. It is the most important subgrid effect near k_c and accounts for most of the energy flow from the resolved scales.

The subgrid spectral eddy viscosity $\nu^{>>}(k)$ and $\nu^{><}(k)$ can be determined from $T^{>>}(k)$ and $T^{><}(k)$ for a given energy spectrum, $E(k)$. Specifically, $\nu^{>>}(k) = -T^{>>}(k)/2k^2E(k)$ and $\nu^{><}(k) = -T^{><}(k)/2k^2E(k)$. Two important features of the quadratic contribution $\nu^{>>}(k)$ should be stressed. First, its positive constant asymptote at small k indicates that the concept of modeling this contribution as an eddy viscosity in analogy to the molecular viscosity is plausible, and second, its value decreases monotonically as k increases toward k_c. This indicates that if we include only the contribution of quadratic velocity products, there is no eddy viscosity cusp at the cutoff k_c. The most important feature of $\nu^{><}(k)$ is the sharp increase at $k \to k_c$.

4.2. Conditional averaging and renormalization group

In a series of papers (McComb and Watt, 1990; McComb, Roberts, and Watt, 1992; McComb and Watt, 1992), an attempt has been made to put the iterative averaging method on a firmer foundation. This has required the development of quite an elaborate formalism, in which the conditional average is defined by reference to a biassed subensemble, with the usual ensemble average being defined by reference to the full, representative ensemble. The conditional average over the $U^>$ (in which the $U^<$ are held constant) is then related to the unconditional, full ensemble average by means of the decomposition

$$U^> = V^> + \Delta^>, \tag{15}$$

where $V^>$ is independent of $U^<$, and hence can be averaged unconditionally, while $\Delta^>$ represents the effect of mode coupling on the conditional average.

This formalism is quite general for problems with mode-mode coupling, and its implementation for any particular physical system needs an ansatz for the relationship between $V^>$ and $U^>$, such that the conditional average of $\Delta^>$ is small. For the case of turbulence, the belief that the energy cascade is local in wavenumber suggests that an appropriate relationship may be obtained by taking $V^>$ to be given by the first-order expansion of $U^>$ in a Taylor series about $k = k_0$, for the first band of wavenumbers eliminated, and generalizing appropriately for subsequent bands.

If we define the scaled effective viscosity $\tilde{\nu}_n(k')$ for the nth band elimination as

$$\nu_n(k_n k') = \alpha^{1/2}\epsilon^{1/3}k_n^{-4/3}\tilde{\nu}_n(k'), \tag{16}$$

the recursion relation for the scaled viscosity may by written as

$$\tilde{\nu}_{n+1}(k') = h^{4/3}\tilde{\nu}_n(hk') + h^{-4/3}\delta\tilde{\nu}_n(k'). \tag{17}$$

Here α is the Kolmogorov spectral constant. Then, introducing a dimensionless measure of bandwidth thickness η, such that $0 \leq \eta \leq 1$, and which is related to the scale parameter h by $h = 1 - \eta$, the increment in the above equation for the recursion relation may be written as

$$\delta\tilde{\nu}_n(k') = \frac{1}{4\pi k'^2}\int d^3j' \frac{L(\mathbf{k}',\mathbf{j}')Q'}{\tilde{\nu}_n(hj')j'^2 + \tilde{\nu}_n(hl')l'^2}, \tag{18}$$

for the wavenumber bands

$$0 \leq k' \leq 1; 1 \leq j', l' \leq h^{-1},$$

where $l' = \mid \mathbf{k}' - \mathbf{j}' \mid$ and Q' is given by

$$Q' = h^{11/3} - \frac{11}{3}h^{14/3}(l' - h^{-1}) + O(\eta^2). \tag{19}$$

Calculations carried out on this basis yield a value of the Kolmogorov spectral constant $\alpha = 1.60 \pm 0.01$, independent of the bandwidth for the range $0.25 \leq \eta \leq 0.45$. The method has also been applied to passive scalar convection (Watt, 1991), where a fixed point is found corresponding to the renormalized scalar diffusivity. In this calculation, the Obukhov-Corrsin spectral constant was found to be $\beta = 1.02 \pm 0.01$ for bandwidth parameters in the range $0.17 \leq \eta \leq 0.33$. The corresponding turbulent Prandtl numbers were found to take values of the order of $0.6 - 0.7$.

Certain quantitative questions remain to be answered. Is the biassed subensemble so unrepresentative that it is impossible to satisfy the statistical criteria for ensemble averaging? Is the turbulent cascade sufficiently local for a first-order truncation of the Taylor series to be justified? Also, in view of the work of Zhou and Vahala (see preceding section), there is some question about our elimination of the triple nonlinearity. Perhaps this is adequate for global quantities (like the Kolmogorov constant) which are evaluated by means of integrals over wavenumber, but does not give the correct local behaviour near the cut-off wavenumber.

Nevertheless, the numbers which we calculate are good, and the qualitative behaviour of our RG theory is generally now satisfactory. These improvements over the previous iterative averaging theory may be attributed to the fact that our mode-elimination procedure is a rational approximation to the Navier-Stokes equation, in the sense that all terms neglected are of order η^2, or higher, where η is the dimensionless bandwidth parameter.

4.3. Lam's phenomenological renormalization group

Lam's approach (Lam, 1992), although belonging to the FNS school of activity, is quite different from the other members of that group. It does not make use of the λ-expansion procedure used by FNS, nor the correspondence principle and the ϵ-expansion procedure used in the YO theory. Starting with the Navier-Stokes equation *without* an extra external random force, it focuses attention on the extra term—called g_{fast} by Lam—which arises in the momentum equations for the resolved modes as a natural consequence of the mode elimination process. In other words, g_{fast} represents the second and third terms on the right hand side of equation (10). Following previous RG theories, Lam represented the low-k part of g_{fast} by a k-independent eddy viscosity term. The *residual* of g_{fast} is not treated analytically— thus side-stepping the triple nonlinearity problem. Instead, it was modelled (actually, its correlation function) by a profile which peaks in the neighborhood of $k = \Lambda$, the ultra-violet cutoff wavenumber (denoted as k_c elsewhere in this paper). In this way, a physical interpretation could be given to the arbitrarily inserted random force of the YO theory (*i.e.* in effect, this would offer an explanation of their correspondence principle), but it was concluded that this new 'force' must be Λ-dependent— in direct contradiction to the implicit assumption by YO that the force in their theory is scale invariant.

Once the scale invariance hypothesis is abandoned, equation (12) can be rewritten as

$$W(k) = \text{dimensionless constant} \times \varepsilon \times k^{-3}\left(\frac{k}{\Lambda}\right)^{4-\epsilon}, \qquad (20)$$

where (for space dimension $d = 3$) we have made the replacement

$y = \epsilon - 1$. Note that ϵ may take on different values on either side of the peak, although only the ϵ associated with the low k side is relevant.

This choice of the Λ-dependent W profile allows the Kolmogorov energy spectrum to be recovered (and to remain dimensionally consistent at all times) without restricting the value of ϵ to be 4 for all k. Hence the ϵ in (20) is available as an adjustable parameter.

Next, the relation between the dissipation rate ε and the physical flow field of interest was provided by an *ad hoc* expression for the renormalized dissipation rate (Lam, 1992), thus:

$$\varepsilon \equiv \lim_{\Lambda \to \infty} 2\nu_T(\Lambda) \int_0^{\Lambda} k^2 E(k) dk, \tag{21}$$

where ν_T is the k-independent eddy viscosity. This proposal is equivalent to equation (14), which was derived from the Navier-Stokes equation for a more general case. With (21) in place, Lam showed that one single adjustable parameter is sufficient to determine eight scaling law amplitudes of engineering interest.

5. Statement and Critique of the Yakhot-Orszag Proposal

The FNS theory of stirred fluids, as outlined in Section 3.2, generated a great deal of interest. In particular, DeDominicis and Martin (1979) recovered essentially the same results, using field-theoretic methods. They also pointed out that if one chose $y = 3$, in the FNS expression for the force covariance (i.e. equation (12)), then the predicted energy spectrum takes the Kolmogorov form. However, the difficulty with this particular proposal is that, for stationarity, any choice of force covariance must result in a form of $W(k)$ which satisfies equation (6), and maintains the dissipation rate ε. As McComb (1990) has pointed out, their choice of $W_0 = \varepsilon$, in order to specify the dissipation rate, when taken in conjunction with equation (12), means that the force spectrum (and hence the energy spectrum) must be bounded by wavenumbers k_{min} and k_{max}, where $k_{max}/k_{min} = 1.083$!

Yakhot and Orszag (1986: referred to as YO), proposed that such a conjecture could be used for practical calculations. Their proposal was enunciated as the *correspondence principle*, to the effect that the numerical results obtained from an *ersatz* inertial range, produced by strong stirring, would be the same as those appropriate to a real turbulent inertial range. Unlike DeDominicis and Martin (1979), they argued that the relation between the stirring forces and the dissipation rate could be obtained from renormalized perturbation theory, and obtained a result equivalent to $y = 3$ and $W_0 = 11.12\varepsilon$. McComb (1990) has commented that this makes the YO proposal a hybrid of RG and RPT, but Yakhot (1991) has replied that YO was later used to derive the RPT result (Dannevik, Yakhot, and Orszag (1987)). Nevertheless, the fact remains that with this particular relationship between the rate of doing work and the rate of energy dissipation, the wavenumbers bounding the stirring forces (and hence the velocity field) are restricted to be in the ratio $k_{max}/k_{min} = 1.007$.

In dealing with the problem of the triple nonlinearity, YO have argued that RG-induced terms are irrelevant to the renormalized Navier-Stokes equation and eddy viscosity when $\epsilon \to 0$. However, Zhou and Vahala (1993a) have pointed out that there is no good reason why these terms would still be irrelevant at $\epsilon = 4$. Indeed, measurements of numerical simulation datasets have demonstrated that these terms are dominant near the cutoff k_c (see Subsection 4.1).

In the course of developing his own theory, Lam has presented quite an extensive critique of YO. In order to make contact between the two approaches, we observe that the form of $W(k)$ used by YO can be recovered from our equation (20) by setting Λ equal to unity (but it should be noted that the dimension of ε is now ϵ-dependent). Under these circumstances, the equivalent of (20) in YO theory is only dimensionally homogeneous when the expansion parameter $\epsilon = 4$. However, the quantitative results obtained by YO using $\epsilon = 4$ and the λ-expansion did not agree with experiment. As a consequence of their scale invariance hypothesis, the YO theory deduced that ϵ is proportional to λ^2. They argued that since the λ-expansion procedure had already been used, ϵ must therefore also be considered as a small parameter in order to be consistent. Accordingly, they advocated and applied the so-called ϵ-expansion procedure. Operationally, this procedure expands all ϵ-dependence about $\epsilon = 0$ normally —keeping only the leading order term— except that all ϵ appearing in exponents are exempt to preserve dimensional consistency (For a discussion, see: Smith and Reynolds, 1992). The mathematics of this *ad hoc* procedure is clearly unconventional, but good numerical results were obtained. The exception is the decay law of homogeneous isotropic turbulence. Smith and Reynolds showed that the original numerical result for the decay law in YO was derived with an algebraic error, and when the error is corrected the result no longer agrees with experiment. A later paper by Yakhot and Smith (1992) revisited this problem, and found it necessary to introduce an infrared cutoff, along with other approximations, in order to yield an acceptable numerical answer. A detailed critique of the work by Yakhot and Smith can be found elsewhere in this volume (Lam, 1993).

We conclude this particular discussion by drawing a clear distinction between YO as an RG theory of turbulence and YO as a *computational model* of turbulence. Turbulence models are mysterious in their origins and inscrutable in terms of the reasons for their successes and failures. When compared to each other, performance is all; and it would appear that the performance of the YO version of the $k - \varepsilon$ model is very good indeed. So this is certainly not the subject of our criticism (although it has to be said that the YO turbulence model already shows signs of the classic malaise of all turbulence models, in that imperfect predictions are to be treated by fine tuning rather than by a sound, fundamental starting point). No, our strictures are directed at YO when presented as an RG theory. These criticisms have not been rebutted by Yakhot and Orszag, who ignore much other work in the field, including some which substantially predates theirs. One purpose of the present paper is to try to rectify this situation and give a more balanced view of the subject.

6. Conclusion

In their review of RG, as applied to critical phenomena, Wilson and Kogut (1974) draw a distinction between those theories which rely on imponderable partial summation procedures (usually known as renormalized perturbation theories) and those which consciously reduce the number of degrees of freedom and thereby qualify as RG. The RG theories of turbulence summarized in Sections 4.1 and 4.2, do involve the conscious reduction of the number of degrees of freedom (Fourier modes); and they both achieve this by a genuine manipulation of the Navier-Stokes equation in the inertial range of wavenumbers. Their performance offers grounds for optimism, but they are now at the stage where there should be a concerted effort to reconcile the differences between them.

The phenomenological theory presented in Section 4.3, offers a completely new approach, which could be the basis of an honest computational model for turbulence, with the introduction of an adjustable constant even in the isotropic case. Certainly our discussions suggested that Lam's model could be interpreted in terms of the more 'microscopic' theories, and we are convinced that an interaction between the three methods is likely to prove fruitful in increasing our understanding in this area.

References

Dannevik, W. P., Yakhot, V. and Orszag, S. A., 1987. "Analytical theories of turbulence and the ϵ expansion," *Phys. Fluids* **30**, p. 2021.

DeDominicis, C. and Martin, P. C., 1979. "Energy spectra of certain randomly-stirred fluids," *Phys. Rev. A* **19**, p. 419.

Forster, D., Nelson, D. R., and Stephen, M. J., 1976. "Long-time tails and the large-eddy behavior of a randomly stirred fluid," *Phys. Rev. Lett.* **36**, p. 867.

Forster, D., Nelson, D. R., and Stephen, M. J., 1977. "Large-distance and long-time properties of a randomly stirred fluids," *Phys. Rev. A* **16**, p. 732.

Kraichnan, R. H., 1976. "Eddy viscosity in two and three dimensions," *J. Atmos. Sci.* **33**, p. 1521.

Lam, S. H., 1992. "On the RNG theory of turbulence," *Phys. Fluids A* **4**, p. 1007.

Lam, S. H., 1993. "On RNG theory and the decay law of homogeneous isotropic turbulence," to appear in this volume.

McComb, W. D., 1982. "Reformulation of the statistical equations for turbulent shear flow," *Phys. Rev. A* **26**, p. 1078.

McComb, W. D., 1986. "Application of renormalization group (RG) methods to the subgrid modeling problems," in *Direct and large eddy simulation of turbulence* (eds U. Schumann and R. Friedrich.) *Notes on numerical fluid mechanics*, Vol. 15. Vieweg, Braunschweig.

McComb, W. D., 1990. *The Physics of Fluid Turbulence*, Oxford University Press.

McComb, W. D., Roberts, W., and Watt, A. G., 1992. "Conditional averaging procedure for problems with mode-mode coupling," *Phys. Rev. A* **45**, p. 3507.

McComb, W. D. and Shanmugasundaram, V., 1983. "Fluid turbulence and the renormalization group: a preliminary calculation of the eddy viscosity," *Phys. Rev. A* **28**, p. 2588.

McComb, W. D. and Watt, A. G., 1990. "Conditional averaging procedure for the elimination of the small-scale modes from incompressible fluid turbulence at high Reynolds numbers," *Phys. Rev. Lett.* **65**, p. 3281.

McComb, W. D. and Watt, A. G., 1992. "Two-field theory of incompressible fluid turbulence," *Phys. Rev. A* **46**, p. 4797.

Rose, H. A., 1977. "Eddy diffusivity, eddy noise and subgrid scale modeling," *J. Fluid Mech.* **81**, p. 719.

Smith, L. M. and Reynolds, W. C., 1992. "On the Yakhot-Orszag renormalization group method for deriving turbulence statistics and models," *Phys. Fluids A* **4**, p. 364.

Watt, A. G., 1991. "A study of isotropic turbulence," Ph.D. thesis, University of Edinburgh.

Wilson, K. G., 1975. "The renormalization group: critical phenomena and the Kondo problem," *Rev. Mod. Phys.* **47**, p. 773.

Wilson, K. G. and Kogut, J., 1974. "The renormalization group and the ϵ expansion," *Phys. Rep.* **12C**, p. 75.

Yakhot, V., 1991. Personal communication to W. D. McComb.

Yakhot, V. and Orszag, S. A., 1986. "Renormalization group analysis of turbulence theory. I. Basic theory," *J. Sci. Comput.* **1**, p. 3.

Yakhot, V. and Smith, L. M., 1992. "The renormalization group, the ϵ expansion, and derivation of turbulence models," *J. Sci. Comput.* **7**, p. 35.

Zhou, Y., 1993a. "Degrees of locality of energy transfer in the inertial range," *Phys. Fluids A* **5**, p. 1092.

Zhou, Y., 1993b. "Interacting scales and energy transfer in isotropic turbulence," *Phys. Fluids A* **5**, p. 2511.

Zhou, Y. and Vahala, G., 1992. "Local interactions in renormalization methods for Navier-Stokes turbulence," *Phys. Rev. A* **46**, p. 1136.

Zhou, Y. and Vahala, G., 1993a. "Reformulation of recursive renormalization group based subgrid modeling of turbulence," *Phys. Rev. E* **47**, p. 2503.

Zhou, Y. and Vahala, G., 1993b. "Renormalization group estimates of transport coefficients in the advection of a passive scalar by incompressible turbulence," ICASE Report 93-1, to appear in *Phys. Rev. E.*

Zhou, Y., Vahala, G., and Hossain, M., 1988. "Renormalization group theory for the eddy viscosity in subgrid modeling," *Phys. Rev. A* **37**, p. 2590.

Zhou, Y., Vahala, G., and Hossain, M., 1989. "Renormalized eddy viscosity and Komogorov's constant in forced Navier-Stokes turbulence," *Phys. Rev. A* **40**, p. 5865.

THE YAKHOT-ORSZAG THEORY AND LOCAL INTERACTIONS

R. Rubinstein

Institute for Computational Mechanics in Propulsion
NASA Lewis Research Center
Cleveland, Ohio 44135

ABSTRACT

The connection between the ϵ expansion and the distant interaction approximation in the Yakhot-Orszag turbulence theory is discussed.

The ϵ expansion of the (Yakhot and Orszag, 1986) turbulence theory has been interpreted (Kraichnan, 1987) as an extrapolation from a theory in which distant interactions, amenable to description in terms of an eddy viscosity, are weakly dominant to the physical case of Kolmogorov turbulence. Recent formulation (Zhou and Vahala, 1992) of a renormalization group theory in which local interactions are not ignored suggests further investigation of the role of distant and local interactions in the Yakhot-Orszag theory.

The DIA response equation for stationary isotropic turbulence is (Kraichnan, 1959)

$$\frac{\partial}{\partial t}G(k,t-s)+\int_s^t dr\ \eta(k,t-r)\ G(k,r-s)=0 \tag{1}$$

where G is the Green's function and the damping function η is defined by

$$\eta(k,\sigma)=\int_{\mathbf{k}=\mathbf{p}+\mathbf{q}}d\mathbf{p}d\mathbf{q}\ \beta(k,p,q)G(p,\sigma)U(q,\sigma) \tag{2}$$

where $U(q,\sigma)$ is the stationary two time correlation function and

$$\beta(k,p,q)=P_{rmn}(\mathbf{k})P_{nrs}(\mathbf{p})P_{ms}(\mathbf{q}). \tag{3}$$

Leslie's evaluation (Leslie, 1972) of the inertial range constants assumes the long time form

$$G(p,\sigma)=e^{-\theta(p)\sigma},\ \theta(p)=C_D\varepsilon^{1/3}p^{2/3} \tag{4}$$

M. Y. Hussaini et al. (eds.), Transition, Turbulence and Combustion, Vol. II, 163–166.

and the fluctuation dissipation relation

$$U(p,\sigma) = G(p,\sigma)C_K\varepsilon^{2/3}p^{-11/3}/2\pi. \tag{5}$$

The failure of this calculation due to the divergence of Eq. (2) for the Kolmogorov forms (4) and (5) when $p \sim 0$ is well known (Kraichnan, 1964; Leslie, 1972). The divergence can be removed (Kraichnan, 1964) by restricting the integration in Eq. (2) to the region $\{p \geq k \text{ or } q \geq k\}$. Let us perform this calculation, replacing (4) and (5) by the general scaling forms

$$\theta(p) = C_D(\epsilon)D^{1/3}p^{2-\epsilon/3} \tag{6}$$

and

$$E(p) = C_K(\epsilon)D^{2/3}p^{1-2\epsilon/3} \tag{7}$$

where D and ϵ are as in the Yakhot-Orszag theory (Yakhot and Orszag, 1986). Following (Leslie, 1972), set $s = 0$ in Eq. (1) and integrate over all t. The result is

$$\begin{aligned} 1 &= \frac{1}{2}\frac{C_K}{C_D^2}k^{-2+\epsilon/3}\int_{p\geq k} dp\,\left\{\frac{\beta(k,p,q)p^{1-2\epsilon/3}}{p^{2-\epsilon/3}+q^{2-\epsilon/3}} + \frac{\beta(k,p,q)q^{1-2\epsilon/3}}{p^{2-\epsilon/3}+q^{2-\epsilon/3}}\right\} \\ &\equiv \frac{C_K}{C_D^2}k^{-2+\epsilon/3}Z(k). \end{aligned} \tag{8}$$

Now expand the integrand in Eq. (8) in powers of k/p; the first term in the expansion is the *distant interaction limit* (Kraichnan, 1987). The result has the form

$$\begin{aligned} Z(k) &= A_1(\epsilon)k^2\int_k^\infty dp\,p^{-1-\epsilon/3} + A_2(\epsilon)k^3\int_k^\infty dp\,p^{-2-\epsilon/3} + \cdots \\ &= k^{2-\epsilon/3}[A_1(\epsilon)\tfrac{3}{\epsilon} + A_2(\epsilon)\tfrac{1}{1+\epsilon/3} + A_3(\epsilon)\tfrac{1}{2+\epsilon/3} + \cdots] \end{aligned} \tag{9}$$

where

$$A_1(\epsilon) = \frac{1}{2}\frac{d^2-d-\epsilon}{d(d+2)} = \frac{3-\epsilon/2}{15} \text{ for } d=3 \tag{10}$$

and $A_n(\epsilon)$ is a polynomial of degree n in ϵ. Substituting Eq. (9) in Eq. (8), there results

$$1 = \frac{C_K}{C_D^2}[A_1(\epsilon)\frac{3}{\epsilon} + A_2(\epsilon)\frac{1}{1+\epsilon/3} + A_3(\epsilon)\frac{1}{2+\epsilon/3} + \cdots]. \tag{11}$$

To lowest order in ϵ, using Eq. (10),

$$1 = \frac{C_K}{C_D^2} A_1(0) \frac{3}{\epsilon} = \frac{C_K}{C_D^2} \frac{3}{5\epsilon} \tag{12}$$

which is equivalent to the calculation of the fixed point value of the renormalized coupling constant in (Yakhot and Orszag, 1986). Eq. (12) has also been stated in (Woodruff, 1992).

Previous speculation (Yakhot and Orszag, 1990) on the ϵ expansion has focused on its role as an expansion in the connectivity type of diagrams, much as it occurs in the theory of critical phenomena. Thus, the lowest order term approximately sums all "bubble" diagrams (Orszag, 1972) obtained by repeatedly replacing propagator lines by single loop corrections in all possible ways. The next order includes contributions from diagrams built up the same way from the lowest order "crossing" diagram (Orszag, 1972, Fig. 6.3) and so on. But Eq. (11) suggests another possibility, that corrections to the distant interaction approximation of the Yakhot-Orszag theory also generate a series expansion in ϵ.

The behavior of Eq. (9) when $\epsilon = 0$ can be understood in terms of the dominance of distant interactions in the logarithmic theory defined by $\epsilon = 0$ (Kraichnan, 1987; Fournier and Frisch, 1983) in which $E(k) \sim k^1$ and $\nu(k) \sim \log(k)$. DIA itself does not suggest bringing this theory into consideration; in fact, for DIA without the regularization of Eq. (8), (Woodruff, 1992) suggests a transition when $\epsilon \to 3$ to a dominant sweeping interaction. Regularization appears to suppress this crossover.

Accepting an expansion about the logarithmic theory with $\epsilon = 0$ as valid for turbulence, there is no doubt that Eq. (12) is the consistent lowest order approximation. The deeper question remains in what sense the sum in Eq. (11) is well approximated by the lowest order term, Eq. (12).

References

Fournier, J. D. and Frisch, U., 1983. "Remarks on the renormalization group in statistical fluid dynamics," *Phys. Rev. A* **28**, p. 1000

Kraichnan, R. H., 1959. "The structure of turbulence at very high Reynolds number," *J. Fluid Mech.* **5**, p. 497.

Kraichnan, R. H., 1964. "Kolmogorov's hypothesis and Eulerian turbulence theory," *Phys. Fluids* **7**, p. 1030.

Kraichnan, R. H., 1987. "An interpretation of the Yakhot-Orszag turbulence theory," *Phys. Fluids* **30**, p. 2400.

Leslie, D. C., 1972. "Modern developments in the theory of turbulence," Oxford University Press.

Orszag, S. A., 1972. "Lectures on the statistical theory of turbulence," in *Fluid Dynamics*, edited by R. Balian and J.-L. Preube (Gordon and Breach, New York).

Woodruff, S. L., 1992. "Dyson equation analysis of inertial range turbulence," *Phys. Fluids A* **4**, p. 1077.

Yakhot, V. and Orszag, S. A., 1986. "Renormalization group analysis of turbulence. I. Basic theory," *J. Sci. Comput.* **1**, p. 3.

Yakhot, V. and Orszag, S. A., 1990. "Analysis of the ϵ expansion in turbulence theory: Approximate renormalization group for diffusion of a passive scalar in a random velocity field," Princeton University preprint.

Zhou, Y., and Vahala, G., 1992. "Local interactions in renormalization methods for Navier-Stokes turbulence," *Phys. Rev. A.* **46**, p. 1136.

THE APPLICATION OF A RECURSION *RNG*-BASED TURBULENCE MODEL FOR SEPARATED FLOWS[1]

Siva Thangam[2], *Ye Zhou, and George Vahala*[3]

Institute for Computer Applications in Science and Engineering
NASA Langley Research Center
Hampton, Virginia 23681-0001

ABSTRACT

Recursion renormalization group theory based Reynolds stress closure models are developed for the prediction of turbulent separated flows. The model uses a finite wavenumber truncation scheme to account for the spectral distribution of energy and incorporates both local and nonlocal interactions. The nonlocal interactions are shown to yield a contribution identical to that from the small perturbation *RNG* (ϵ-RNG) model, while the local interactions introduce higher-order dispersive effects. The development of the model is summarized and its ability to accurately predict separated flows is analyzed. Turbulent flow past a backward facing step is chosen as a test case and the results obtained based on detailed computations demonstrate the viability of the proposed recursion - *RNG* model.

1. Introduction

Turbulent flows of practical interest involve a broad spectrum of length and time scales and require some type of modeling for Reynolds stresses. The most widely used are the two-equation turbulence models based on the transport equations for parameters that involve the length and the time scales and require the simplest level of Reynolds stress closure that do not depend specifically on the flow geometry. In its standard form the two-equation Reynolds stress turbulence models involve the turbulence kinetic energy and dissipation based on a Boussinesq type approximation involving isotropic eddy

---

[1]This research was supported by the National Aeronautics and Space Administration under NASA Contract No. NAS1-19480 while the authors were in residence at the Institute for Computer Applications in Science and Engineering (ICASE), NASA Langley Research Center, Hampton, VA 23681.

[2]Permanent address: Stevens Institute of Technology, Hoboken, NJ 07030.

[3]Permanent address: The College of William and Mary, Williamsburg, VA 23187.

M. Y. Hussaini et al. (eds.), Transition, Turbulence and Combustion, Vol. II, 167–178.

viscosity. Such a representation of turbulence is often not effective from both theoretical as well as phenomenological point of view. To over come this, models that are nonlinear (i.e., quadratic) in the mean strain rate were proposed in the form of a constitutive relation. These models are capable of predicting the anisotropy in the Reynolds stresses in complex flows but require empirical evaluation of the model constants (Speziale, 1991).

The present study addresses the need for a more effective approach in the development of two-equation turbulence models, and in this context the renormalization group (RNG) theory based models are considered. These models fall into two distinct categories: (a) $\epsilon - RNG$, where a small parameter ϵ is introduced into the exponent of the forcing correlation function (with the forcing function being introduced into the momentum equation), and the theory is then developed for $\epsilon \ll 1$ (Yakhot and Orszag, 1986); and (b) recursion-RNG, which does not rely on an ϵ-expansion, and treats explicitly the cubic nonlinearities introduced into the renormalized momentum equation (Zhou, et al. 1988). It should be noted that in the $\epsilon - RNG$ while all constants generated are evaluated in the limit $\epsilon \ll 1$, at the same time, all exponents that are ϵ-dependent are evaluated at $\epsilon = 4$ (Lam, 1992). In fact, $\epsilon = 4$ is required in the $\epsilon - RNG$ to recover the Kolmogorov energy spectrum in the inertial range (Lam, 1992). In addition, $\epsilon - RNG$ theory can only take into account non-local interactions (Smith and Reynolds, 1992).

On the other hand, recursion-RNG does not rely on an ϵ-expansion, and treats explicitly the cubic nonlinearities introduced into the renormalized momentum equation. Moreover, recursion-RNG can handle both local and non-local interactions. Effects such as the cusp behavior in the transport coefficients (Zhou and Vahala, 1993a) which are due to local interactions and the cubic nonlinearities are recovered in these theories. As in $\epsilon - RNG$, the eddy viscosity is readily determined from the solution of a relatively simple differential equation (Zhou and Vahala, 1993a). However, unlike $\epsilon - RNG$, the transport coefficients are determined over the whole resolvable scales and not just in the wavenumber limit $k \to 0$.

In the present work, the recursion RNG based model is utilized for the analysis of turbulent separated flows. The model introduces two additional terms arising from the local interactions effects that are of higher order than those obtained using the $\epsilon - RNG$ method. The first of these two terms as well as those from the conventional turbulence models are shown to be a part of the integrity basis used in the representation of the anisotropic part of the Reynolds stress tensor (Gatski and Speziale, 1993). The second term which arises from pressure-strain coupling is quartic in strain rate.

The turbulent flow past a backward-facing step which has played a central role in benchmarking the performance of turbulence models for separated flows is analyzed using the recursion-RNG model. During the past decade - beginning with the 1980/81 Stanford conference on complex turbulent flows - various two-equation turbulence models have been tested and compared with the experimental data of Eaton and Johnston (1981) for the backstep problem. Initial results indicated that the standard $K-\varepsilon$ model, with wall functions, underpredicted the reattachment point by a substantial amount on the order of 20-25%. Several independent studies have been subsequently published using alternative forms of the $K-\varepsilon$ model wherein a variety of conflicting results have been reported. Considering the need to accurately predict separated turbulent flows the proposed model is applied for the backstep problem. The computations based on a sufficiently resolved finite-volume algorithm show that the proposed model based on the recursive application of the renormalization group theory (developed independently without any *ad hoc* empiricism) can yield a prediction for the reattachment point that is within a few percent of the experimental result.

2. Formulation of the Physical Problem

The turbulent motion of viscous, incompressible fluids are governed by the Navier-Stokes equations which may be expressed using the following Fourier representation:

$$(\partial_t + \nu_0 k^2)u_i(\boldsymbol{k},t) = M_{i\alpha\beta}(k)\int d^3p\, u_\alpha(\boldsymbol{p},t)u_\beta(\boldsymbol{k}-\boldsymbol{p},t) \qquad (1)$$

$$k_\alpha u_\alpha(\boldsymbol{k},t) = 0. \qquad (2)$$

Here, u_i is the velocity field in the Fourier space, ν_0 is the molecular viscosity, k and p are the wavenumbers and t is the time. It is important to note here that *no* random forcing is introduced unlike in the $\epsilon - RNG$ theories where it plays a critical role in introducing the small parameter ϵ. The non-linear coupling coefficient

$$M_{i\alpha\beta}(k) = k_\beta D_{i\alpha}(k) + k_\alpha D_{i\beta}(k) \qquad (3)$$

with $D_{\alpha\beta}$ being the projection operator defined by

$$D_{\alpha\beta}(k) = \delta_{\alpha\beta} - k_\alpha k_\beta / k^2. \tag{4}$$

A scale factor h is now introduced to partition the wavenumber space to N segments. The velocity field is then decomposed in the Fourier space such that $u_i = u_i^{>}\theta(k - k_h) + u_i^{<}\theta(k_h - k)$ (wherein θ is the Heavyside unit step function, $k_{h(h=1,N)}$ is the local cut-off wavenumber, the superscript $<$ represents large resolvable scale quantities while $>$ represents small unresolved scales with respect to the local cut-off wavenumber). The evolution of these components can be directly obtained from (1). The contributions to the Reynolds stress tensor τ_{ij} from the conventional time-averaging of the equations of motion are due to the nonlinear coupling term and its interaction with the velocity field (Zhou et al., 1993). In particular, the Reynolds stress tensor, $\tau_{ij} = \tau_{ij}^{\gg} + \tau_{ij}^{><}$, where the $\tau_{ij}^{\gg}$ - part arises from the infrared limit of $k \to 0$ and is due to the $u_i^{>} - u_j^{>}$ distant interaction limit while the $\tau_{ij}^{><}$ - part arises from the $0 < k \leq k_c$ spectrum and is due to the $u_i^{>} - u_j^{<}$ local interaction limit.

It should be noted that in $\epsilon - RNG$ approach, one is forced into taking the large-scale infrared limit $k \to 0$. In essence, this forces a spectral gap between the resolvable part of the flow field and the small unresolved scales. If this spectral gap were somehow present initially, it would be quickly populated in just a few eddy turn over times. Thus, retaining only the distant interactions may not be appropriate. In fact, it has been shown (Zhou and Vahala, 1993a) that the energy transfer function that corresponds to local interactions accounts for most of the energy flow out of the resolvable scales. It is, therefore, important to retain both local and nonlocal interactions in the modeling of the Reynolds stress and this can be readily achieved by recursion-RNG. But in the $\epsilon - RNG$ model, the Reynolds stress $\tau_{ij} = \tau_{ij}^{>>}$ and is obtained purely from the $u_i^{>} - u_j^{>}$ interaction in the small unresolved scale momentum equation.

Now, if one follows standard recursion-RNG procedures (Zhou et al., 1993), the relevant part of the small scale velocity field that contributes to $\tau_{ij}^{><}$ is obtained in the wavenumber space. This could be mapped to the physical space and if the resulting kernels are approximated by considering the flow parameters (Zhou et al., 1993) the following algebraic representation is obtained.

$$\tau_{ij}^{><}(\boldsymbol{x}) = C_{R1}\frac{K^4}{\varepsilon^3}[\partial_\alpha U_i \partial_\beta U_j \partial_\alpha U_\beta + i \leftrightarrow j] + C_{R2}\frac{K^7}{\varepsilon^5}[(\partial_\alpha U_i)\partial_j(\partial_\gamma U_\beta \partial_\alpha \partial_\beta U_\gamma) + i \leftrightarrow j] \tag{5}$$

where U_i is the time-averaged mean velocity (and the second term inside the square brackets is obtained by switching the indices). However, for most flows of interest, k_c varies considerably throughout the flow domain. Thus, if one wishes to use the local approximation for the nonlocal kernels by the use of delta functions, as well as retain $k_c \sim O(1)$ effects so as to obtain an algebraic form for $\tau_{ij}^{><}(\boldsymbol{x})$, then the coefficients C_{R1} and C_{R2} will be function of the flow quantities. As a first attempt at applying this recursion-RNG model we make a lowest order approximation that these coefficients are constants. In particular, for the present analysis, C_{R1} and C_{R2} are taken to be 0.025 and 0.342×10^{-3}, respectively (Zhou et al., 1993).

We note here that Rubinstein and Barton (1990) have derived a Reynolds stress model using the $\epsilon - RNG$ method (which corresponds to the infrared limit of $k \to 0$) of the following form:

$$\tau_{ij}^{\gg} = -\frac{2}{3}K\delta_{ij} + v_T(\partial_j U_i + \partial_i U_j) - \frac{K^3}{\varepsilon^2}[C_{\tau 1}(\partial_\alpha U_i \partial_\alpha U_j)^* + C_{\tau 2}(\partial_\alpha U_i \partial_i U_\alpha + \partial_\alpha U_j \partial_i U_\alpha)^* + C_{\tau 3}(\partial_i U_\alpha \partial_j U_\alpha)^*]. \tag{6}$$

The constants $C_{\tau 1} = 0.034$, $C_{\tau 2} = 0.104$ and $C_{\tau 3} = -0.014$, and $(\ldots)^*$ denotes the deviatoric part of the expression within the parenthesis. The first two terms correspond to the linear model and $v_T = C_\mu K^2/\varepsilon$ is the isotropic eddy-viscosity, where ε is the turbulence dissipation and $C_\mu \approx 0.09$ based on empirical data from equilibrium boundary layer flows. The above model is quadratic in mean strain rate, includes the effect of convection and diffusion and is qualitatively similar to other second order models (Speziale, 1991).

Combining (5) - (6) a formal expression for Reynolds stress which includes both the local and nonlocal interactions may be obtained. It is of some interest to note that integrity basis representations are commonly employed to represent the anisotropic part of the Reynolds stress tensor for three dimensional turbulent flows based on a systematic derivation from a hierarchy of second-order closure models

(Gatski and Speziale, 1993). It can readily be shown that the tensors that constitute the integrity basis are recovered for most part when the proposed recursion-RNG model is recast appropriately (Zhou et al., 1993).

The above expression for Reynolds stress (5,6) are to be used along with the equations of motion by specifying turbulent kinetic energy and dissipation. In two-equation turbulence models, this closure is achieved through the development of transport equations for the turbulent kinetic energy and dissipation - quantities that are directly related to the length and time scales - of the following general form:

$$\partial_t K + U_j \partial_j K = \mathcal{P} - \varepsilon + \partial_i[(v_0 + v_T/\alpha_K)\partial_i K] \tag{7}$$

$$\partial_t \varepsilon + U_j \partial_j \varepsilon = C_{\varepsilon 1}\mathcal{P}\varepsilon/K - C_{\varepsilon 2}\varepsilon^2/K + \partial_i[(v_0 + v_T/\alpha_\varepsilon)\partial_i \varepsilon] - \mathcal{R} \tag{8}$$

where, $\nu = \nu_0 + \nu_T$ is the total viscosity, $\mathcal{P} = -\tau_{ij}(\partial U_i/\partial x_j)$ is the turbulence production, ε is the scalar turbulent dissipation rate, and $\mathcal{R} = 2\nu_0 S_{ij}\overline{(\partial u_j/\partial x_i)(\partial u_i/\partial x_j)}$ is the turbulent strain rate correlation term. The quantities $C_{\varepsilon 1}$, $C_{\varepsilon 2}$, α_K, α_ε are dimensionless and taken to be 1.44, 1.92, 1.0 and 1.3, respectively, consistent with the standard form of the two-equation $K - \varepsilon$ model (based on empirical data obtained from equilibrium boundary layer flows). It should be noted here that these are calculated explicitly in the $\varepsilon - RNG$-based formulations to be 1.42, 1.68, 0.719 and 0.719, respectively (Yakhot and Orszag, 1986). In addition, it is customary to neglect the strain rate correlation term $\mathcal{R}$ in the standard $K - \varepsilon$ model as well as in some RNG-based models since R must be modeled in order to achieve closure. Special forms of approximations have been recently employed in the case of some of the $\epsilon - RNG$-based models (Yakhot et al., 1991). To avoid ambiguity (and to avoid introducing *ad hoc* assumptions required to model $\mathcal{R}$) the contribution from the turbulent strain rate correlation term is neglected in the present work.

3. Turbulent Flow Past a Backward-Facing Step: A Case Study

Turbulent flow over a backward facing step (Figure 1) is one of the standard test cases used to evaluate the performance of turbulence models in separated flows. Computations are performed for the configuration wherein the expansion ratio, E is 1:3 (i.e., step height to outlet channel height, $H : H_2$) and the Reynolds number $Re = 132{,}000$ (based on the inlet centerline mean velocity and outlet channel height). This configuration was selected since it is a standard test case and because of the available experimental data (Eaton and Johnston, 1981).

The mean turbulence equations are solved using a finite-volume scheme with a 200×100 nonuniform mesh which is known to yield results that are essentially grid independent (Thangam and Speziale, 1991). Inlet profiles are specified five step heights upstream of the step corner based on a separate set of calculations for developing turbulent flow in a channel that matches the experimental conditions of Eaton and Johnston (1981). Conservative extrapolated outflow conditions are applied thirty step heights downstream of the step corner. Along the upper and the lower walls as well along the inside of the step, the standard law of the wall conditions are imposed (Thangam and Speziale, 1991). The computed solution was assumed to have converged to its steady state when the magnitude of the relative average difference between successive iterations for all the flow quantities and the location of the reattachment point was less than 10^{-4}. The computational results are presented in Figures 2-6 and compared with the experimental data of Eaton and Johnston (1981). Approximately 2000 iterations were needed for the convergence of the standard $K-\varepsilon$ model; this corresponds to approximately 20 minutes of CPU time in a partially vectorized mode on the Cray-YMP supercomputer using 64-bit precision. The recursive RNG-based $K-\varepsilon$ model requires approximately 33% more CPU time due to the fact that the additional terms in the RNG-based $K-\varepsilon$ model have to be evaluated during each iteration.

In Figure 2 the computed streamlines are for the standard $K-\varepsilon$ model are shown, indicating reattachment at $X_r/H \approx 6.12$ - a result which is about 15% underprediction of the experimental reattachment point of $X_r/H \approx 7.1$. Now, it will be demonstrated that the use of the proposed recursive RNG-based model can yield a more significant improvement in the results. The computed streamlines for the flow field shown in Figure 3a have a mean reattachment point, $X_r/H \approx 6.72$, a result which is about 5% lower than the experimental value. The corresponding mean velocity profiles shown in Figure 3b indicate very good agreement with the experimental results. It should be noted that computations performed based on the $\epsilon-RNG$-based model for the same flow conditions and model constants yield a mean reattachment point, $X_r/H \approx 6.42$ (not shown herein). This difference in the size of the separated flow region is clearly due to the contributions from the Reynolds stress terms representing the local interactions effects in the recursion RNG model.

Furthermore, the overall agreement between the turbulence intensity and the shear stresses with the experimental data shown in Figures 4 and 5 are also good. In Figure 6, the pressure coefficient (based on the reference pressure and the inlet centerline velocity) obtained from the standard and recursion RNG-based $K - \varepsilon$ models at the bottom wall are compared with the experimental data of Eaton and Johnston (1981). As can be seen, the recursion-RNG-based $K - \varepsilon$ model performs comparably well in reproducing the experimental trends.

4. Conclusions

The recursion renormalization group (RNG) theory is utilized to develop Reynolds stress closure models for the prediction of turbulent separated flows. Unlike the small parameter $\epsilon - RNG$ model that only consider the interaction of the long wavelength modes, the recursion-RNG model includes both the local and nonlocal interaction of all the relevant resolvable scales. The ability of the proposed model to accurately predict separated flows is analyzed from a combined theoretical and computational standpoint by considering turbulent flow past a backward facing step as a test case. The results obtained based on detailed computations demonstrate that the proposed recursion-RNG model can yield very good predictions for the turbulent flow of an incompressible viscous fluid over a backward-facing step. Thus, in spite of its well known the deficiencies, when the anisotropy of the turbulent stresses are properly accounted for, the two-equation turbulence models can be quite effective for the prediction of turbulent separated flows.

Acknowledgments

The authors are indebted to Drs. T. B. Gatski and M. Y. Hussaini for their helpful comments and encouragement during the course of this work.

References

Eaton, J. and Johnston, J. P., 1981. "Turbulent flow reattachment: An experimental study of the flow and structure behind a backward-facing step," Technical Report MD-39, Stanford University, CA.

Gatski, T. B. and Speziale, C. G., 1993. NASA-ICASE Report 92-58, to appear in *J. Fluid Mech.*

Lam, S. H., 1992. "On the RNG theory of turbulence," *Phys. Fluids A* **4**, p. 1007.

McComb, W. D., 1990. *The Physics of Fluid Turbulence*, Clarendon Press.

Rubinstein, R. and Barton, J. M., 1990. "Nonlinear Reynolds stress models and the renormalization group," *Phys. Fluids A* **2**, p. 1472.

Smith, L. M. and Reynolds, W. C., 1992. "On the Yakhot-Orszag renormalization group method for deriving turbulence statistics and models," *Phys. Fluids A* **4**, p. 364.

Speziale, C. G., 1991. "Analytical methods for the developments of Reynolds stress closure," *Ann. Rev. Fluid Mech.* **23**, p. 107.

Thangam, S. and Speziale, C. G., 1992. "Turbulent separated flow past a backward-facing step: A critical evaluation of two-equation turbulence models," *AIAA Journal* **30**, p. 1314.

Yakhot, V. and Orszag, S. A., 1986. "Renormalization group analysis of turbulence. I. Basic Theory," *J. Sci. Comput.* **1**, p. 3.

Yakhot, V., Orszag, S. A., Thangam, S., Gatski, T. B., and Speziale, C. G., 1992. "Development of turbulence models for shear flows by a double-expansion technique," *Phys. Fluids A* **4**, p. 1510.

Zhou, Y., Vahala, G., and Hossain, M., 1988. "Renormalization-group theory for the eddy viscosity in subgrid modeling," *Phys. Rev. A* **37**, p. 2590.

Zhou, Y. and Vahala, G., 1993a. "Reformulation of recursive-renormalization-group-based subgrid modeling of turbulence," *Phys. Rev. E* **47**, p. 2503.

Zhou, Y. and Vahala, G., 1993b. "Renormalization group estimates of transport coefficients in the advection of a passive scalar by incompressible turbulence," *Phys. Rev. E*, submitted; also available as ICASE Report 93-1.

Zhou, Y., Vahala, G., and Thangam, S., 1993. "Development of a recursion-RNG based turbulence model," *Phys. Rev. E*, submitted; also available as ICASE Report 93-51.

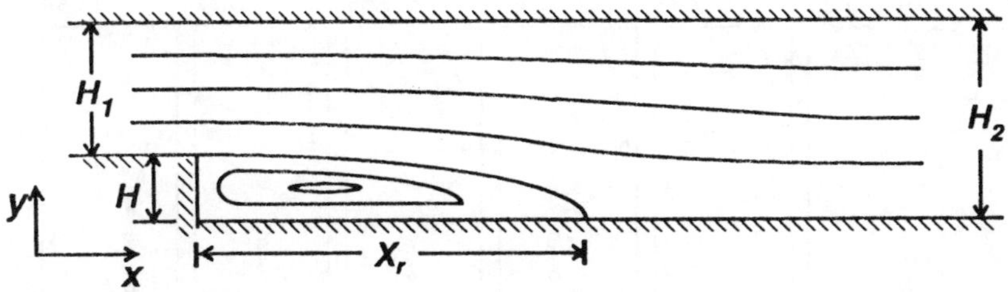

Figure 1. Schematic of turbulent flow over a backward facing step

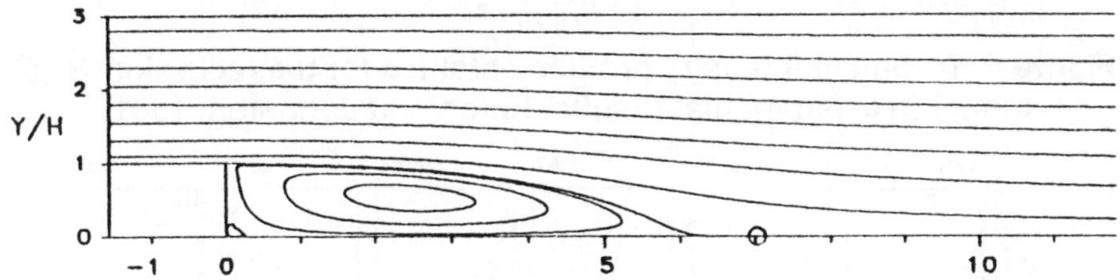

Figure 2. Mean flow streamlines obtained from the standard *K*-ε model (o experimental mean reattachment point)

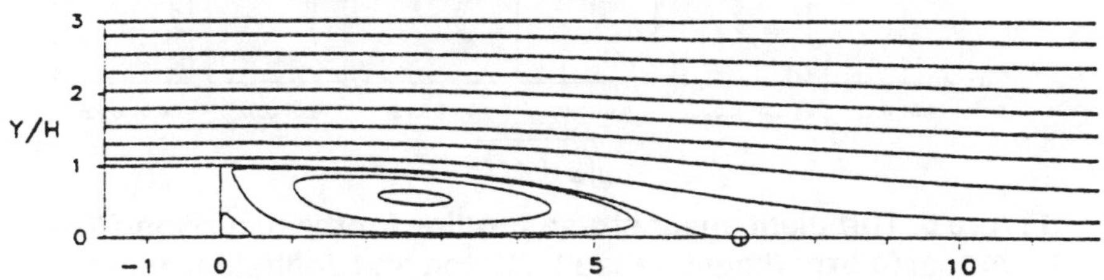

Figure 3a. Mean flow streamlines obtained for the recursion-*RNG* model (o experimental mean reattachment point)

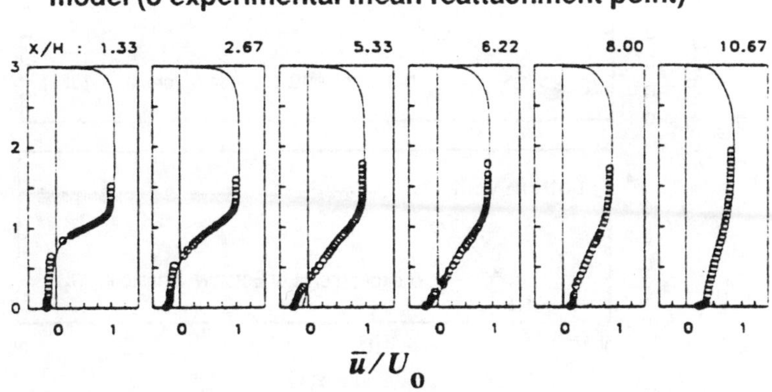

Figure 3b. Mean flow velocity profiles for the recursion-*RNG* model (o experimental results, Eaton and Johnston, 1980)

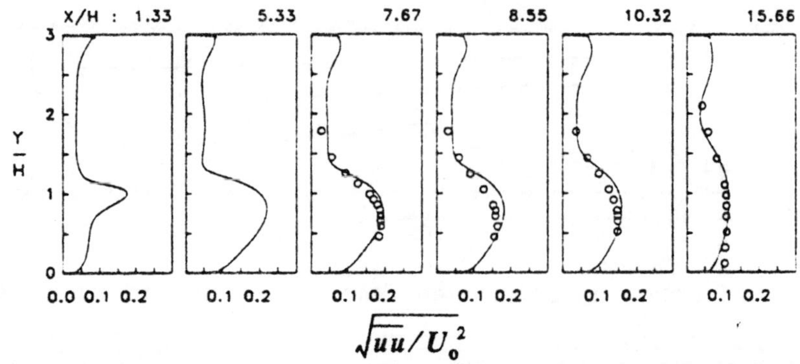

Figure 4. Turbulent intensity profiles obtained for the recursion-*RNG* model (o experimental results, Eaton and Johnston, 1981)

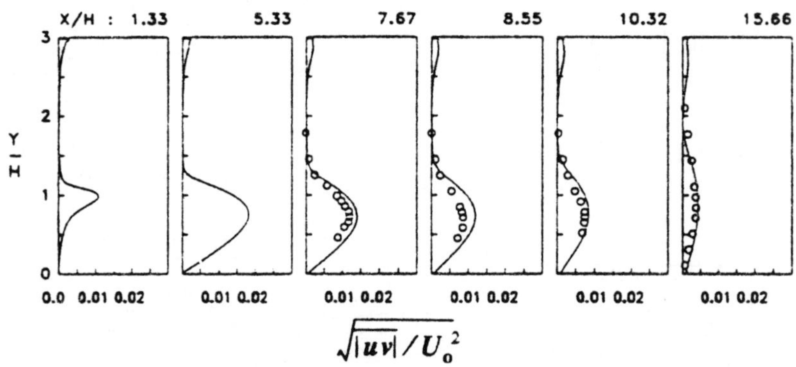

Figure 5. Turbulent shear stress profiles for the recursion-*RNG* model (o experimental results, Eaton and Johnston, 1981)

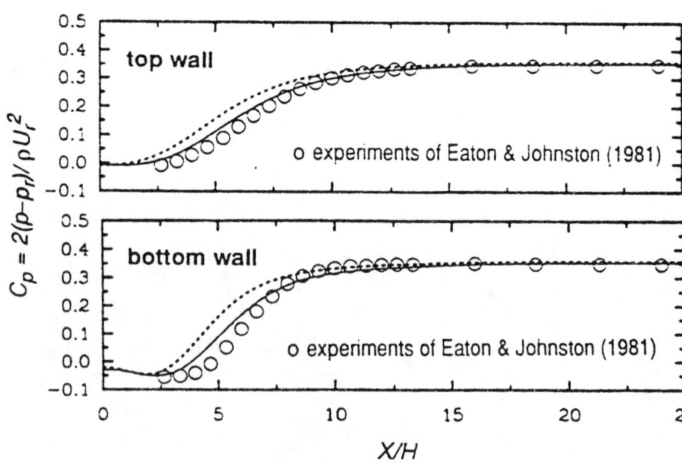

Figure 6. Pressure coefficient along the top and bottom walls (– – standard *K*-ε model; —— recursion-*RNG* model)

AN OVERVIEW OF RNG METHODS IN TURBULENCE MODELING: PANEL DISCUSSION SUMMARY[1]

Ye Zhou

Institute for Computer Applications in Science and Engineering
NASA Langley Research Center
Hampton, Virginia 23681-0001

Charles G. Speziale

Department of Aerospace and Mechanical Engineering
Boston University
Boston, Massachusetts 02215

ABSTRACT

In an effort to foster increased research activities in RNG turbulence modeling, a special panel discussion, organized by Y. Zhou, was held during the ICASE/LaRC Workshop on Transition, Turbulence and Combustion on 29 June, 1993. Its primary goal was to provide an open discussion of the fundamentals of renormalization group approaches and to review the status of RNG methods for turbulence modeling. S. A. Orszag and V. Yakhot (Princeton University) came specifically to ICASE on June 29 to participate in this panel discussion. The panel discussion consisted of presentations by Y. Zhou (ICASE), V. Yakhot (Princeton University), S. A. Orszag (Princeton University), and G. Vahala (College of William and Mary) along with a general discussion moderated by W. D. McComb (University of Edinburgh, UK). Unfortunately, S. H. Lam (Princeton University), who was an active participant in the workshop, could not attend the panel discussion due to a prior commitment.

1. Introduction

Turbulence has been characterized as one of the last unsolved problems in classical physics despite the fact that the governing equations of turbulence (the Navier Stokes equations) have been known

[1]This work was supported by the National Aeronautics and Space Administration under NASA Contract No. NAS1-19480 while the authors were in residence at the Institute for Computer Applications in Science and Engineering (ICASE), NASA Langley Research Center, Hampton, VA 23681-0001.

M. Y. Hussaini et al. (eds.), Transition, Turbulence and Combustion, Vol. II, 179–196.

for more than a century. At high Reynolds numbers (R_e), a wide range of length and time scales in turbulent flows become excited. Direct numerical simulations (DNS) – which are three dimensional, time dependent numerical solutions where all significant scales of motion are computed without a turbulence model – are presently restricted to low turbulence Reynolds numbers (c.f., Reynolds, 1990).

Although DNS can be extremely useful in many areas related to the study of turbulence physics and the assessment of theories (Reynolds, 1990), virtually all scientific and engineering calculations of nontrivial turbulent flows, at high R_e, are based on some type of modeling (Speziale, 1991). Most modeling efforts have been focused on subgrid scale models for large-eddy simulations and Reynolds stress models – the two forms of modeling that will be discussed in this summary.

Large-eddy simulations (LES) are a logical 'modeling extension' of DNS. They are based on the observation that the small scales are more universal in character than the large, energy containing scales of motion that are primarily responsible for turbulent transport. In LES, the three-dimensional time-dependent motion of these large scales are computed directly while the small scales are modeled (the so called 'subgrid scale modeling' problem; see Rogallo and Moin (1984) and Reynolds (1990)). LES is useful in the study of turbulence physics at high Reynolds numbers that are unattainable by DNS (Reynolds, 1990; Zhou, 1993b). Unlike DNS at low Reynolds numbers, LES at high R_e with subgrid scale models must be interpreted with great caution. Separating physics from numerical effects is very important because of the limited computational domain (Zhou, 1993b).

LES is also intended to be useful in the development of turbulence models for the prediction of the complex flows of technical interest where simpler modeling approaches fail (Reynolds, 1990). However, all applications of LES have been in simple geometries where Van Driest damping could be used – an empirical approach that generally does not work well when there is flow separation (Speziale, 1991).

LES and DNS are direct products of the rapid increase in computer capacity. The Reynolds stress modeling concept, however, is one of great age; it was introduced by Osborne Reynolds in 1895. Reynolds stress models remain the most important tool for the highly repetitive engineering calculations associated with design (Reynolds, 1990). The goal of Reynolds stress modeling is to provide reliable information about first and second one-point moments (e.g., the mean velocity, mean pressure, and turbulence intensity), which is usually all that is needed for design purposes (Speziale, 1991).

2. General Remarks on RNG in Turbulence Modeling

The technique of renormalization group theory has achieved great success in the field of critical phenomena (Wilson, 1975; Wilson and Kogut, 1974). Recently, the methodology of RNG has attracted considerable attention as a systematic approach to subgrid scale modeling in LES and Reynolds stress modeling. Indeed, the RNG removal of only the smallest scales generates a subgrid scale model while Reynolds stress models are obtained as successively larger scales are removed. RNG theory is a natural tool for both subgrid scale modeling and Reynolds stress modeling. It is important to realize that these calculation fall into two distinct categories: (a) ϵ-RNG, pioneered by Forster *et al.* (1977), and (b) recursive-RNG, pioneered by Rose (1977). These techniques have been critiqued and compared (Kraichnan, 1987; Smith and Reynolds, 1991; Lam, 1992; and Zhou and Vahala, 1992). It should be noted that McComb's conditional averaged RNG (McComb and Watt, 1992) has not been applied to turbulence modeling and so it will not be discussed further here.

In ϵ-RNG (Yakhot and Orszag, 1986; Yakhot and Smith, 1992; Yakhot *et al.*, 1992), one invariably introduces a zero-mean Gaussian random forcing term into the Navier-Stokes equation. This white noise forcing is determined by its correlation function, which is assumed to obey a subgrid wavenumber power law spectrum. A small parameter ϵ is introduced into the exponent of this power-law. One then calculates the effect on the Navier-Stokes equation of removing a small wavenumber shell. One of the resulting effects is that the molecular viscosity is replaced by the turbulent viscosity. Another effect appears in the introduction of triple nonlinearities into the renormalized equation. By recourse to the small ϵ-expansion, it can be shown that these triple nonlinearities are, in 'RNG' terminology (Wilson and Kogut, 1974), 'irrelevant'. To recover the Kolmogorov energy spectrum, one must choose $\epsilon = 4$ while assuming that the triple nonlinearities can be neglected.

Independently, the recursive RNG (Zhou *et al.*, 1988, 1989, 1993; Zhou and Vahala, 1992, 1993a,b) has been applied to subgrid scale modeling and Reynolds stress modeling. The major differences between the two RNG procedures are that in the recursive RNG:

- the ϵ expansion is not applied,
- the turbulent transport coefficients are determined for the whole range of resolvable wavenumber scales, and
- triple nonlinearities are generated and retained in the renormalized Navier-Stokes equation.

The discussion of these issues is presented in more detail in the sections to follow.

3. Synopses of Presentations

3.1. Turbulence modeling by recursive RNG

Ye Zhou, ICASE

The first part of the talk focused on the kind of basic physics that RNG procedures must preserve in turbulence modeling. A prime objective of the RNG method is to reduce, step by step, the degrees of freedom without compromising the underlying physics. Two important basic physical features should be retained in the decimated system:

- The energy transfer budget
- Triad interactions including both local and nonlocal interactions.

One must first reduce the degrees of freedom and then determine what the decimated system looks like. This point can be demonstrated by examining how the first subgrid scale shell is removed in the Navier-Stokes equation. We denote k_1 as the cut-off wavenumber that separates the resolvable scales from the small scales and introduce the usual notation

$$u_\alpha(\mathbf{k},t) = \begin{cases} u_\alpha^>(\mathbf{k},t) & \text{if } k_1 < k < k_0 \ ; \\ u_\alpha^<(\mathbf{k},t) & \text{if } k < k_1, \end{cases}$$

where k_0 is of the order of the Kolmogorov wavenumber. For $k < k_1$, the resolvable scale Navier-Stokes equation is

$$\left[\frac{\partial}{\partial t} + \nu_0 k^2\right] u_\alpha^<(\mathbf{k},t) = M_{\alpha\beta\gamma}(k) \int d^3j [u_\beta^<(\mathbf{j},t) u_\gamma^<(\mathbf{k}-\mathbf{j},t)$$

$$+2u_\beta^>(\mathbf{j},t)u_\gamma^<(\mathbf{k}-\mathbf{j},t)+u_\beta^>(\mathbf{j},t)u_\gamma^>(\mathbf{k}-\mathbf{j},t)], \tag{1}$$

where $M_{\alpha\beta\gamma}(k) = k_\beta D_{\alpha\gamma}(k)+k_\gamma D_{\alpha\beta}(k)$ and $D_{\alpha\beta}(k) = \delta_{\alpha\beta}-k_\alpha k_\beta/k^2$.

The RNG procedure eliminates the small scale modes by solving the corresponding small scale equation and then substituting it into the resolvable scale equation. This step introduces a turbulent eddy viscosity (from the last term on the RHS of (1)) and a triple nonlinear term (from the second term on the RHS of (1)). Recursive RNG keeps the triple nonlinear term and argues that this type of higher order interactions is not only a natural by-product of RNG in equilibrium problems but is, in fact, essential for obtaining a correct theory. On the other hand, in ϵ-RNG, the triple nonlinear term ($\epsilon^{3/2}$) is neglected by appealing to the limit of $\epsilon \to 0$.

To shed some light on this issue, we examine the energy transfer equation which can be readily derived from (1)

$$\left[\frac{\partial}{\partial t}+2\nu_0 k^2\right]E(k,t) = T^{<<}(k)+T^{><}(k)+T^{>>}(k) \tag{2}$$

where $k < k_1$. $T^{<<}(k)$ gives the rate of energy transfer to mode k from interactions between the resolvable scale velocity fields; $T^{><}(k)$ is the energy transfer rate to mode k from the interactions between resolvable scale and small scale modes; and $T^{>>}(k)$ is the transfer rate to mode k from the interactions between small scale modes. How the transfer spectra are affected by the various modal interactions can be determined from LES and DNS data sets. $T^{<<}(k)$ moves the energy to higher wavenumbers – essentially to the last octave. $T^{>>}(k)$ removes energy throughout the resolvable scales in a manner consistent with the concept of eddy viscosity. This term represents the energy transfer due to nonlocal interactions. The term in dispute here, $T^{><}(k)$, removes energy from the last resolvable octave which was transferred there by $T^{<<}(k)$, and, thus, it is a local flow of energy through k_1 (Zhou and Vahala, 1993a). A detailed study of local and nonlocal interactions in isotropic turbulence can be found in Zhou (1993a,b).

The spectral eddy viscosities (Kraichnan, 1976) can be defined as

$$\nu^{>>}(k) = -T^{>>}(k)/2k^2E(k),$$

and

$$\nu^{><}(k) = -T^{><}(k)/2k^2E(k).$$

We stress that the information contained in the spectral eddy viscosity is equivalent to that of the energy transfer. DNS and LES data indicate that $\nu^{>>}(k)$ is the most important contribution to the cusp-up behavior predicted by Kraichnan (1976) in his test field model and by Chollet and Lesieur (1981) in their EDQNM model.

The role of the triple nonlinear term can be summarized as follows:

- maintaining an energy balance,
- incorporating both local and nonlocal interactions, and
- maintaining the Kolmogorov spectrum for the whole range of resolvable scales.

We mention in passing that the higher order nonlinear interactions have also been generated in other completely different physical systems such as the Ising model of Wilson (Wilson, 1975; Wilson and Kogut, 1974) and in the study of bifurcations of low order dynamical systems (Berkooz and Holmes, private communication, 1993). A comparison between the ϵ-RNG and recursive RNG is presented in Table 1.

This numerical study is one example of how DNS/LES can help in turbulence modeling. It is clear that one should retain both local and nonlocal interactions in turbulence modeling. For example, the Reynolds stress, $\tau_{ij} = \tau_{ij}^{>>} + \tau_{ij}^{><}$, has two components. The $\tau_{ij}^{>>}$ part is due to the $u^{>} - u^{>}$ distant interactions limit while the $\tau_{ij}^{><}$ part is due to the $u^{>} - u^{<}$ local interactions (Zhou et al., 1993).

A Reynolds stress closure model based on the recursive renormalization group theory was recently derived for the prediction of turbulent separated flows. The model uses a finite wavenumber truncation to account for the spectral distribution of energy. In particular, the model incorporates effects of both local and nonlocal interactions. The nonlocal interactions are shown to yield a contribution identical to that obtained from ϵ-RNG (which corresponds to the distant interaction limit $k \to 0$; see Rubinstein and Barton, 1990). The first two terms correspond to the isotropic eddy viscosity. On the other hand, the third term, which is quadratic in the mean strain rate, incorporates the effect of convection and diffusion and is qualitatively similar to other second order models. The local interaction term introduces higher order dispersive effects. The model can be shown to be qualitatively similar to those recently derived from a hierarchy of second order closure models (Gatski and Speziale, 1993).

•Methods	ϵ-RNG	Recursive RNG
•Interactions (Kraichnan; Smith & Reynolds)	Nonlocal $(k \to 0)$	Nonlocal and local $(k \in [0, k_0])$
•Results of RNG procedure	$\nu_T(k \to 0 \mid k_0)$	$\nu_t(k \mid k_0)$ and uuu
•ϵ-expansion (Lam)	Yes Coefficient $(\epsilon \to 0) \exp[(\epsilon \to 4)]$	No
•Reference energy spectrum (Kraichnan)	$E(k) \sim k(\epsilon \to 0)$ $E(k) \sim k^{-5/3}(\epsilon \to 4)$	$E(k) \sim k^{-5/3}$ (input)

Table 1. Comparison of RNG Methods

The ability of the model to accurately predict separated flows was analyzed from a computational standpoint by considering turbulent flow past a backward facing step as a test case. Results obtained from finite difference computations demonstrate that the model yields good predictions for this separated turbulent flow. The details of this work can be found in Zhou *et al.* (1993).

3.2. RNG in turbulence modeling, Part I

V. Yakhot, Princeton University

First I have few remarks on Dr. Zhou's presentation:

(1) The spectral eddy viscosity is nonsense. People use it because it was proposed by Kraichnan (1976). For example, in two-dimensional turbulence, the eddy viscosity is negative. One should use the self-energy concept instead. Take the problem of the mean velocity profile in two-dimensional fully-developed channel flow, for example, where no-slip boundary conditions are assumed at the walls. According to your definition, in 2D the nonlinear interaction between fluctuating modes is equivalent at large enough scales to a *negative* eddy viscosity. So you are left with a linear problem where there is a negative coefficient in front of the Laplacian. There is no solution to this problem and the positive $O(\nabla^4)$ dissipation does not help. I think that this model is good for the description of mean energy transfer in homogeneous flows. The relevance of the cusp in the effective viscosity, derived from the mean energy transfer relations, is not at all clear. I am not saying that it cannot be right, but the model has to be obtained from the only correct definition of viscosity (eddy or molecular):

$$T_{ij} = 2\nu S_{ij}$$

in the limit $S_{ij} \to 0$.

(2) The stirring force and ϵ parameter are introduced into the Navier-Stokes equation as

$$< ff >= \delta(k)\mathcal{E}\delta(\mathbf{k}+\mathbf{k}') \tag{3}$$

where

$$\delta(k) = \lim_{\epsilon\to 4}(4-\epsilon)k^{1-\epsilon} \;\; for\; k \to 0 \tag{4}$$

and $\mathcal{E}$ is the dissipation rate.

Leonard (private communication, 1993) has calculated the isotropic decay rate – with a resolution of 512^3 grid points for over a thousand eddy turn over times using Delta. The calculated decay rate is in agreement with the RNG prediction.

Returning to turbulence modeling, the RNG dissipation equation (Yakhot *et al.*, 1992) is

$$D_t\varepsilon = 1.42(\varepsilon/K)\tau_{ij}\partial_j u_i - 1.68\varepsilon^2/K + \partial_i(\alpha\nu_T\partial_i\varepsilon) - \mathcal{R}. \tag{5}$$

We have applied the RNG $K - \varepsilon$ model to the following cases:

- backward facing step
- U bend with separation bubble predicted
- constricted tube with separation bubble size better predicted
- flow over a cylinder with periodic vortex shedding.
- unsteady flow over a model car with the correct pressure, lift and drag distributions computed
- external flows: multiple time scales where RNG can fail
- rotating flows; swirl flows in combustion
- compressible flows with shocks.

Surprisingly good results have been obtained in almost every case. In particular, time-dependent vortex shedding is properly predicted behind a circular cylinder. Further tests of the RNG $K - \varepsilon$ model are currently underway using the FLUENT computer code.

3.3. RNG in turbulence modeling, Part II

S. A. Orszag, Princeton University

I will talk about several other RNG applications:

- near wall turbulence
- narrow band velocity of Kraichnan
- compressibility: $\varepsilon \to \varepsilon(1 + M^2)$ – Sarkar's correction. Correspondingly, for the eddy viscosity, $\nu_T \to \nu_T/(1 + M^2)$ where M is the turbulence Mach number
- $K - \varepsilon$ model: modified to include combustion and chemical reaction.

Insofar as the first point is concerned, the RNG $K - \varepsilon$ model can be integrated directly to solid boundaries without the need for *ad hoc* wall damping functions. In regard to the last two points, the compressible version of the RNG $K - \varepsilon$ model yields results for turbulent flows with combustion and chemical reactions that are far superior to the commonly used variable density extensions of the standard $K - \varepsilon$ model. More complex tests are currently underway with FLUENT.

3.4. Comments on ϵ-RNG, phenomenological RNG, and recursive RNG

G. Vahala, College of William and Mary

We have heard a lot of applications. Let us return to two basic questions and see how ϵ-RNG, phenomenological RNG, and recursive RNG handle them.

I. Form invariance of Navier-Stokes under decimation of modes:

- ϵ-RNG: introduces 'stirring forces' and a parameter ϵ into the Navier-Stokes equation.
- phenomenological RNG (Lam, 1992): role of g^{fast} – not handled self-consistently
- recursive RNG: introduction of the triple velocity products uuu, closure problem as in ϵ-RNG.

II. Role of the parameter ϵ:

- ϵ-RNG: is ϵ treated consistently or in an *ad hoc* fashion?
- phenomenological RNG: ϵ is a free parameter
- recursive RNG: ϵ is not introduced.

The correlation function of the external forcing is

$$< f_i(\mathbf{k},\omega) f_j(\mathbf{k}',\omega') > \equiv 2D_o^*(\epsilon,k)k^{-3}(2\pi)^4[\delta_{ij} - k_ik_j/k^2] \cdot \delta(\mathbf{k}+\mathbf{k}')\delta(\omega+\omega') \tag{6}$$

Lam pointed out that the energy dissipation rate $\mathcal{E}$, assumed independent of k, has the following dimensional form

$$\mathcal{E} \equiv \Pi_3 D_o^*(\epsilon, k_c) k_c^{4-\epsilon}, \tag{7}$$

where Π_3 is a constant. As a result, if $\epsilon \neq 4$, the stirring force is not scale invariant.

Following Lam, we rewrite the Navier-Stokes equation as

$$[\partial/\partial t - \nu_T \nabla^2] u^< = M^< u^< u^< + g^{fast} \tag{8}$$

where

$$g^{fast} \equiv -(1/\rho)\nabla p^> + \nabla \cdot (u^< u^< - uu) - L[u^>; \nu] + (\nu - \nu_T)\nabla^2 u^< \tag{9}$$

and $L = [\partial/\partial t - \nu_T \nabla^2]$.

The problem now is how to handle g^{fast}. The methods are different in various RNG approaches:

- ϵ-RNG: try to show $g^{fast} \sim O(\epsilon^{3/2})$ as $\epsilon \to 0$.
- conditional averaged RNG of McComb: define conditional averaging so as to eliminate g^{fast}
- phenomenological RNG: model g^{fast} for proper behavior as $k \to 0$ and $k \to k_c$.
- recursive RNG: attempts to handle g^{fast} perturbatively.

Inconsistencies in the ϵ-RNG approach has been discussed in detail by Lam (1992). In brief, the fixed point solution of the effective coupling coefficient ($\bar{\lambda}_*^2$) is

$$3A_3(\epsilon)\bar{\lambda}_*^2 = \epsilon \tag{10}$$

Another term, which $\sim dD_0(\epsilon, k_c)/dk_c$, has been neglected above in ϵ-RNG which actually can be done so only for $\epsilon = 4$. Lam argued that the ϵ-RNG treatment is actually an *ad hoc* procedure: in all coefficients $\epsilon \to 0$ ($A_3(\epsilon)$, LHS) while $\epsilon \to 4$ for all exponents (RHS). In the mean time, $\bar{\lambda}_*^2 \sim O(\epsilon)$ must be small in order to argue that $g^{fast} \sim O(\epsilon^{3/2})$.

In phenomenological RNG, the stirring force is not introduced. Furthermore, g^{fast} is not calculated explicitly. Instead, based on physical arguments, Lam argued that $g^{fast} \to 0$ as $k \to 0$ but it becomes important as $k \to k_c$. In Lam's formulation, ϵ is a free parameter that is not necessarily small.

A connection between g^{fast} in Lam's phenomenological RNG and uuu in recursive RNG can be found (Zhou and Vahala, unpublished,

1993). The most important term in g^{fast} is the first one and it can be rewritten as

$$u^{<}u^{<} - uu = u^{<}u^{<} - [u^{<}u^{<} + 2u^{>}u^{<} + u^{>}u^{>}]$$
$$= -[2u^{>}u^{<} + u^{>}u^{>}].$$

Indeed, it is precisely the interaction represented by the first term on the RHS that gives rise to the cubic nonlinearity and local interactions in the recursive RNG. The second term on the RHS contributes to the turbulent eddy viscosity. Note that recursive RNG can also recover ϵ-RNG coefficients in the limit as $k \to 0$ [nonlocal interactions].

Finally, I would like to say a few words about the RNG $K - \varepsilon$ model. Much of the success of this model rests on the $\mathcal{R}$-term which results in a non-closure problem which cannot be handled by RNG. The treatment of Yakhot *et al.* (1992) claims to use Padé approximations to determine this $\mathcal{R}$-term. I disagree. What is done is nothing like the Padé approximations that I have learned from Bender and Orszag (1978), especially since Yakhot *et al.* (1992) do not even know the coefficient in the Taylor expansion for $\mathcal{R}$ and the whole scheme behind Padé approximations is the manipulation of those Taylor coefficients.

4. Discussion during the Presentations

Yakhot: (to Zhou) Your triple nonlinear term will destroy Galilean invariance.

Zhou: No. Even with the triple nonlinear term, Galilean invariance is preserved. The details have been presented in ICASE report 93-1 (Zhou and Vahala, 1993b).

Erlebacher (*ICASE*): (to Yakhot) Since you are talking about physics, your ϵ-RNG method is not the same RNG as that applied to critical phenomena, is it?

Yakhot: No, it is not. Why should it be? The RNG in critical phenomena deals with thermodynamic equilibrium. This is a static theory. The RNG for turbulence is based on the dynamic renormalization group developed for the Navier-Stokes equation.

Zhou: (to Yakhot) The comparison between the decay rate of isotropic turbulence simulations and that of ϵ-RNG predictions is not appropriate. As pointed out by Lam (1993), the ϵ-RNG coefficient is calculated from an input spectrum which has a sharp jump between small wavenumbers and the inertial range spectra.

Yahkot: What is the big deal with a jump?

Zhou: It is unphysical.[2]

5. General Discussion

5.1. Moderator's opening statement

The moderator, W. D. McComb, opened the general discussion with the following slide:

The basic issue is 'the correspondence principle'.

How do you get from a theory of a fluid, stirred so weakly that the inertial cascade does not take place (which is valid only in the limit of infinitely long wavelengths) to a theory of isotropic, stationary turbulence, at Reynolds numbers large enough for the Kolmogorov $k^{-5/3}$ spectrum to exist.

Now, one divides the velocity field into the resolvable and small scales. Substituting the decomposition into the Navier-Stokes equation and eliminating $u^>$ from $u^<$, we find the following problems:

(1) the generation of the new terms $u^<u^<u^<$.

(2) averages such as $< u^<u^>u^> > \approx u^< < u^>u^> >$.

However, for stirred fluids, one finds

$u^<u^<u^< \to 0$ as $k \to 0$

$< u^<u^>u^> > = u^< < u^>u^> >$.

5.2. Summary of principal discussions

Yakhot: The way to view the correspondence principle is starting from the $\delta(k)$ definition in my presentation.

One way to see how the ϵ-expansion works is by looking at the simple example

$$F(\epsilon) = \epsilon^{1/2}/(1+\epsilon^2). \tag{11}$$

The ϵ expansion for this model problem is

$$F \approx \underbrace{\sqrt{\epsilon}}_{\epsilon=4}\underbrace{(1-\epsilon^2+....)}_{\epsilon=0}. \tag{12}$$

[2]Note added by YZ and CGS: for a detailed discussion of this subject, see the contribution by S. H. Lam and the reply by V. Yakhot in this volume.

Orszag: The ϵ expansion cannot be proved for Navier-Stokes turbulence. However, for a model problem, we have shown that it works well.

Zhou: The spectral eddy viscosity is defined by the energy transfer function. Regardless of whether you like or dislike the spectral eddy viscosity concept, the correct energy balance must be preserved. If one computes the turbulent viscosity from the term $< u^> u^> >$ alone, the DNS and LES indicate that you cannot have the correct inflow and outflow. The problem with this approximation can be seen from the triad interaction point of view. From the scaling interaction analysis, ϵ-RNG kept only eddy viscosity type interactions while neglecting the straining type of interactions.

Orszag: (to Zhou) You may be interested in this preprint on the application of RNG to Rossby wave turbulence where we have also computed the spectral eddy viscosity.

Vahala: Two-dimensional flow is not a good example for RNG. Liang and Diamond (1993) proved that there is no fixed point in two dimensional flow.

Orszag: We have proven that a fixed point exists.

Vahala: No, I disagree. I have seen a recent preprint on the application of ϵ-RNG to Rossby wave turbulence and these authors categorically say that they do *not* prove the existence of a fixed point – but merely assume that it exists. In two-dimensional turbulence that is very dangerous due to the inverse cascade, as discussed by Diamond and his student. So I totally disagree with Yakhot's example of eddy viscosity in 2D.

McComb: Let us stop bringing up 2D turbulence as an example, since we all know it is pathological.

Speziale: In regard to Sarkar's compressible correction, it should be noted that, while it works fairly well for homogeneous turbulence, it breaks down in the log-layer of turbulent boundary layers.

Younis (*City University of London*): (to Yakhot) One problem with two-equation models is the Boussinesq stress-strain relationship which is particularly inadequate when turbulence is distorted by body forces or extra rates of strain. This relationship represents some kind of a bottle neck through which the models must pass, with penalty, irrespective of how well you handle the epsilon equation. Are you therefore hopeful that RNG can provide useful answers for complex flows?

Yakhot: No problem. The model we have works very well for cases involving strong streamline curvature, vortex shedding and swirling flows as in hydrocyclones.

Demuren (*Old Dominion University*): (to Yakhot) If I use your RNG $K-\varepsilon$ model equation published in *Phys. Fluids A*, and apply it to the cases you have just shown, can I reproduce all of your results without changing any coefficients?

Yakhot: No.

Younis: (to Yakhot) Have you tested your model against well-documented two-dimensional free shear flows of the type generally accepted as necessary test cases for model verification?

Yakhot: Yes.

Younis: Did you get the right answers for both the plane and axi-symmetric free jets?

Yakhot: Yes.

Younis: I have bad news for you: I tried the RNG model for those flows and found that the predictions are much worse than with the standard $K-\varepsilon$ model.

Yakhot: You did it wrong. Maybe you used the wrong equations.

Younis: I did this with Charles Speziale and he should know the right equations.

Speziale: We used the model given in the Physics of Fluids paper and I guess that is different from the RNG model that you currently use.

Yakhot: Correct. The renormalization group, as we developed it, gives us a calculus – a consistent set of prescriptions to derive effective equations of motion for flows in external fields such as rotation, stratification, magnetic fields, shock, etc. As a result, we have obtained a rather general two-equation model with all these fields included. The equations can be used in a wide variety of cases and the contributions corresponding to external fields turn themselves on automatically when and where they are dynamically important. In its entire form, the model has not yet been published since we first wanted to test it on as many cases as possible.

The renormalization group for turbulence is not yet a mathematically sound construction. There is no point in arguing whose approach is better since none of them is really good enough from a theoretical viewpoint. This is a classical example of a pudding: the proof is in the eating. The main reason why it took us so many years to come up with publications was that we wanted to derive as many results as possible: dimensionless constants, the detailed description of various complex flows, etc. There is a simple way to show that another approach is better than ours: calculate as many flows as we did and demonstrate superior results.

Speziale: In response to the earlier comments of George Vahala concerning the term $\mathcal{R}$ in the RNG $K-\varepsilon$ model, it should be noted that such a term can be systematically derived from the transport equation for the tensor dissipation rate when the standard equilibrium hypothesis is invoked for homogeneous turbulence. This is the same hypothesis by which $K-\varepsilon$ type models with an anisotropic eddy viscosity can be formally derived from a second moment of the Navier-Stokes equation (see Gatski and Speziale (1993)).

Acknowledgments

The authors thank Dr. M. Y. Hussaini and Ms. E. Todd for their assistance in organizing the Panel Discussion. We are also grateful to Drs. T. B. Gastki, S. Thangam, and B. A. Younis for providing tapes and notes.

References

Bender, C. M. and Orszag, S. A., 1978. *Advanced Mathematical Methods for Scientists and Engineers*, Academic, NY.

Chollet, J. P. and Lesieur, 1981. "Parameterization of small scales of three-dimensional isotropic turbulence utilizing spectral closures," *J. Atmos. Sci.* **38**, p. 2747.

Forster, D., Nelson, D., and Stephen, M., 1977. "Large-distance and long-time properties of a randomly stirred fluid," *Phys. Rev. A* **16**, p. 732.

Gatski, T. B. and Speziale, C. G., 1993. "On explicit algebraic stress models for complex turbulent flows," *J. Fluid Mech.* **254**, p. 59.

Kraichnan, R. H., 1976. "Eddy viscosity in two and three dimensions," *J. Atmos. Sci.* **33**, p. 1521.

Kraichnan, R. H., 1987. "An interpretation of the Yakhot-Orszag turbulence theory," *Phys. Fluids* **30**, p. 2400.

Lam, S. H., 1992. "On the RNG theory of turbulence," *Phys. Fluids A* **4**, p. 1007.

Lam, S. H., 1993. "On RNG theory and the decay law of homogeneous isotropic turbulence," to appear in this volume.

Liang, W. Z. and Diamond, P. H., 1993. "A renormalization group analysis of two-dimensional magnetohydrodynamic turbulence," *Phys. Fluids B* **5**, p. 63.

McComb, W. D. and Watt, A. G., 1992. "Two-field theory of incompressible fluid turbulence," *Phys. Rev. A* **46**, p. 4797.

Reynolds, W. C., 1990. "The potential and limitations of direct and large-eddy simulations," in *Whither Turbulence*, edited by J. Lumley, p. 313, (Springer, Berlin, 1990).

Rose, H. A., 1977. "Eddy diffusivity, eddy noise and subgrid scale modelling," *J. Fluid Mech.* **81**, p. 719.

Rubinstein, R. and Barton, J. M., 1990. "Nonlinear Reynolds stress models and the renormalization group," *Phys. Fluids A* **2**, p. 1472.

Smith, L. M. and Reynolds, W. C., 1992. "On the Yakhot-Orszag renormalization group method for deriving turbulence statistics and models," *Phys. Fluids A* **4**, p. 364.

Speziale, C. G., 1991. "Analytical methods for the development of Reynolds stress closures in turbulence," *Ann. Rev. Fluid Mech.* **23**, p. 107.

Wilson, K. G., 1975. "The renormalization group: critical phenomena and the Kondo problem," *Rev. Mod. Phys.* **47**, p. 773.

Wilson, K. G. and Kogut, J., 1974. "The renormalization group and the ϵ expansion," *Phys. Rep.* **12C**, p. 75.

Yakhot, V. and Orszag, S. A., 1986. "Renormalization group analysis of turbulence. I. Basic theory," *J. Sci. Comput.* **1**, p. 3.

Yakhot, V. and Smith, L. M., 1992. "The renormalization group, the ϵ expansion, and derivation of turbulence models," *J. Sci. Comput.* **3**, p. 35.

Yakhot, V., Orszag, S. A., Thangam, S., Gatski, T. B., and Speziale, C. G., 1992. "Development of turbulence models for shear flows by a double expansion technique," *Phys. Fluids A* **4**, p. 1510.

Zhou, Y., 1993a. "Degrees of locality of energy transfer in the inertial range," *Phys. Fluids A* **5**, p. 1092.

Zhou, Y., 1993b. "Interacting scales and energy transfer in isotropic turbulence," *Phys. Fluids A* **5**, p. 2511.

Zhou, Y. and Vahala, G., 1992. "Local interactions in renormalization methods for Navier-Stokes turbulence," *Phys. Rev. A* **46**, p. 1136.

Zhou, Y. and Vahala, G., 1993a. "Reformulation of recursive renormalization group based subgrid modeling of turbulence," *Phys. Rev. E* **47**, p. 2503.

Zhou, Y. and Vahala, G., 1993b. "Renormalization group estimates of transport coefficients in the advection of a passive scalar by incompressible turbulence," ICASE Report 93-1, to appear in *Phys. Rev. E*.

Zhou, Y., Vahala, G., and Hossain, M., 1988. "Renormalization group theory for the eddy viscosity in subgrid modeling," *Phys. Rev. A*, **37**, p. 2590.

Zhou, Y., Vahala, G., and Hossain, M., 1989. "Renormalized eddy viscosity and Komogorov's constant in forced Navier-Stokes turbulence," *Phys. Rev. A* **40**, p. 5865.

Zhou, Y., Vahala, G., and Thangam, S., 1993. "Development of a recursive RNG based turbulence model," ICASE Report 93-51.

COMBUSTION

COUNTER-JET DIFFUSION FLAMES

GROUP SUMMARY: COUNTER-JET DIFFUSION FLAMES

Phil Drummond

NASA Langley Research Center
Hampton, Virginia 23681-0001

The design of many propulsion devices requires a detailed knowledge of the structure and the mechanism of extinction of the flame in order to understand the effects of aerodynamic strain on combustion chemistry. The complex nature of the problem necessitates a combined experimental, numerical, and analytical approach. Because well controlled laboratory experiments can only be performed over a limited parameter range, numerical models are needed to predict the structure of these flames at conditions for which experimental measurements are not available. For situations where complex chemistry is required, analytical methods can provide insights into the behavior of the flow and often lead to closed form expressions useful for correlation. Ultimately, however, analytical models must be tested by comparison with numerical predictions and experiments.

The workshop participants hosted by the Counter-Jet Diffusion Flame Group were John Buckmaster (University of Illinois); Phil Drummond, Gerry Pellett, and Bill Roberts (NASA Langley Research Center); K. M. Isaac (University of Missouri-Rolla); Tom Jackson (ICASE); Moshe Matalon (Northwestern University); Bob Pitz and Tom Brown (Vanderbilt University); Kal Seshadri (University of California-San Diego); and Mitch Smooke (Yale University). The Group chose to examine several critical issues of flame extinction and structure using the three-tiered approach as reflected by the following four papers. The model for these studies was the counter-flow burner. Pitz, Brown, and Nandula experimentally studied a counter-jet diffusion flame using nonintrusive laser diagnostics and began collecting data for comparison with numerical and analytical studies. Jackson and Matalon examined the stability of twin premixed flames in a counter-flow configuration using an asymptotic method and characterized the stability in terms of several critical parameters. Seshadri studied the effect of contaminants on the structure of nonpremixed flames using an asymptotic method and compared those results with experiments. Finally, Isaac used a one-dimensional numerical model resulting from a similarity transformation of the multidimensional governing equations to examine the flame structure and extinction characteristics of a counter-jet diffusion flame.

M. Y. Hussaini et al. (eds.), Transition, Turbulence and Combustion, Vol. II, 201.

CHARACTERISTICS OF STRETCHED HYDROGEN-AIR DIFFUSION FLAMES AT HIGH PRESSURES

K. M. Isaac[1]

The University of Missouri-Rolla
Mechanical & Aerospace Engineering Department
Rolla, Missouri 65401

ABSTRACT

A one-dimensional model based on similarity transformation for the H_2-air opposed-jet counter flow diffusion flame (CFDF) has been developed and numerically solved to study flame structure and extinction limits at a pressure of ten atmospheres. The property profiles for this case are very similar to those of an earlier study in which the pressure was one atm. For the pure hydrogen-air case, the flame extinction takes place at a strain rate of 140,000 s^{-1}.

Nomenclature

a	$= \partial u/\partial r$, air side strain rate $(1/s)$
C	$(\rho\mu)/(\rho_e\mu_e)$
D	Jet diameter (cm)
D_{ij}	Binary diffusion coefficient (cm^2/s)
D_i^T	Thermal diffusion coefficient (cm^2/s)
f	Similarity function
h	Enthalpy per unit mass (erg/g)
h_i°	Formation enthalpy per unit mass of species i (erg/g)
J	Number of nodes in the computational domain
K	Number of species; equilibrium constant
Le_i	$= \frac{\rho D_i \overline{C_p}}{k}$, Lewis number
Le_i^T	$= D_i^T/D_i$, Thermal Lewis number of species i
n	Index for flow type, n=0 for two-dimensional flow n=1 for axisymmetric flow;

[1]Research was supported by the National Aeronautics and Space Administration Under NASA Contract No. NAS1-19480 while the author was in residence at the Institute for Computer Applications in Science and Engineering (ICASE), NASA Langley Research Center, Hampton, VA 23681-0001. Financial support for this work was provided by NASA Langley Research Center under the Grant, NAG-1-861, G. L. Pellett, technical monitor.

M. Y. Hussaini et al. (eds.), Transition, Turbulence and Combustion, Vol. II, 203–216.

	superscript for integration time level
$\overline{R}$	Universal gas constant ($erg/g/K$)
r	Radial coordinate
T	Temperature (K)
t	Time
u	Velocity component in r direction (cm/s)
V	Diffusion velocity (cm/s)
v	Velocity component in z direction (cm/s)
X_i	Mole fraction of species i
Y_i	Mass fraction of species i
z	Axial coordinate
η	Similarity variable
$\dot{w}_i$	Net production rate of species i ($g/cm^3/s$)
μ	Coefficient of dynamic viscosity ($g/cm/s$)
ρ	Density (g/cm^3)
θ	Normalized temperature

Subscripts

ext	extinction condition
i	Species i
j	Node j in computation domain
e	Boundary layer edge condition; right hand side control surface
η	Derivative with respect to η

1. Introduction

Ignition and extinction are of considerable importance in combustion and re-entry problems. Because of the constraints under which a jet engine has to operate, it is nearly impossible to achieve mixing at the molecular level before combustion and, therefore, diffusion of fuel and oxidant into each other is the basic mechanism by which combustion takes place. Any analysis of practical systems involving diffusion flames must include realistic models for turbulence and finite-rate chemistry. The structure of stretched laminar diffusion flames is being recognized as an important building block in turbulence modeling. A laminar flamelet model capable of handling finite rate chemistry has been proposed (Peters, 1984) for non-premixed turbulent flames. The dynamics of stretched flames has been reviewed recently in Law (1988). In this approach the turbulent diffusion flame is conceived as an ensemble of thin laminar flamelets stretched and distorted by the turbulent flowfield. The implementation of the flamelet model requires detailed knowledge of stretched laminar flames. The counter flow configuration has been extensively used in the past to study the behavior of stretched diffusion flames.

The objective of the present analysis is to examine how the flame behavior is affected at elevated pressures by flow velocity and composition of the fuel and air streams. Determination of the critical values of the flowfield and chemical parameters for flame extinction was another important objective of the present study.

2. Flame Model

We assume the flow is laminar in the stagnation region. For the nonreacting case, analytical solution for two incompressible jets impinging on one another exists. However, as the term 'diffusion flames' suggests, the present problem is dominated by viscous and chemical effects; the fuel contained in one jet will diffuse into the other jet and the oxidant will do likewise in the opposite direction, thus establishing a narrow zone in which fuel and oxidant coexist. If the mixture is ignited, a thin flame zone will be established. The temperature will rise sharply at the flame. This sharp temperature peak will affect the density of the gases and the flow characteristics will now be different from those of the nonreacting flow. This flow must be determined from the simultaneous solution of the equations of motion coupled with the equations of the conservation of energy and species. Since the effect of the flame on the flowfield is restricted to a narrow region, the problem is similar to a boundary layer type flow in which the viscous effects are restricted to the boundary layer when compared to the entire flowfield. This analogy, together with the assumption that the jets are infinitely wide compared to the zone of interest, suggests that the equations representing the counter flow diffusion flame may be transformed into a set of ordinary differential equations of the boundary-value type with the use of a similarity transformation.

A complete description of the flowfield is given by the simultaneous partial differential equations of the conservation of momentum, energy and individual chemical species together with the appropriate boundary conditions. Details of the procedure used to transform these equations to the one-dimensional form are given in Ho et al. (1989) and Ho et al. (1991) which closely follows the formulation of Fay and Riddell (1958). The final form of the equations and the boundary conditions are given below. Subscript η in these equations stands for derivative with respect to η. The equation for the similarity function f is of third order

$$(Cf_{\eta\eta})_\eta + ff_{\eta\eta} + \frac{1}{n+1}\left(\frac{\rho_e}{\rho} - f_\eta^2\right) = 0. \tag{1}$$

The index $n = 0$ for two-dimensional flow and $= 1$ for axisymmetric flow.

The equations for energy and mass fractions are of second order

$$\left(\frac{C}{P_r}h_\eta\right)_\eta + (fh)_\eta - f_\eta h + \left\{\frac{C}{P_r}\sum_{i=1}^{K} h_i\left[(Le_i - 1)Y_{i_\eta} + \frac{Le_i^T T_\eta}{T}Y_i\right]\right\}_\eta = 0 \tag{2}$$

for $i = 1, ..., K-1$:

$$\left(\frac{CLe_i}{P_r}Y_{i\eta}\right)_\eta + (fY_i)_\eta - f_\eta Y_i + \left(\frac{CLe_i^T T_\eta}{P_r T}Y_i\right)_\eta + \frac{1}{(n+1)}\frac{\dot{w}_i}{a\rho} = 0. \tag{3}$$

For a system with K species, Eq. 3 represents K - 1 equations. The Kth species mass fraction is obtained from the relation

$$Y_K = 1 - \sum_{i=1}^{K-1} Y_i \tag{4}$$

2.1. Boundary conditions

From a consideration of momentum balance the following boundary condition for the function f can be written.

$$f_\eta(-\infty) = \frac{u_{-\infty}}{u_\infty} = \sqrt{\frac{\rho_\infty}{\rho_{-\infty}}} \tag{5}$$

The following additional conditions are obtained from the radial velocity at the edge of the solution domain and the zero axial velocity at the stagnation point.

$$f_\eta(\infty) = 1 \quad (6)$$
$$f(0) = 0. \quad (7)$$

Energy equation

$$h(-\infty) = \sum_{i=1}^{K} Y_{i,-\infty}(h_{i,-\infty} - h^\circ_{i,-\infty}) \quad (8)$$
$$h(\infty) = \sum_{i=1}^{K} Y_{i,\infty}(h_{i,\infty} - h^\circ_{i,\infty}) \quad (9)$$

Species equations
for $i = 1, ..., K-1$:

$$Y_i(-\infty) = Y_{i,-\infty} \quad (10)$$
$$Y_i(\infty) = Y_{i,\infty}. \quad (11)$$

Note that the coefficients appearing in the governing equations consist of non-dimensional groups involving the mixture density and the transport properties of viscosity, thermal conductivity and diffusion coefficient. Species diffusion velocities were calculated using the multicomponent formulation

$$V_i = \frac{1}{X_i \overline{M}} \sum_{j \neq i}^{K} M_j D_{ij} \frac{dX_j}{dx} - \frac{D_i^T}{\rho Y_i} \frac{1}{T} \frac{dT}{dx} \quad (12)$$

in which the first term on the right hand side represents ordinary diffusion and the second term represents thermal diffusion. $\overline{M}$ stands for the mean molecular weight of the gas mixture. Thermal diffusion was considered in the trace, light component limit. The transport properties as described above were computed using the CHEMKIN routines (Kee and Miller, 1986) and the non-dimensional parameters, C, P_r, Le_i and Le_i^T appearing in Eqs. 1, 2, 3 were calculated using these transport properties. The derivatives appearing in the governing equations as coefficients were evaluated using a curve-fitting routine.

3. Numerical Algorithm

It is well known that the equations involving chemical reactions are stiff (mathematically, the coefficient matrix of the locally linearized representation of stiff equations gives rise to eigenvalues that have negative real parts which differ by several orders of magnitude). In chemically reacting systems, stiffness arises because of the widely varying reaction rates in the chemistry model. Thus, the degree of stiffness of the hydrogen-air system is much higher than the hydrocarbon-air system. These observations led to the choice of a time-dependent, finite-volume technique (Patankar, 1980) for the numerical solution of the equations. The method is based on conservation of variables in the control volumes formed by the grid in the solution domain. The degree of variation of each of the terms in the equations was carefully examined to determine whether to choose an implicit or explicit form for integration. This choice was critical for the species production term $\dot{w}_i$, since the stiffness associated with the present problem is reflected in this term. Since the term $\dot{w}_i$ is the net production rate due to the forward and the backward steps in the reaction set, this term can be split into a creation term and a destruction term and handled separately to improve numerical stability (Patankar, 1980). Our experience showed that the finite-volume technique results in more stable solutions compared to finite-difference techniques which were also attempted in the present work. A method based on arc length continuation has been used by other workers (Smooke, 1980; Puri et al., 1987) to solve the CFDF problem. One of the items that requires special attention is the choice of the initial profiles for the time integration of the equations. We have tried initial profiles obtained using different techniques and found that some work satisfactorily and the others do not. The initial profiles obtained from analytical expressions of the flame sheet model were not particularly useful in getting steady state solutions. Subsequently, calculations using a one-step model for the chemical reaction were attempted and the temperature and major species initial profiles from the one-step model were used as initial profiles for the calculations with detailed chemistry. The solution was very insensitive to the minor species profiles and, therefore, they were chosen somewhat arbitrarily. This two-step procedure gave stable solutions without difficulty. For the global time-marching procedure, we start with an initial guess of profiles for both the temperature and the species mass fractions obtained from the one-step model. These initial profiles are then smoothed by a commonly used curve-fitting routine. At each level in the time integration, the $L2$ norm of the mass fraction residuals is calculated as follows:

$$L2 \quad = \quad \sqrt{\frac{1}{(J-2)K} \sum_{j=2}^{J-1} \sum_{i=1}^{K} \left(\frac{Y_{i,j}^{n+1} - Y_{i,j}^{n}}{Y_{i,j}^{n}} \right)^2} . \tag{13}$$

Note that j denotes computational node, n time level, and i species, in Eq. 13. If the $L2$ norm is greater than a specified tolerance ϵ, a new temperature distribution will be calculated from the enthalpy using the interval-halving method, the coefficients updated, and the time marching continued. Other variables (f, h) were not considered in the calculation of the $L2$ norm because they were found to converge faster than the mass fractions. The solution obtained for one case would be used as the initial guesses for other cases which lie in its neighborhood; a series of solutions might be obtained in this manner, each time varying the parameters only slightly. This procedure greatly reduces the computer time and effort involved in running a case.

4. Results and Discussion

Several cases were run by varying the strain rate ranging from low to high values. The general trends are very similar to the results from the one atm case. The temperature, velocity, H_2O mass fraction, and H_2O production rate profiles for a range of values of 'a' are presented in Figures 1-4. The flame location and flame width are given in Figures 5 and 6, respectively. Figures 7 and 8 show respectively, the maximum mass fraction and maximum production rates of H_2O plotted against the strain rate. Flame extinction for this elevated pressure case occurred at 'a' = 140,000 s^{-1}. Similar results for the H_2-air flame using the arc length continuation method are reported in Guthiel and Williams (1989).

The trends seen in the temperature profiles of Figure 1 are very similar to those for the one atm case. When the flame stretch is low, the flame is thicker and the maximum temperature is close to, but lower than, the adiabatic flame temperature. As the flame stretch increases, the maximum temperature drops and the flame shifts towards the stagnation plane. An explanation for this migration of the flame towards the stagnation plane with increasing flame stretch is given in an earlier work (Isaac et al., 1994). As a precursor to extinction, the flame becomes thinner at higher strain rates. The maximum temperature as the flame approaches extinction is about 1600K. This may be compared to a maximum flame temperature of 1350K for the counterpart flame at one atm pressure. Extensive experimental results for the H_2-air CFDF has been reported in Guerra et al. (1987), Pellett et al. (1989), and Pellett et al. (1991). Comparisons with this data at 1 atm have been given for the present method in Ho et al. (1991) and Isaac et al. (1994).

The velocity profiles are given Figure 2. Notice the steep gradients as the strain rate increases.

The H_2O mass fraction profiles are given in Figure 3. These profiles show a few interesting features. The maximum value is about 0.24 for a wide range of strain rates, as is the case in the earlier work for a pressure of one atm. However, at smaller strain rates, the profiles are wider. The reason for the existence of this plateau region can be explained as follows: At smaller values of 'a', there is a region in which equilibrium conditions exist in the flame. As the strain rate increases, this equilibrium region diminishes in width, and finally disappears altogether. The species production rates profiles shown in Figure 4 support this explanation. For small values of 'a', these profiles are flatter and have smaller peaks; note that in the state of complete equilibrium, the productions rates should be zero.

Flame location and flame width are given in Figures 5 and 6. Both these variables show two distinct trends. At smaller strain rates, there is a rapid fall-off in the flame location and flame width. In the higher range of strain rates, the flame location and flame width do not decrease appreciably with increasing strain rates.

5. Conclusions

1. The general trends observed in the numerical simulation of the hydrogen-air diffusion flame (CFDF) at an elevated pressure of ten atm are similar to those at one atm.

2. Flame extinction at this elevated pressure takes place at a strain rate of 140,000 s^{-1} as compared to the one atm case in which extinction takes place at 8200 s^{-1} (Isaac et al., 1994).

3. The profiles of temperature and H_2O mass fraction exhibit the equilibrium nature of the flame at small values of flame stretch. As flame stretch increases, non-equilibrium processes become dominant as indicated by the higher values of the net production rates of H_2O.

Acknowledgements

Some of the results presented in this paper are based on computer runs made by Jun Zhao, graduate student, UM-Rolla.

References

Fay, J. A. and Riddell, F. R., 1958. "Theory of stagnation point heat transfer in dissociated air," *J. Aeronautical Sciences* **25**, pp. 73-85.

Guerra, R., Pellett, G. L., Wilson, L. G., and Northam, G. B., 1987. "Opposed jet burner studies of effects of CO, CO_2 and N_2 air-contaminants on hydrogen-air diffusion flames", AIAA Paper No. AIAA-87-1960.

Guthiel, E. and Williams, F. A., 1989. "Numerical and asymptotic investigation of structures of hydrogen-air diffusion flames," Western States Section Meeting of the Combustion Institute, Livermore, CA.

Ho, Y. H. and Isaac, K. M., 1989. "Analysis of opposed-jet hydrogen-air counter flow diffusion flame," Technical Memorandum MAE TM-25, The University of Missouri-Rolla.

Ho, Y. H., Isaac, K. M., Pellett, G. L., and Northam, G. B., 1991. "Analysis of hydrogen-air counter flow diffusion flames," AIAA 29th Aerospace Sciences Meeting, Reno, AIAA-91-0582.

Isaac, K. M., Ho, Y. H., Zhao, J., Pellett, G. L., and Northam, G. B., 1994. "Global characteristics and structure of hydrogen-air counter flow diffusion flame: A one-dimensional model," AIAA Paper 94-0680, 32nd Aerospace Sciences Meeting, Reno, to appear.

Kee, R. J. and Miller, J. A., 1986. "A structured approach to the computational modeling of chemical kinetics and molecular transport of flowing systems," Sandia Report, SAND86-8841.

Law, C. K., 1988. "Dynamics of stretched flames," *Proceedings of the Twenty-second Symp. (Int.) on Combustion*, The Combustion Institute, pp. 1381-1402.

Patankar, S. V., 1980. *Numerical Heat Transfer and Fluid Flow*, Hemisphere, Washington D.C.

Pellett, G. L., Northam, G. B., Guerra, R., Wilson, L. G., Jarrett Jr., O., Antcliff, R. R., Dancey, C. L., and Wang, J. A., 1989. "Opposed jet diffusion flames of nitrogen-diluted hydrogen vs. air: Axial LDA and CARS surveys; Fuel/air strain rates at extinction," AIAA Paper No. AIAA-89-2522.

Pellett, G. L., Northam, G. B., and Wilson, L., 1991. "Counter flow diffusion flames of hydrogen plus methane, ethylene, propane and silane vs. air: Strain rate at extinction," AIAA-91-0370, AIAA Aerospace Sciences Meeting, Reno, Nevada.

Peters, N., 1984. *Prog. Energy Combust. Sci.* **10**.

Puri, I., Seshadri, K., Smooke, M. D., and Keys, D., 1987. "A comparison between numerical calculations and experimental measurements of the structure of a counterflow diffusion flame burning diluted methane in diluted air," *Combustion Science and Technology* **56**, p. 1.

Smooke, M. D., 1980. "Solution of burner-stabilized premixed laminar flames by boundary value methods," *J. Comp. Phys.* **48**, p. 72.

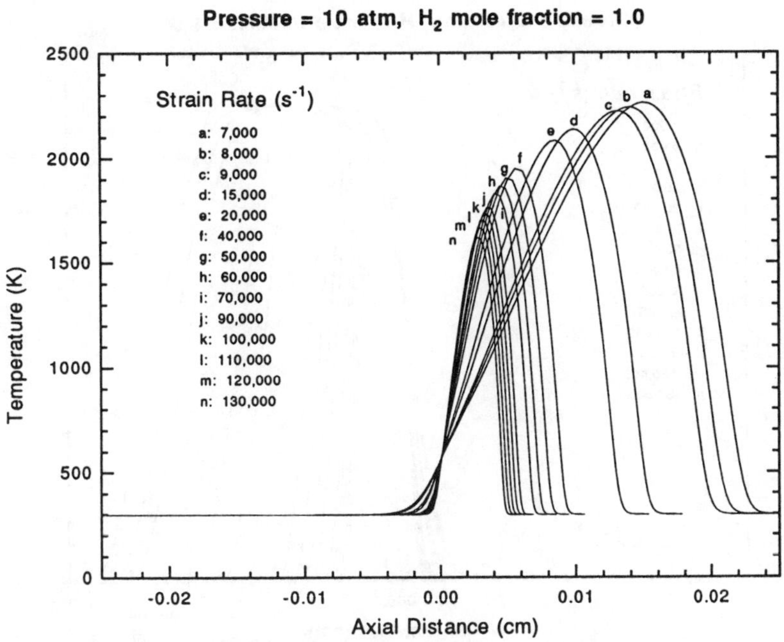

Figure 1. Temperature Profiles

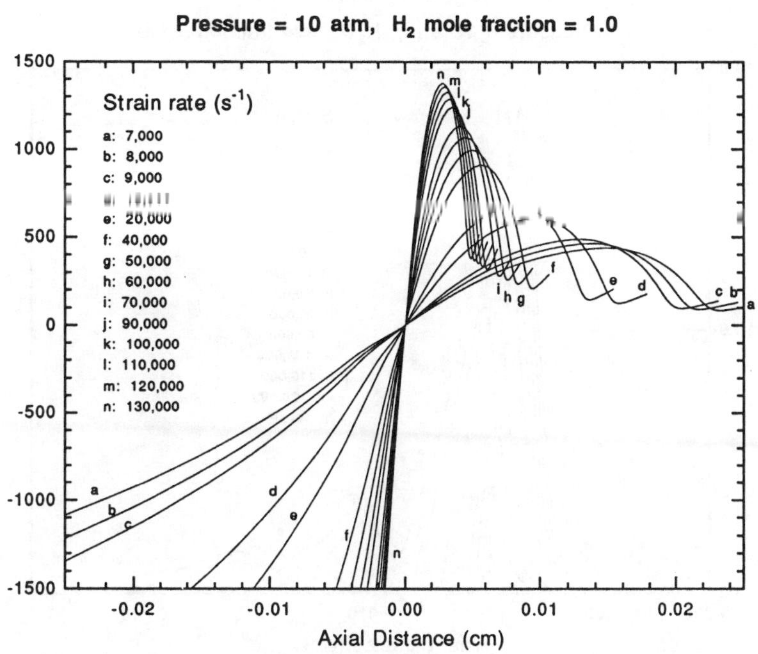

Figure 2. Velocity Profiles

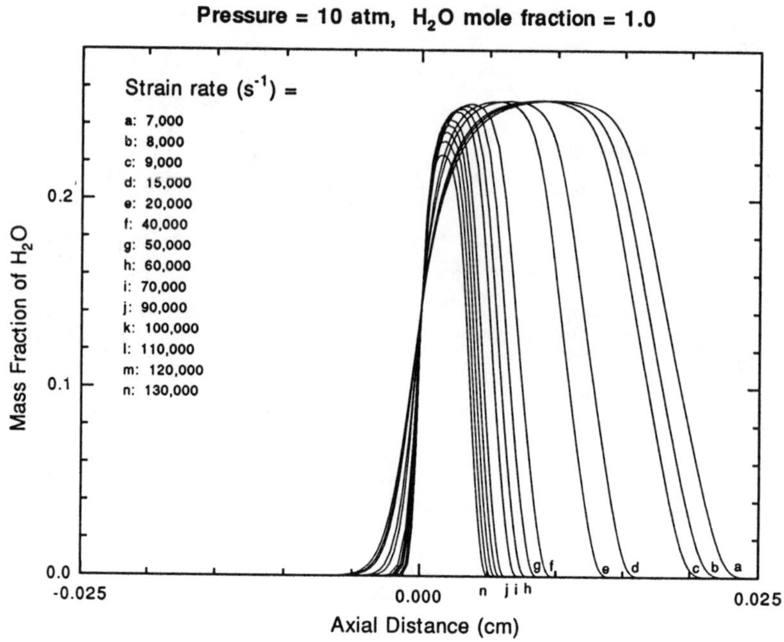

Figure 3. H_2O Mass Fraction Profiles

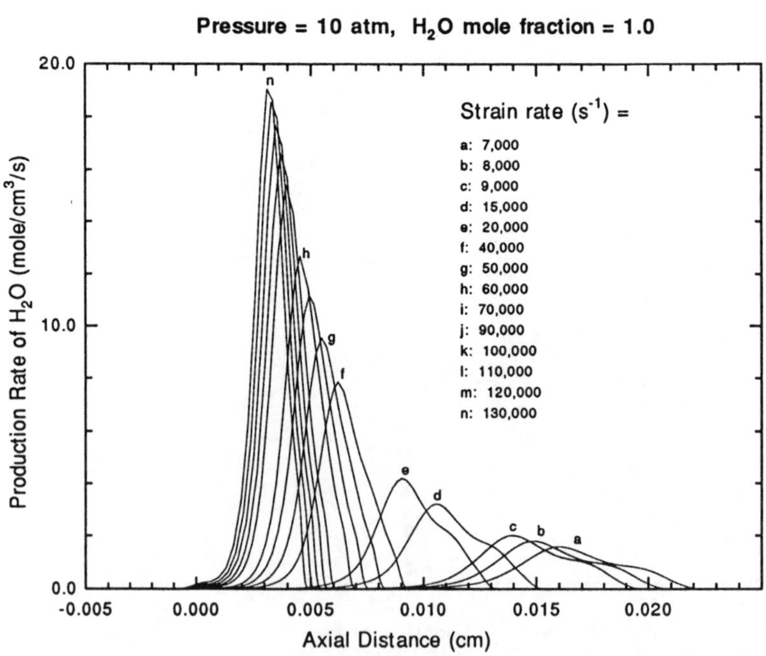

Figure 4. H_2O Production Rate Profiles

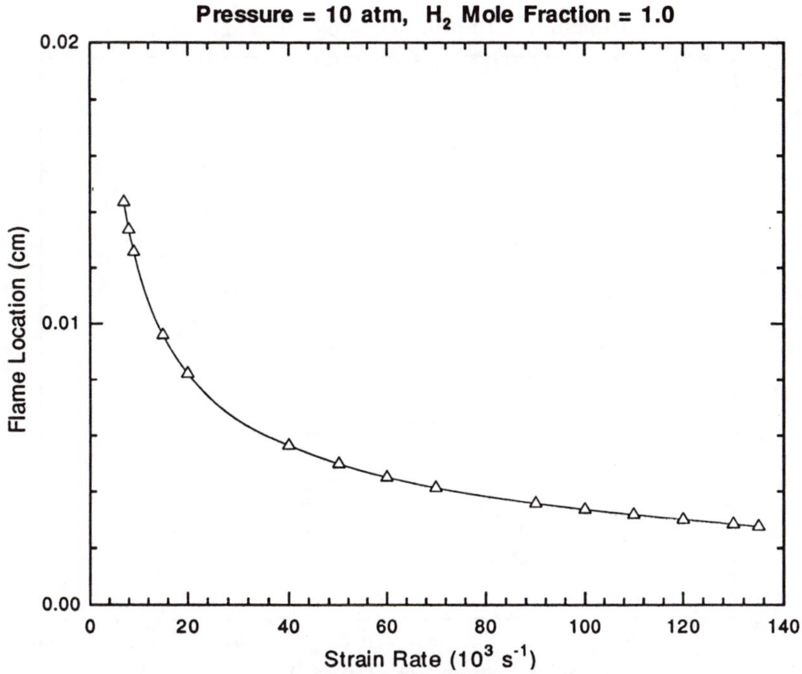

Figure 5. Flame Location vs. Strain Rate

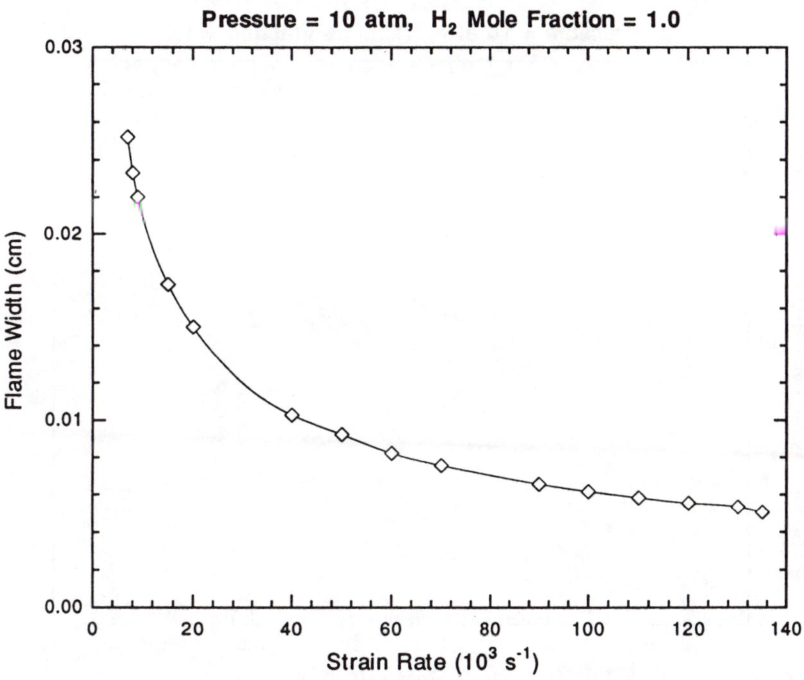

Figure 6. Flame Width vs. Strain Rate

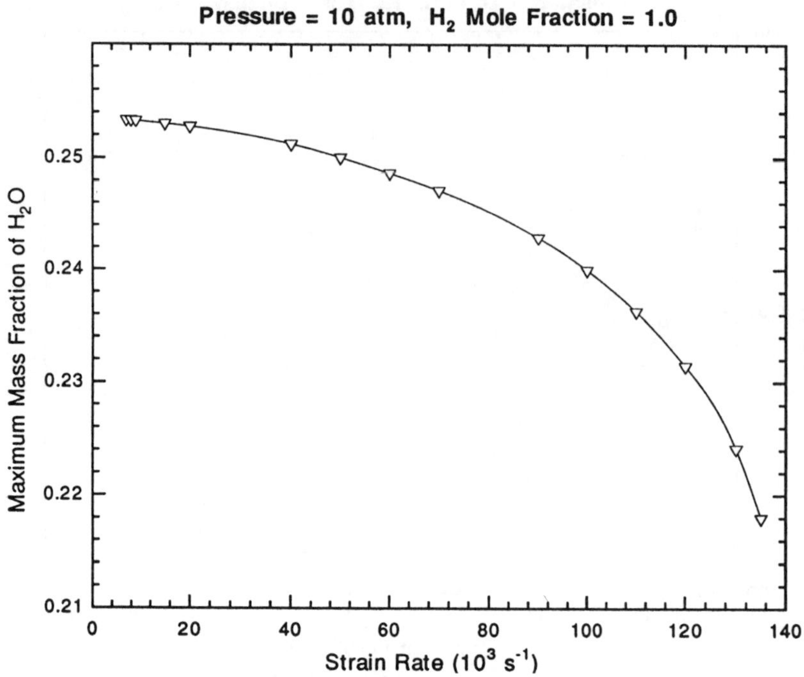

Figure 7. Maximum H_2O Mass Fraction vs. Strain Rate

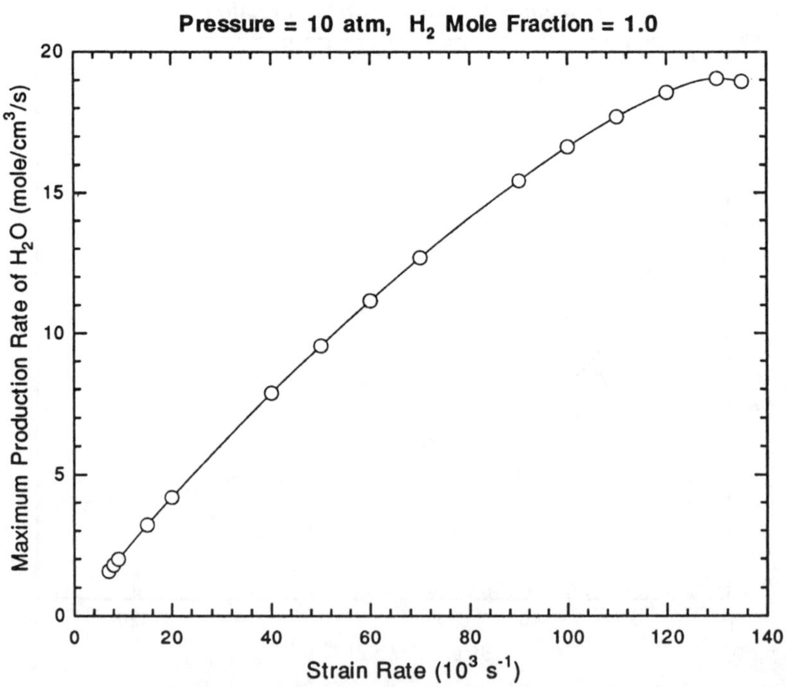

Figure 8. Maximum H_2O Production Rate vs. Strain Rate

LINEAR RESPONSE OF TWIN PREMIXED FLAMES IN A COUNTERFLOW TO SMALL DISTURBANCES[1]

T. L. Jackson

Institute for Computer Applications in Science and Engineering
NASA Langley Research
Hampton, Virginia 23681-0001

and

M. Matalon

Engineering Sciences and Applied Mathematics
McCormick School of Engineering and Applied Science
Northwestern University
Evanston, Illinois 60208

ABSTRACT

In this study the dynamics of twin premixed flames in a counterflow configuration is analyzed. The problem is of fundamental importance because this configuration is often used in laboratory experiments and because of its relevance to weakly turbulent premixed flames. The stability characteristics have been delineated in terms of all relevant parameters and in particular in terms of the strain rate. A comparison is also made between the two cases in which the corrugated flames are in or out of phase with respect to each other. It is shown that the difference is significant only in a limited range of moderate wavelength.

1. Introduction

A premixed flame held in the stagnation-point of the flow against an infinite plate or of the flow of two opposed jets, is a configuration often used in laboratory experiments. The flow field in such a config-

[1]This work was supported by the National Aeronautics and Space Administration under NASA Contract No. NAS1-19480 while in residence at the Institute for Computer Applications in Science and Engineering (ICASE), NASA Langley Research, Hampton, VA 23681-0001 and while attending the ICASE-NASA Langley Research Center workshop on Transition, Turbulence and Combustion held during June 7 - July 2, 1993. M.M also acknowledges support from the National Science Foundation.

M. Y. Hussaini et al. (eds.), Transition, Turbulence and Combustion, Vol. II, 217–226.

uration imposes on the bulk a well defined strain rate controlled by the mass flow rate of the incoming fresh mixture. It has been noticed that the flame in such circumstances is markedly more stable than an unstretched premixed flame. At low values of the strain rate the flame becomes unstable and various patterns can be observed on its surface. A premixed flame held in a stagnation-point flow is also a prototype of a premixed laminar flamelet and can therefore be used to study the dynamics of weakly turbulent flame fronts. For these reasons, the stability properties of strained premixed flames and the degree of instability is of primary importance.

It is well known that the stability of a premixed flame is significantly affected by disturbances of wavelength long compared to its thickness (Landau, 1944; Darrieus, 1946). Whereas diffusion often introduces stabilizing influences, long wavelength disturbances are hydrodynamically unstable. Stretch, however, could have an important effect in suppressing the hydrodynamic instability. Theoretical analyses of hydrodynamic stability in the presence of stretch encounter formidable difficulties for analytical considerations. Even for the simplest situation, the basic unperturbed flow is non parallel, with velocities varying normal and tangential to the flame front. As a consequence the linearized disturbance equations for the stability problem are *not* separable and the equations cannot be resolved using normal modes in both transverse directions to the flame front. Hence Kim and Matalon (1990) restricted attention to the case of perturbations with periodic waviness in the cross-stream direction only, with no corrugations in the direction of flow divergence. Even then, the disturbance equations in the region of burned gas, which reduce to a system of ordinary differential equations with variable coefficients, had to be solved numerically. More recently, Jackson and Matalon (1993) have reexamined the problem allowing for a more general class of three dimensional disturbances. The various modes, represented by the index n, which corresponds to the wavenumber of the corrugations in the direction of flow divergence, are all coupled so that the disturbance equations in the region of burned gas reduce to an infinite set of coupled systems of ordinary differential equations. The resulting problem was solved numerically after truncating the series systematically at some $n = N$, where N represents the smallest allowable length scale of corrugations in the direction of stretch.

There have been also attempts to gain qualitative understanding in the nature of the instability by examining the case of small heat release (Sivashinsky et al., 1982), for which density variations induced by thermal expansion are neglected, or adopting a potential model (Joulin and Sivashinsky, 1993), where the vorticity generated at the corrugated flame front is ignored. Because of their simplicity, these models yield analytical expressions for the growth rate and also permit the treatment of more complex situations. For example, Joulin and Sivashinsky (1993) considered, among others, the case of twin flames in a counterflow configuration. This problem, which has implications to weakly turbulent flames, is reconsidered here. The method used by Jackson and Matalon (1993) has been adopted where general three-dimensional disturbances have been considered without recourse to simplifying assumptions.

2. Problem Formulation

Two opposed jets of a combustible mixture are impinging toward each other. The flow field near the stagnation-point supports a planar premixed flame. Thus, two flames, located on either side of the stagnation point $x = 0$, separate the burned combustion products from the upstream regions of fresh unburned mixture. Far upstream, the flow field is the classical stagnation point flow characterized by the strain rate e. The stability characteristics of this twin-flames configuration depends on the strain rate as well as on the mixture properties. The latter is characterized by the Lewis number $Le = D_T/D_M$, where D_T is the thermal diffusivity of the mixture and D_M the mass diffusivity of the deficient reactant in the mixture, and by the thermal expansion parameter σ defined as the ratio of the burned to unburned gas densities.

The mathematical formulation is identical to the previous publications by Kim and Matalon (1990) and Jackson and Matalon (1993), which should be consulted for further details. Disturbances characterized by the length scale L, much larger than the flame thickness are considered. The latter is measured in units of the diffusion length $L_D = D_T/S_f$ where S_f is the laminar flame speed. Thus $\delta \equiv L_D/L \ll 1$ and, in the limit as $\delta \to 0$, the flame shrinks to a surface. The flow field on either side of the flame front is determined by the Euler's equations for an incompressible inviscid fluid. The flame structure, resolved on the $O(\delta)$ scale, yields jump conditions for the fluid variables across the flame as well as an evolution equation that determines the instantaneous shape and location of the flame front.

In the following, the length L has been used as a unit of distance, the laminar flame speed as a unit of speed and L/S_f as a unit of time. The dimensionless strain rate is $\epsilon = (L/S_f)e$. The state of the fresh mixture $\rho_\infty, P_\infty, T_\infty$ has been used to nondimensionalize the mixture density, pressure and temperature, respectively. All mixture properties have been assumed constant.

3. Basic State

The basic unperturbed state corresponds to two flat flames located at $x = \pm d$. The temperature is a piecewise-constant function with the values 1 and σ in the unburned/burned gas respectively. The density $\rho = 1/T$. It is assumed that all the reactant is consumed at the flame. The velocity field is given by

$$\mathbf{v} = (\ F(x),\ -yF'(x),\ 0) \tag{1}$$

with the function $F(x)$ expanded in the form

$$F(x) = F_0(x) + \delta F_1(x) + \cdots$$

where

$$F_0(x) = \begin{cases} -\epsilon(x + a_0) & -\infty < x < -d \\ -\frac{\sigma^{3/2}}{\sqrt{\sigma - 1}} \sin qx & -d < x < \ \ d \\ -\epsilon(x - a_0) & \ \ d < x < \ \ \infty \end{cases}$$

and

$$F_1(x) = \begin{cases} -\epsilon a_1 & -\infty < x < -d \\ \epsilon\sigma\frac{\gamma\ell}{2}\sqrt{\frac{\sigma}{\sigma-1}}(\sin qx - qx\cos qx) & -d < x < \ \ d \\ \epsilon a_1 & \ \ d < x < \ \ \infty. \end{cases}$$

Here

$$q = \frac{\epsilon\sqrt{\sigma - 1}}{\sigma}, \quad \gamma = \int_1^\sigma \frac{\ln \xi}{\xi - 1}\, d\xi, \quad \ell = \frac{E}{RT_\infty}\ \frac{Le - 1}{\sigma^2}$$

where E is the overall activation energy of the chemical reaction and R the gas constant. The flame positions $\pm d$ and the displacement distances $\pm a$ are expressed as $d = d_0 + \delta d_1 + \cdots$ and $a = a_0 + \delta a_1 + \cdots$, where

$$d_0 = \eta/\epsilon, \qquad d_1 = -\beta - \gamma\ell\eta/2$$
$$a_0 = d_0 - 1/\epsilon, \qquad a_1 = d_1 + \beta + \gamma\ell/2$$
$$\eta = \frac{\sigma\tan^{-1}\sqrt{\sigma-1}}{\sqrt{\sigma-1}}, \qquad \beta = \frac{\sigma\ln\sigma}{\sigma-1}.$$

The dynamic pressure is calculated directly from the momentum equation, if needed.

4. Disturbances

The perturbed flame fronts are now expressed in the form

$$x = -d + \phi(y)e^{ikz+\omega t}, \quad x = d + \psi(y)e^{ikz+\omega t}.$$

Similarly the velocity field perturbation, which is to be superimposed to (1), is given by

$$\tilde{\mathbf{v}} = (u(x,y),\, v(x,y),\, w(x,y))\, e^{ikz+\omega t}$$

When implemented in the governing equations, an eigenvalue problem results for the determination of the temporal growth rate ω and its dependence on the physico-chemical parameters $(\epsilon, \sigma, \ell, \delta)$ and on the form of the disturbance. The latter is characterized by the wavenumber k associated with wrinkles in the z-direction and an appropriate parameter that identifies the form of the corrugations in the y-direction.

If only wrinkles in the z-direction are of interest, ϕ and ψ are constants and the governing equations admit a self-similar solution in which all variables vary along the axial direction with the exception of v which increases linearly with y. The eigenvalue problem in this case reduces to a system of ordinary differential equations (Kim and Matalon, 1990).

In general, when wrinkles in the y-direction are also present the solution is expected to increase with distance from the axis as y^n. Consequently,

$$u = \sum_{n=0}^{\infty} U_n(x)\, y^n, \quad v = \sum_{n=0}^{\infty} V_n(x)\, y^n, \quad w = \sum_{n=0}^{\infty} ikW_n(x)\, y^n$$

and the various modes, corresponding to different indices n, are all coupled. The eigenvalue problem, thus, consists of an infinite set of coupled systems of ordinary differential equations. A solution can be found by truncating the series at $n = N$, as in Jackson and Matalon (1993). The growth rate then depends on N which represents the smallest allowable length scale for the corrugations in the y-direction. Note that the self-similar solution corresponds to $N = 1$.

As commented earlier, the model adopted here is asymptotically valid when $\delta \ll 1$. Hence the growth rate ω is expressed in the form

$$\omega = \epsilon\alpha_0(\sigma, \kappa, N) + \delta\epsilon^2\alpha_1(\sigma, \ell, \kappa, N) \tag{2}$$

where $\kappa = k/\epsilon$. The leading term, α_0 represents the modification of the Darrieus-Landau instability for this configuration. The second term incorporates the effects due to diffusion.

5. Results

It is convenient to distinguish between the following two cases (see also Figure 1):

1. $\phi\psi > 0$ everywhere, so that the twin flames are in phase with respect to each other, and
2. $\phi\psi < 0$ everywhere, so that the twin flames are out of phase with respect to each other.

In the first case the location of the stagnation point remains at $x = 0$ so that the axial velocity must vanish there. The problem is therefore identical to that of a flame standing in the front stagnation-point flow generated when a combustible jet impinges against a flat wall. The results, reported below, are therefore deduced from Jackson and Matalon (1993). In the second case the location of the stagnation-point is slightly shifted; but symmetry considerations requires that the axial velocity gradient vanishes at the unperturbed position $x = 0$. The calculation for this case were performed in a similar way as for the previous case. In both cases, the least stable mode corresponds to disturbances which are self-similar, i.e. with $N = 1$. These are the only one shown in Figures 2 and 3.

In Figure 2, the leading term α_0 is plotted against the scaled wavenumber κ. It is also instructive to write down the explicit expression for α_0, obtained by adopting a *potential model*:

$$\alpha_0 = -1 + \frac{1}{2}\kappa(\sigma - 1)\left(1 \pm e^{-2\kappa\sigma}\right), \tag{3}$$

where the $+/-$ correspond to the two cases in which the twin flames are in phase and out of phase, respectively. This expression agrees qualitatively with the numerical calculations which fully account for the vorticity in the burned gas, both in the basic state (1) and in the perturbed flow. The first term in (3) indicates the stabilizing effect of stretch on the long wavelength disturbances. The second term in (3) is the Darrieus-Landau instability. The last term in (3) is the damping effect resulting from the proximity of the two flames. This damping effect is significant for long wavelength disturbances only and is more effective when the twin flames are out of phase with respect to each other.

In Figure 3 the growth rate $\alpha = \omega/\epsilon$, which include the combined effects of hydrodynamics and diffusional-thermal, is plotted against κ for selected values of $\epsilon\delta$. Clearly, diffusion has a stabilizing influence on the short wavelength disturbances. The results here indicate that there exist a band of wavenumbers for which twin flames out of phase are more stable than twin flames in phase depending on $\epsilon\delta$. For very long or very short waves the relative position of the flames is of no particular importance.

References

Darrieus, G., 1946. *Unpublished manuscript* (1938). *Sixth International Congress of Applied Mechanics.*

Jackson, T. L. and Matalon, M., 1993. *Combustion Science and Technology*, **90**, 385.

Joulin, G. and Sivashinsky, G. I., 1992. *Twenty-Fourth Symposium (International) on Combustion*, The Combustion Institute, Pittsburgh, PA, p. 37.

Kim, Y. D. and Matalon, M., 1990. *Combustion Science and Technology* **69**(4-6), p. 85.

Landau, L. D., 1944. *Acta Physicochimica URSS* **19**, p. 77.

Sivashinsky, G. I., Law, C. K., and Joulin, G., 1982. *Combustion Science and Technology* **28**, p. 155.

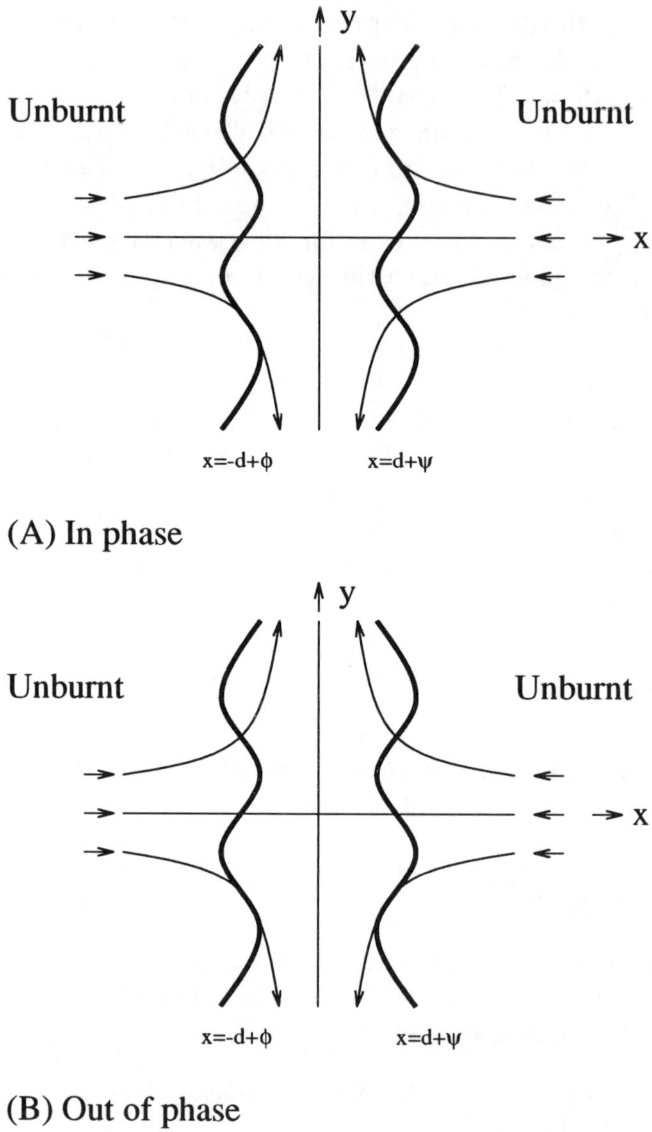

Figure 1. Schematic representation of premixed twin flames in a counterflow.

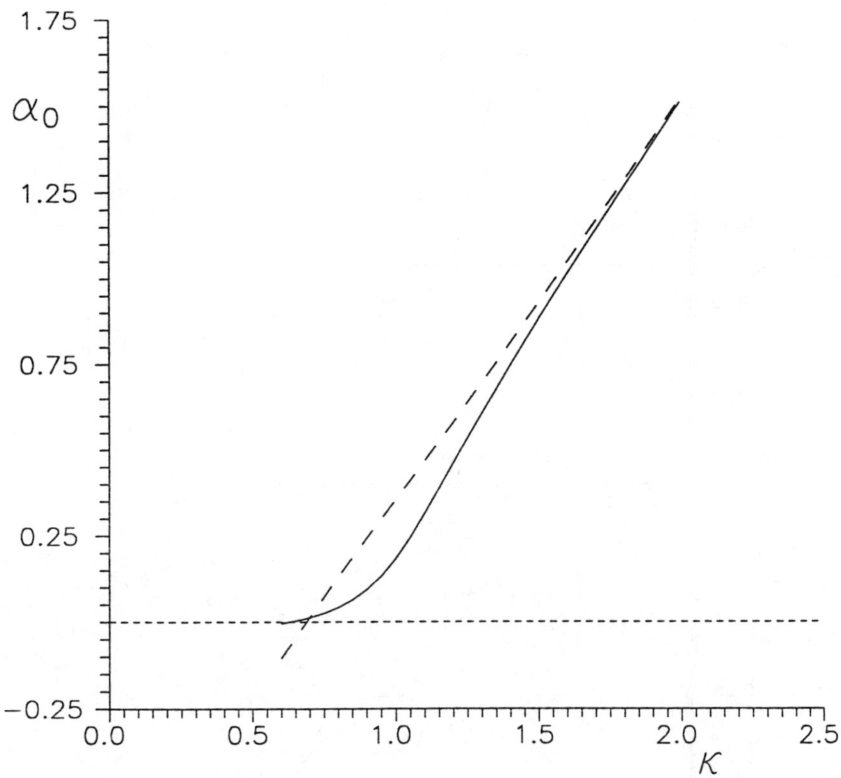

Figure 2. The growth rate α_0 plotted against the scaled wavenumber κ for $N = 1$.

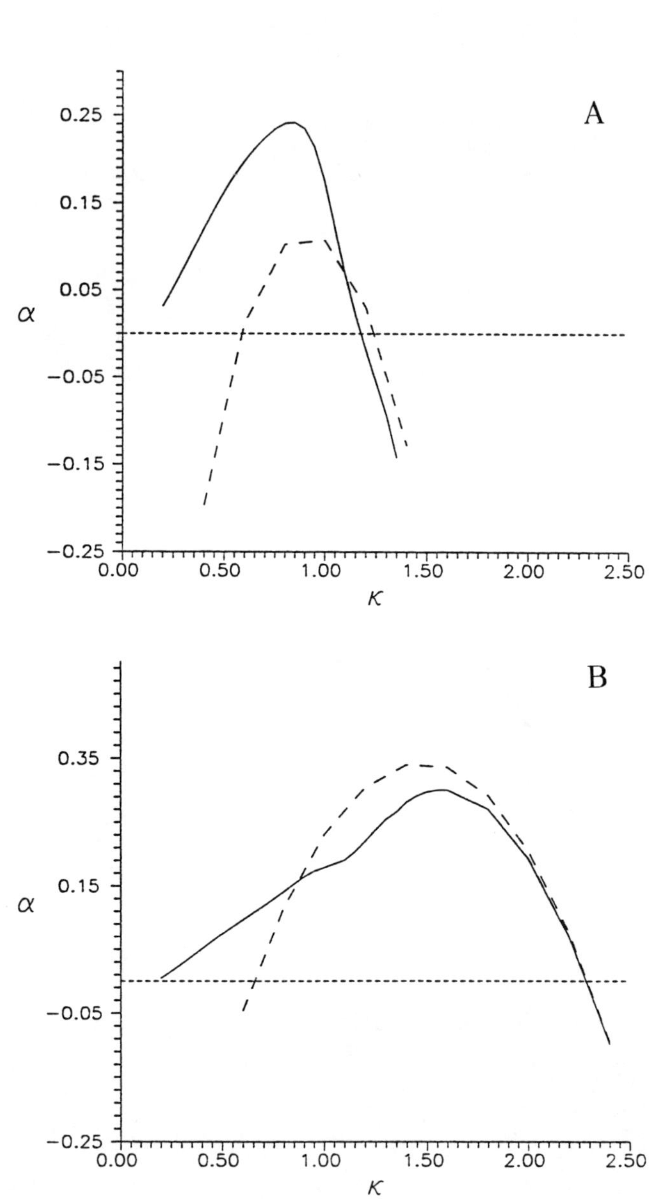

Figure 3. The growth rate $\alpha = \omega/\epsilon$ plotted against the scaled wavenumber κ for $N = 1$, $\ell = -0.45$ and (a) $\epsilon\delta = 0.8$ and (b) $\epsilon\delta = 0.4$.

RAMAN SCATTERING IMAGING OF OPPOSED JET DIFFUSION FLAMES[1]

R. W. Pitz, T. M. Brown, and S. P. Nandula

Department of Mechanical Engineering
Vanderbilt University
Nashville, Tennessee 37235

ABSTRACT

In this work, the structure of stretched one-dimensional (1D) opposed flow laminar diffusion flames is studied with accurate nonintrusive laser diagnostics. In recent turbulent flame models, turbulent diffusion flames are predicted by modeling the reaction zones as a statistical ensemble of laminar flamelets undergoing different values of stretch. The laminar flamelets are described by the structure of a laminar opposed flow diffusion flame which is calculated with detailed transport, complex chemistry models. These models however have not been verified by nonintrusive laser measurements that include species concentrations and temperature. This paper describes preliminary work to measure species concentrations and temperature using line Raman scattering imaging from a KrF excimer laser. Measurements are made in an opposed jet diffusion flame of hydrogen and air.

1. Introduction

One-dimensional, stretched opposed flow laminar flames have been studied extensively for a number of years as they provide a simple experimental configuration to understand the effect of strain rate on the chemistry (e.g. Tsuji, 1982; Law, 1988). The theoretical model of the chemistry and transport in the stretched flame can be reduced to a simple two-point boundary value problem where the calculation of the complex chemistry and detailed transport can be more easily implemented (Smooke et al., 1985). The motivation of many of the laminar flamelet studies has been the promise that these

[1]This work was supported in part by ICASE through a fellowship for TMB and summer support for RWP while in residence at NASA Langley Research Center. The authors also wish to acknowledge support from the National Science Foundation (PYI Grant CTS-8657130 and equipment Grant CTS-9007519).

M. Y. Hussaini et al. (eds.), Transition, Turbulence and Combustion, Vol. II, 227–237.

laminar flamelets can be used as a model of turbulent reaction zones. Laminar flamelet models of turbulent combustion have been developed that describe turbulent flame fronts as a ensemble of stretched laminar flame fronts (Peters, 1984, 1985; Peters and Williams, 1983, Rogg et al., 1986).

The flamelet libraries used in turbulent combustion codes are calculated from detailed transport and complex chemistry calculations that have not been verified by nonintrusive laser diagnostics that accurately measure the chemical species and temperature in the reaction zones. These predictions have been compared to probe measurements of species concentrations and temperature (e.g. Puri et al., 1987; Smooke et al., 1988). Probe measurements can effect the flow and can be biased by sample line reactions. Also minor species such as OH that are sensitive to the chemistry are not measured in these studies. The temperature is measured by thermocouples which have to be corrected for radiation loss and will not survive at high temperatures (above 2000 K).

Recently a UV Raman technique has been developed at Vanderbilt University that instantaneously and simultaneously measures major species (H_2, O_2, N_2, and H_2O), minor species (OH), and temperature in a turbulent flame (Wehrmeyer et al., 1992, Cheng et al., 1992). The ultraviolet (UV) radiation from a narrowband KrF excimer laser produces an enhanced Raman signal that is used to measure the major species concentrations and temperature. The OH concentration is determined from laser-induced predissociative fluorescence (LIPF) of OH from the same laser pulse.

Mixing and finite-rate chemistry have been studied by UV Raman scattering in a number of turbulent flames including lifted jet diffusion flames (Cheng et al., 1992, Pitz et al., 1992), swirled jet diffusion flames (March et al., 1991), and supersonic jet diffusion flames (Cheng et al., 1991). All these measurements were made at a single point using photomultiplier tubes to record the Raman signals. The flame was scanned to obtain concentration and temperature profiles.

The UV Raman system has been recently improved to make measurements along a line using a digital camera imaging system (Nandula et al., 1992). Raman spectra from individual points along the line are interpreted to give the chemical composition and temperature at each point. From the gradients of temperature and composition, the scalar dissipation can be estimated in the flame zone (Nandula et al., 1993).

In this work, the line Raman method is applied to an opposed jet diffusion flame. The line Raman method is described and preliminary measurements are presented for the opposed jet diffusion flame at different values of stretch.

2. Experimental Measurements

UV Raman/LIPF measurements

A schematic of the UV Raman imaging system is shown in Fig. 1. A narrowband KrF excimer laser is focussed by a 3 m lens into the flame. The narrowband KrF excimer laser has a pulse energy of 200 mJ and a pulse length of 17 ns. The scattered light is collected by a reflective Cassegrain optics and focussed onto the entrance slit of 1/2 m spectrometer. A butyl acetate filter rejects the laser light but passes the Stokes Raman lines.

The spectrometer disperses the spectra across the face of a Princeton Instruments intensified charge coupled device (ICCD) camera. The image from the ICCD camera is digitized and stored in a personal computer. The 4.8 mm line image recorded by the camera in a flame is depicted in Fig. 2. A typical single-pulse spectral image is shown in Fig. 3a.

For species concentration measurement, the integrated ICCD camera counts (Q_i) from the ith species Raman line is given by the equation:

$$Q_i = K_i Q_l [N_i] f_i(T)$$

where $[N_i]$ is the number density of species i, Q_l is the incident laser energy, and K_i is a proportionality constant dependent on the vibrational Raman cross section, geometry, and optical efficiency. The bandwidth factor, $f_i(T)$, accounts for a number of the temperature-dependent factors including the distribution of molecules in their allowed quantum states, spectral cutoff due to narrow exit slits, and spatial non-uniformities in detector response. The bandwidth factor can be theoretically calculated for diatomic molecules; however, the calculation is less reliable for polyatomic molecules. Hence, the terms K_i and $f_i(T)$ are combined to give the calibration factor $C_i = 1/[K_i f_i(T)]$ which is obtained from calibration in a Hencken burner (Cheng et al., 1992). The temperature is calculated from the ideal gas law using the measured total number density of all the chemical species and the known pressure.

For the OH concentration measurement, the laser-induced predissociative fluorescence (LIPF) process is employed (Cheng et al., 1992). The 5% unlocked broadband part of the narrowband laser excites OH molecules from $v'' = 0$ to $v' = 3$ in the $A^2\Sigma \leftarrow X^2\Pi$ system. The $v' = 3$ state is predissociated due to a crossing of the $a^4\Sigma^-$ state. The measured OH fluorescence intensity is related to the OH number density as follows:

$$I_F = C\frac{A}{A+Q+P}I_L[N_{OH}]f_B(T)$$

where C is a constant dependent on optical and detector efficiency, A is the fluorescence rate, Q is the collisional quenching rate, P is the predissociating rate, I_L is the incident laser intensity, $[N_{OH}]$ is the OH number density, and $f_B(T)$ is the temperature dependent Boltzmann population fraction. Since P is $\sim$ 20 times greater than A plus Q, the above equation can be simplified,

$$I_F = C\frac{A}{P}I_L[N_{OH}]f_B(T).$$

Thus the fluorescence is independent of the collisional quenching – a very important result. Other LIF techniques must correct the OH fluorescence for collisional quenching (i.e. Barlow et al., 1990). From our calculations (Cheng et al, 1992), the OH fluorescence is temperature insensitive ($< 10\%$) from 1500 to 3000 K. Thus the OH LIPF signal is linearly proportional to the OH concentration. The OH LIPF system is calibrated in the Hencken burner.

Using the theory described above and measuring signals at the known high temperature conditions in the Hencken burner, the species concentrations and temperature measurement system is calibrated.

The line Raman system has been applied to measure the chemistry of a turbulent flame (Nandula et al., 1993). A 4.8 mm Raman line image is measured through the turbulent flame front as sketched in Fig. 2. The line Raman spectra produced by a 248 nm KrF excimer laser are shown in Fig. 3a. Here 24 Raman spectra are shown for the different radial positions through the turbulent hydrogen/air diffusion flame. The Raman peaks for H_2 (276 nm) can be seen in the center of the jet which is at the bottom of the figure. The Raman peaks for O_2 (258 nm) and N_2 (263 nm) are on the air side of the jet. The H_2O Raman line is in between the air and fuel. From these Raman scattering signals, the mole fractions of the species and the temperature can be determined as shown in Fig. 3b for the same laser shot. The spatial resolution of this measurement is 0.4 mm. The line Raman system has also been applied to the turbulent jet diffusion flame to measure dissipation rate and differential diffusion effects in the turbulent reaction zones (Nandula et al., 1993).

Opposed jet UV Raman/LIPF measurements

Preliminary line Raman and LIPF measurements have been made in the opposed jet diffusion flame sketched in Fig. 4. The nozzle diameter (D) is 7.2 mm and the opposed jets are spaced 1D apart. The KrF excimer laser beam is projected down the axis of the burner. Line Raman images taken at low and moderate stretch conditions are shown in Fig. 5. The axial dimension of the line Raman image is 4.8 mm. The count levels of the images are proportional to the species number density (#/cc) and thus inversely proportional to the temperature. The hydrogen fuel (277 nm) is at the top of the image and the air (O_2: 257 nm, N_2: 263 nm) at the bottom. Since the hydrogen fuel is diluted with nitrogen, there is nitrogen throughout the opposed flow flame. The N_2 signal decreases in the flame due to the increased temperature. The H_2O (273 nm) and OH LIPF (297 nm) signals can be seen at the center of the flame sheet. In the higher stretch flame (Fig. 5b), the flame is thinner, the gradients are steeper, and the OH number densities are smaller.

Since this is a laminar flame, multiple images can be recorded and averaged. With the increased signals the spatial resolution of the averaged image can be reduced to about 0.2 mm. For low stretched flames (typically a few millimeters thick), this resolution is very good.

Using the calibration measurements, the images shown in Fig. 5 will be interpreted as discussed earlier. The Raman signals will give the major species concentrations and temperature. The OH LIPF fluorescence will give the concentration of OH in the flame.

3. Conclusions

A line Raman/LIPF imaging method has been described that measures species concentrations and temperatures nonintrusively in a flame zone. Line Raman images have been recorded in an opposed jet laminar diffusion flames. The line Raman/LIPF images qualitatively show the concentrations of H_2, O_2, N_2, H_2O and OH in the opposed jet flame. As expected, at higher values of stretch, the flame zone is found to be thinner and the OH concentrations are reduced. The line Raman images are currently being analyzed to give quantitative axial concentrations and temperature profiles.

Acknowledgements

The technical contribution of G. Pellett and L. Wilson of NASA Langley Research Center in the design and construction of the opposed jet burner is gratefully acknowledged.

References

Barlow, R. S., Dibble, R. W., Chen, J.-Y., and Lucht, R. P., 1990. "Effect of Damköhler number on superequilibrium OH concentrations in turbulent nonpremixed jet flames," *Comb. Flame* **82**, p. 235.

Cheng, T. S., Wehrmeyer, J. A., and Pitz, R. W., 1992. "Simultaneous temperature and multi-species measurement in a lifted hydrogen diffusion flame," *Comb. Flame* **91**, p. 323.

Cheng, T. S., Wehrmeyer, J. A., Pitz, R. W., Jarrett, Jr., O., and Northam, G. B., 1991. "Finite-rate chemistry effects in a Mach 2 reacting flow," AIAA/SAE/ASME/ASEE 27th Joint Propulsion Conference, AIAA-91-2320.

Law, C. K., 1988. "Dynamics of stretched flames," 22nd Symposium (Int.) on Combustion, The Combustion Institute, Pittsburgh, p. 1381.

March, S. R., Cheng, T. S., Pitz, R. W., Wehrmeyer, J. A., and Debelak, K. A., 1991. "Temperature and species concentration measurements in a swirled hydrogen diffusion flame," AIAA 29th Aerospace Sciences Meeting, AIAA-91-0480.

Nandula, S. P., Brown, T. M., Cole, W. A., and Pitz, R. W., 1992. "Simultaneous multi-species multi-point measurements in H_2-air flames using a narrowband KrF excimer laser," AIAA/SAE/ASME/ASEE 28th Joint Propulsion Conference, AIAA-92-3348.

Nandula, S., Brown, T. M., and Pitz, R. W., 1993. "Measurements of differential diffusion and scalar dissipation in reaction zones by Raman scattering," Western States Section of the Combustion Institute, Paper #93-064, Fall Meeting, Menlo Park, CA, October.

Peters, N., 1984. "Laminar diffusion flamelet models in non-premixed turbulent combustion," *Prog. Energy Comb. Sci.* **10**, p. 319.

Peters, N., 1984. "Partially premixed diffusion flamelets in non-premixed turbulent combustion," 20th Symposium (Int.) on Combustion, The Combustion Institute, p. 353.

Peters, N. and Williams, F. A., 1983. "Liftoff characteristics of turbulent jet diffusion flames," *AIAA J.* **21**, p. 423.

Pitz, R. W., Nandula, S., and Brown, T. M., 1992. "Comparison of reaction zones in turbulent lifted diffusion flames to stretched laminar flamelets," AIAA/SAE/ASME/ASEE 28th Joint Propulsion Conference, AIAA-92-3349.

Puri, I. K., Seshadri, K., Smooke, M. D., Keyes, D. E., 1987. "A comparison between numerical calculations and experimental measurements of the structure of a counterflow methane-air diffusion flame," *Combust. Sci. and Tech.* **56**, pp. 1-22.

Rogg, B., Behrendt, F., and Warnatz, J., 1986. "Turbulent non-premixed combustion in partially premixed diffusion flamelets with detailed chemistry," 21st Symposium (Int.) on Combustion, The Combustion Institute, p. 1533.

Smooke, M. D., Miller, J. A. and Kee, R. J., 1985. "Solution of premixed and counterflow diffusion flame problems by adaptive boundary value methods," in *Numerical Boundary Value ODEs*, U. M. Ascher and R. D. Russell, Eds., Birkhauser, Boston, p. 303.

Smooke, M. D., Seshadri, K., and Puri, I. K., 1988. "The structure and extinction of partially premixed flames burning methane in air," 22nd Symposium (Int.) on Combustion, The Combustion Institute, p. 1555.

Tsuji, H., 1982. "Counterflow diffusion flames," *Prog. Energy Comb. Sci.* **8**, pp. 93-119.

Wehrmeyer, J. A., Cheng, T. S., and Pitz, R. W., 1992. "Raman scattering measurements in flames using a tunable KrF laser," *Applied Optics* **31**, pp. 1495-1504.

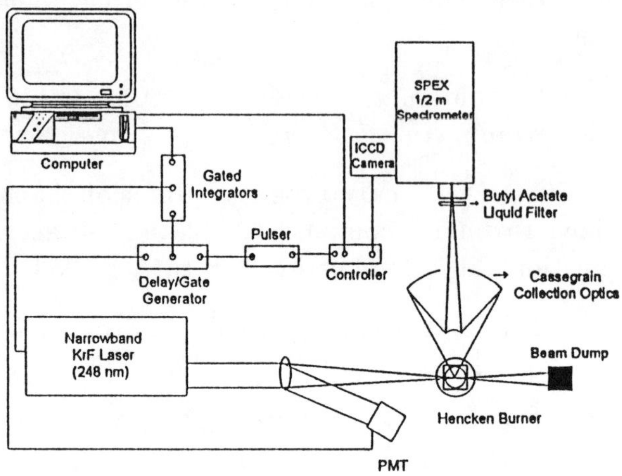

Fig. 1. Schematic of UV Raman imaging system.

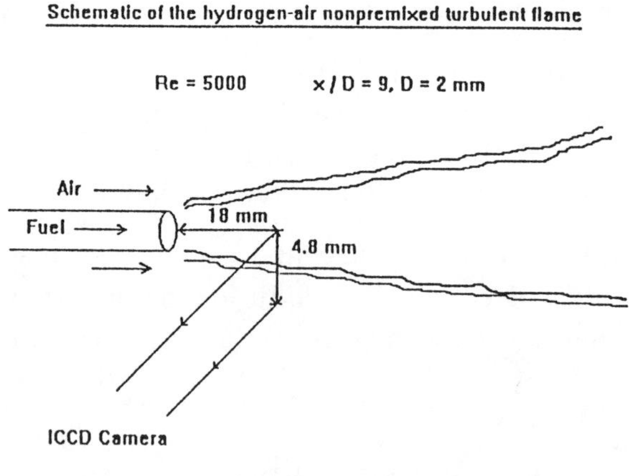

Fig. 2. Schematic of a turbulent jet diffusion flame. The 5 mm vertical line indicates the line imaged by the laser Raman system.

a)

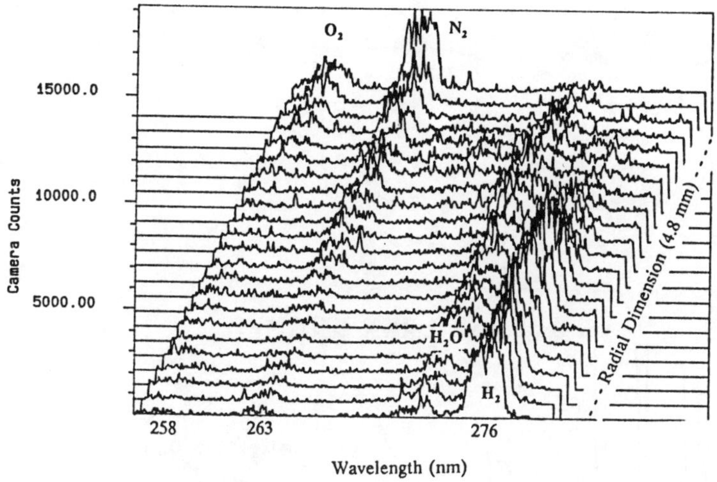

b)

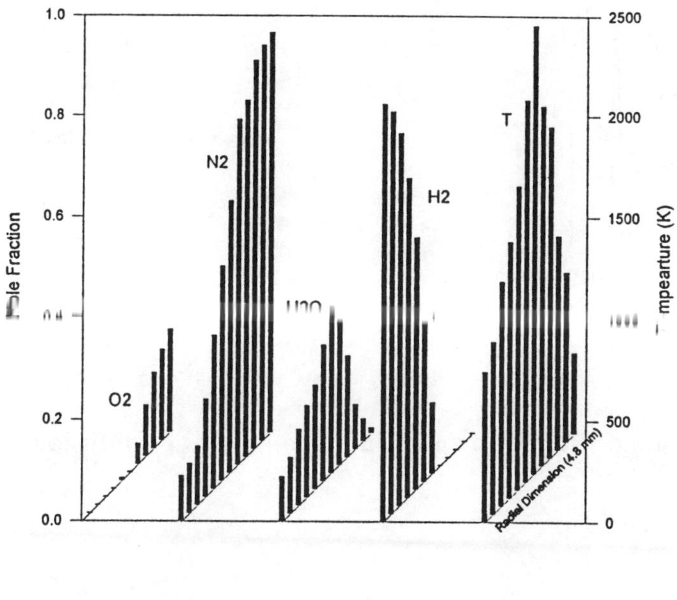

Fig. 3. Single-pulse line Raman imaging in a turbulent H_2/air diffusion flame (Re = 5000, x/D = 9, D = 2 mm). Both images are calculated from the same camera image from a 4.8 mm line. a) Single-pulse line Raman spectra (24 radial locations) b) Single-pulse line Raman mole fraction and temperature image (12 radial locations).

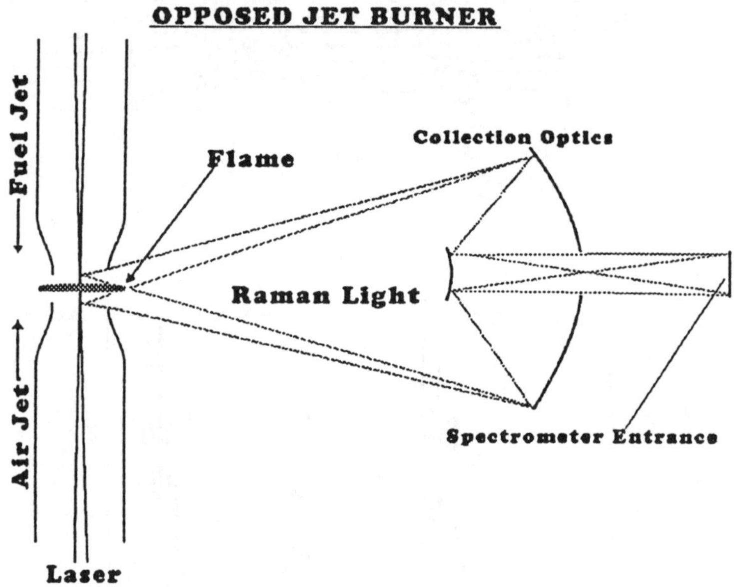

Fig. 4. Line Raman imaging in the opposed jet diffusion flame.

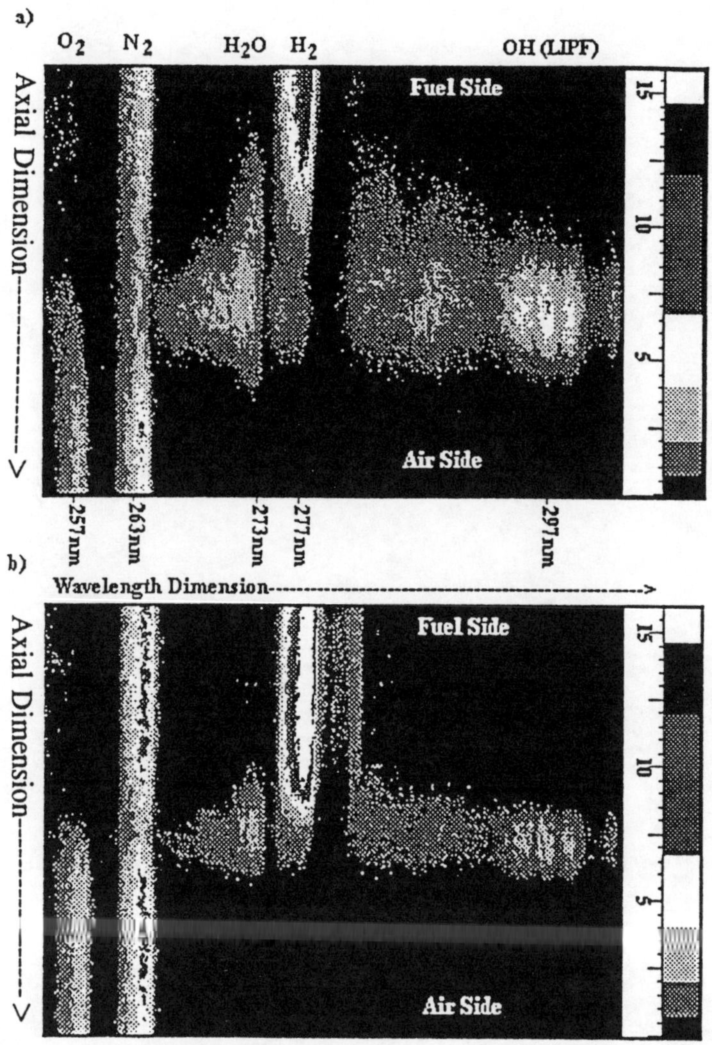

Fig. 5. Line Raman images in H_2-air opposed jet diffusion flames. The fuel is diluted with N_2. a) Low stretch ($a = 135s^{-1}$), fuel: 50% H_2 50% N_2. b) Moderate stretch ($a = 1350s^{-1}$), 40% H_2 60% N_2.

THE INFLUENCE OF CONTAMINANTS ON THE STRUCTURE OF NONPREMIXED HYDROGEN-AIR FLAMES

K. Seshadri

Center for Energy and Combustion Science
University of California, San Diego
La Jolla, California 92093-0310

ABSTRACT

The influence of contaminants on the structure and critical conditions of extinction of counterflow hydrogen/oxygen/nitrogen diffusion flames were examined. The contaminants considered were H_2O and CO_2. The results are anticipated to be useful for relating the performance of supersonic, air-breathing, hydrogen-fueled engines determined from ground-based tests, wherein the air used contains these contaminants, to their performance during flight. Experimental data showing the influence of addition of H_2O and CO_2 to the oxidizer stream of counterflow hydrogen/oxygen/nitrogen diffusion flames, on its critical conditions of extinction are available. Systematic interpretation of these experimental results show that these contaminants increase the value of the strain rate at extinction thus making the flame harder to extinguish. Results of previous asymptotic analyses were used to explain these observations. It was found that addition of the contaminants increased the rate of the recombination reaction $H + O_2 + M \rightarrow HO_2 + M$ which is responsible for the major fraction of the heat released in the flame. The predictions of the asymptotic analyses were compared with the measurements and were found to agree reasonably well.

1. Introduction

There is considerable interest in clarifying the combustion of hydrogen because this compound is considered as a fuel for propulsion of supersonic aircraft. To evaluate the performance of supersonic, air-breathing, hydrogen-fueled engines during flight, it is necessary to perform ground-based tests. Such tests are currently being conducted at NASA Langley Research Center. These ground-based studies on combustion of hydrogen in high speed flows require high

M. Y. Hussaini et al. (eds.), Transition, Turbulence and Combustion, Vol. II, 239–248.

enthalpy air streams. The high enthalpy air flows employed in these studies are generally obtained by use of arc heaters or by vitiation processes using hydrogen or various gaseous hydrocarbon compounds as fuels. The air obtained under these conditions is generally contaminated with H_2O, CO_2 and smaller amounts of CO and NO. Since the composition of the air used in the test facilities is different from that encountered by the aircraft during flight, it is necessary to evaluate the influence of these contaminants on the structure and critical conditions of extinction of hydrogen-air flames. Otherwise results obtained from ground-based tests cannot be confidently used to evaluate the performance of these engines during flight.

Extensive experiments have been performed by Guerra et al. (1987a), Guerra et al. (1987b), and Pellett et al. (1989) to clarify the influence of the contaminants H_2O, CO_2, CO and NO on the critical conditions of extinction of counterflow $H_2/O_2/N_2$ diffusion flames. Flame extinction is referred to as "blowoff" by these investigators (Guerra et al., 1987a, 1987b, and Pellett et al., 1989). These experiments were performed on diffusion flames stabilized in the mixing layer formed between counterflowing streams of H_2 diluted with N_2 and air. The burner employed in these experiments consisted of two opposing tubes with exit diameters of 2.7 mm, 5 mm, and 7.0 mm and were typically placed two to three diameters apart. Thus, the value of aspect ratio, defined as the ratio between the tube diameter and the separation distance between the tubes is typically between 0.5 and 0.33. The characteristic length scale in these experiments was assumed to be equal to the diameter of the tube and the strain rate at extinction a_q was assumed to be proportional to the ratio of the injection velocity of the reactant at the exit of the duct, to its diameter.

Although the previous research summarized here on $H_2/O_2/N_2$ diffusion flames clarified a number of issues concerning the influence of contaminants on the critical conditions of flame extinction, there is need for more research in this area. For example, the thermal and chemical-kinetic aspects associated with the influence of the contaminants on the flame structure is unclear. Therefore, as a first step toward understanding the role of the contaminants on the structure and critical conditions of extinction of counterflow diffusion flames of hydrogen and air, the experimental data of Guerra et al. (1987a), Guerra et al. (1987b), and Pellett et al. (1989) were interpreted using asymptotic theories.

2. Interpretation of Experimental Results

Figures 1 and 2 respectively show the measured final mole percent of H_2O and CO_2 in the oxidizer stream as the function of its ejection velocity U_{air} at blowoff (Pellett et al., 1989). The three different lines in these figures refer to different initial mole percent of O_2 in the oxidizer stream prior to the addition of the contaminants. Both figures show the value of U_{air} to decrease with increasing concentration of the contaminant, which implies that the flame is becoming weaker or less resistant to aerodynamic strain. This is due to the decrease in the calculated value of the adiabatic flame temperature T_{st} with increasing concentration of the contaminant. However further interpretation of the experiment data, shows that for constant values of the calculated adiabatic flame temperature T_{st} shown in Figs. 1 and 2, the value of U_{air} increases with increasing concentration of the contaminant. The value of T_{st} was calculated, assuming that H_2 and O_2 react completely to form H_2O, equilibrium dissociation of the product was neglected and the contaminants H_2O and CO_2 were considered to be inerts. Because the maximum flame temperature in the reaction zone at conditions close to extinction is expected to be low, the value of T_{st} calculated, neglecting dissociation is a proper representation of the overall enthalpy in the reaction zone. This clearly implies that for the same overall enthalpy in the reaction zone, the contaminants make the flame stronger or more resistant to aerodynamic strain. The values of U_{air} and T_{st} at the intersection between the lines of constant T_{st} and the lines where the initial concentration of oxygen in the oxidizer stream prior to addition of the contaminant is a constant are shown in Tables 1a and 1b, again demonstrating that for constant values of T_{st}, the value of U_{air} increases with increasing mole percent of the contaminant in the reaction zone.

The following is a brief illustration of the use of asymptotic results to predict the influence of H_2O and CO_2 on the critical conditions of extinction of counterflow $H_2/O_2/N_2$ diffusion flames. The results obtained from asymptotic analyses of $CH_4/O_2/N_2$ diffusion flames (Seshadri and Peters, 1988, and Yang and Seshadri, 1992), will be used for this purpose, because the functional dependence of the value of a_q to various parameters are expected to be similar for $CH_4/O_2/N_2$ flames and $H_2/O_2/N_2$ (Seshadri and Peters, 1988; Seshadri and Williams, 1993; Seshadri, Peters, and Williams, 1993).

The structure of premixed and nonpremixed $H_2/O_2/N_2$ flames have been calculated previously using a reduced two-step mechanism (Mauss et al., 1993; Trevino and Mauss, 1993) which can be written as

$$3H_2 + O_2 = 2H + 2H_2O, \tag{I}$$

$$2H + M = H_2 + M, \tag{II}$$

where M denotes a third body. The reaction rates for the overall reactions I and II can be related to the rates of the elementary reactions describing the chemical-kinetics of oxidation of hydrogen. The overall reaction I represents the chain-branching steps and the principal rate of this global reaction is the rate of the elementary step

$$H + O_2 = OH + O. \tag{2}$$

The overall step II represents the three body recombination steps and is responsible for a major fraction of the heat released in the flame. The principal rate of this global reaction is the rate of the elementary step

$$H + O_2 + M = HO_2 + M. \tag{3}$$

The reaction rate of the elementary step 3 is proportional to the concentration of the third body C_M. The quantity C_M is proportional to the sum of the products of the mole fraction of species i, X_i, and its chaperon efficiency η_i. Thus

$$C_M \;\alpha\; \sum_{i=1}^{N} \eta_i X_i, \tag{4}$$

where N represents the total number of species. The chaperon efficiency for the various species is roughly H_2 = 1.0, O_2 = 0.4, N_2 = 0.4, CO = 0.75, H_2O = 6.5 and CO_2 = 1.5 (Peters and Rogg, 1993, and Smooke, 1991). The influence of the contaminants H_2O and CO_2 is expected to arise primarily from their effect on the recombination reaction 3. Since their chaperon efficiency is higher than that of N_2, addition of these contaminants to $H_2/O_2/N_2$ systems at constant values of the adiabatic flame temperature T_{st} can be expected to make the flame stronger or more resistant to aerodynamic strain. Tables 1a and 1b show the influence of H_2O on the critical conditions of extinction of the flame to be greater than that of CO_2, because the chaperon efficiency for H_2O is higher than that of CO_2.

For diffusion flames, asymptotic analyses provide explicit algebraic expressions for the scalar dissipation rate at extinction χ_q which is defined as (Seshadri and Williams, 1993)

$$\chi_q = 2\frac{\lambda}{\rho c_p}\left(\nabla\xi\right)^2, \tag{5}$$

where, λ is the thermal conductivity, and c_p is the heat capacity. The mixture fraction ξ is defined such that $\xi = 1$ in the ambient fuel stream and $\xi = 0$ in the ambient oxidizer stream. Results of asymptotic analyses for $CH_4/O_2/N_2$ diffusion flames (Seshadri and Peters, 1988, and Yang and Seshadri, 1992) show that the value of χ_q is proportional to the product of the rate constant for the elementary step 3 denoted by k_3, the concentration of the third body C_M and the square of the stoichiometric mixture fraction ξ_{st}. Thus

$$\chi_q \ \alpha \ k_3 C_M \left(\xi_{st}\right)^2. \tag{6}$$

The quantity χ_q deduced from asymptotic analysis also depends on the value of T_{st}. If the Lewis numbers of species are presumed to be constant, then following the analysis of Peters (1984) the stoichiometric mixture fraction can be written as

$$\xi_{st} = \frac{\nu Y_{O,2}}{\nu Y_{O,2} + Y_{F,1}}, \tag{7}$$

where ν is the ratio of the mass of hydrogen to the mass of oxygen required for stoichiometric combustion, $Y_{O,2}$ denotes the mass fraction of oxygen in the ambient oxidizer stream, and $Y_{F,1}$ denotes the mass fraction of fuel in the ambient fuel stream. Donnerhack and Peters (1984) have shown that the relation between χ_q and the strain rate at extinction a_q can be written as

$$\chi_q = \frac{a_q}{\pi}\exp\left\{-2\left[\operatorname{erfc}^{-1}\left(2\xi_{st}\right)\right]^2\right\}, \tag{8}$$

where erfc^{-1} represents the inverse of the complementary error function and not its reciprocal. At constant values of T_{st}, using Eqs. 6 and 8, the strain rate at extinction for diffusion flames with an inert contaminant added to the oxidizer stream a_q, can be related to the strain rate at extinction for a diffusion flame without the contaminant $(a_q)X_{\mathrm{contam},2} = 0$, as

$$\frac{a_q}{(a_q)_{X_{\mathrm{contam},2}=0}} = \frac{C_M\,(\xi_{st})^2 \exp\left\{2\left[\mathrm{erfc}^{-1}\,(2\xi_{st})\right]^2\right\}}{\left[C_M\,(\xi_{st})^2 \exp\left\{2\left[\mathrm{erfc}^{-1}\,(2\xi_{st})\right]^2\right\}\right]_{X_{\mathrm{contam},2}=0}} \tag{9}$$

In Tables 1a and 1b the results for the ratio of strain rates at extinction predicted using Eq. 9 for addition of H_2O and CO_2 calculated at conditions similar to those used in the experiment are shown. The asymptotic results predict the measured increase in the strain rate at extinction with increasing amounts of the contaminants in the oxidizer stream. Tables 1a and 1b also show the calculated values of T_{st} and values of ξ_{st}. The results of the asymptotic analysis predict the influence of CO_2 rather accurately, but overpredict the influence of H_2O. This is primarily due to the fact that the value of ξ_{st} calculated using Eq. 7, assuming that the Lewis number of H_2 is unity, is inaccurate. Since there are nonnegligible variations in the value of ξ_{st} with addition of H_2O, these inaccuracies can introduce noticeable errors in the predicted values of a_q. However, the variations in the value of ξ_{st} with addition of CO_2 are small, therefore the predictions of the asymptotic analyses are in better agreement with the measurements.

3. Summary and Conclusions

The influence of contaminants on the critical conditions of extinction of counterflow $H_2/O_2/N_2$ were examined. The contaminants considered were H_2O and CO_2. The following remarks summarize the major conclusions

1) At fixed values of the overall enthalpy in the flame, or at fixed values of the adiabatic flame temperature, experimental measurements show that the contaminants increase the value of the strain rate at extinction, or make the flame harder to extinguish.

2) The recombination reaction $H + O_2 + M \rightarrow HO_2 + M$, which is responsible for the major fraction of the heat released in the flame plays a central role in determining the influence of the contaminants. Addition of the contaminants increases the rate of this reaction and thus makes the flame harder to extinguish.

There is a need for more research in this area. In particular, the experiments must be repeated in burners with large aspect ratio where the structure of the flame is nearly one-dimensional. This will allow careful interpretations of the results using asymptotic theories.

References

Donnerhack, S. and Peters, N., 1984. "Stabilization heights in lifted methane-air jet diffusion flames diluted with nitrogen," *Combustion Science and Technology* **41**, p 101.

Guerra, R., Pellet, G. L., Wilson, L. G., and Northam, G. B., 1987a. "Opposed jet burner studies of effects of CO, CO2, and N2 air-contaminants on hydrogen-air diffusion flames," AIAA-87-1960, AIAA/SAE/ASME/ASEE 23rd Joint Propulsion Conference, June 29-July 2, 1987, San Diego, California.

Guerra, R., Pellet, G. L., Wilson, L. G., and Northam, G. B., 1987b. "Opposed jet burner studies of hydrogen combustion with pure and N2, NO contaminated air," AIAA-87-0090, AIAA 25th Aerospace Sciences Meeting, January 12-15, 1987, Reno, Nevada.

Mauss, F., Peters, N., Rogg, B., and Williams, F. A., 1993. "Reduced kinetic mechanisms for premixed hydrogen flames," in *Reduced Kinetic Mechanisms for Applications in Combustion Systems* (N. Peters and B. Rogg, Eds.), Lecture Notes in Physics **m15**, Springer-Verlag, Berlin Heidelberg, p. 29.

Pellet, G. L., Wilson, L. G., Northam, G. B., and Guerra, R., 1989. "Effects of H_2O, CO_2, and N_2 air contaminants on critical air side strain rates for extinction of hydrogen-air counterflow diffusion flames," 26th JANNAF Combustion Meeting, October 23-27, 1989.

Peters, N., 1984. "Laminar diffusion flamelets in nonpremixed turbulent combustion," *Progress in Energy Combustion Science* **10**, p. 319.

Peters, N. and Rogg, B. (Eds.), 1993. *Reduced Kinetic Mechanisms for Applications in Combustion Systems*, Lecture Notes in Physics, **m 15**, Springer Verlag, Heidelberg.

Seshadri, K. and Peters, N., 1988. "Asymptotic structure and extinction of methane-air diffusion flames," *Combustion and Flame* **73**, p. 23.

Seshadri, K., Peters, N., and Williams, F. A., 1993. "Asymptotic analysis of stoichiometric and lean hydrogen-air flames," to appear in *Combustion and Flame.*

Seshadri, K. and Williams, F. A., 1993. "Reduced chemical systems and their application in turbulent combustion," to appear in *Turbulent Reacting Flows* (Eds. P. A. Libby and F. A. Williams), Academic Press.

Smooke, M. D. (Ed.), 1991. *Reduced Kinetic Mechanisms and Asymptotic Approximations for Methane-Air Flames*, Lecture Notes in Physics **384**, Springer-Verlag, Heidelberg.

Trevino, C. and Mauss, F., 1993. "Structure and extinction of non-diluted hydrogen-air diffusion flames," in *Reduced Kinetic Mechanisms for Applications in Combustion Systems* (N. Peters and B. Rogg, Eds.), Lecture Notes in Physics **m15**, Springer-Verlag, Berlin Heidelberg, p. 159.

Yang, B. and Seshadri, K., 1992. "Asymptotic analysis of the structure of nonpremixed methane air flames using reduced chemistry," *Combustion Science and Technology* **88**, p. 115.

Table 1a) Influence of addition of steam to the oxidizer stream of $H_2/O_2/N_2$ diffusion flames.

Initial Mole % O_2 $X_{O_2,1}$	Mole % O_2 in Oxidizer Stream $X_{O_2,2}$	Mole % H_2O in Oxidizer Stream $X_{H_2O,2}$	Adiabatic Flame Temp. T_{ad} K	Stoich. Mixture Fraction Z_{st}	U_{air} Expt. cm/s	$a/(a)_{X_{H_2O,2}=0}$	
						Expt	Asym
20.95	20.95	0.00	2583	0.02824	491	1.0	1.0
25.0	21.86	12.54	2583	0.0307	542	1.1	1.29
29.0	22.56	22.21	2583	0.0327	609	1.24	1.55

Table 1b) Influence of addition of CO_2 to the oxidizer stream of $H_2/O_2/N_2$ diffusion flames.

Initial Mole % O_2 $X_{O_2,1}$	Mole % O_2 in Oxidizer Stream $X_{O_2,2}$	Mole % CO_2 in Oxidizer Stream $X_{CO_2,2}$	Adiabatic Flame Temp. T_{ad} K	Stoich. Mixture Fraction Z_{st}	U_{air} Expt. cm/s	$a/(a)_{X_{CO_2,2}=0}$	
						Expt	Asym
20.95	20.95	0.00	2583	0.02824	491	1.0	1.0
25.0	22.39	10.45	2583	0.02846	510	1.04	1.08
29.0	23.53	18.84	2583	0.02862	539	1.1	1.15

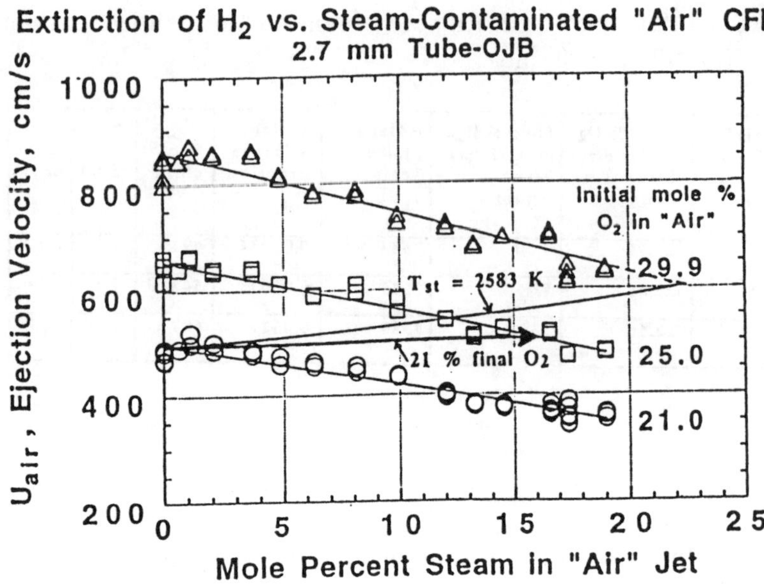

Fig. 1. Measured oxidizer jet velocity limits versus final concentration of steam contaminant (reproduced from Pellett et al. (1989)).

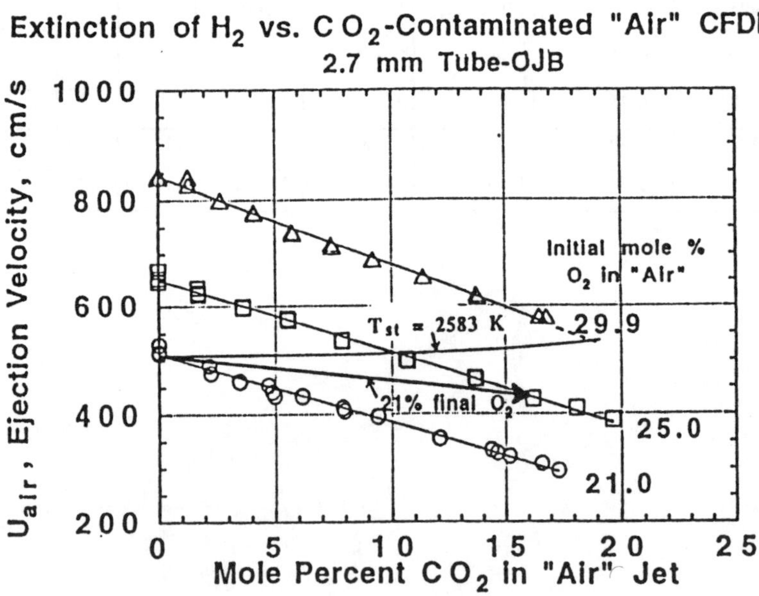

Fig. 2. Measured oxidizer jet velocity limits versus final concentration of CO_2 contaminant (reproduced from Pellett et al. (1989)).

TURBULENT COMBUSTION

GROUP SUMMARY: TURBULENT COMBUSTION

T.L. Jackson
Institute for Computer Applications in Science and Engineering
NASA Langley Research
Hampton, VA 23681-0001

The focus of this group was the modeling of non-premixed, chemically reacting turbulent flow at low speeds. The workshop participants hosted by this group were: Sharath Girimaji (ICASE), Peyman Givi (SUNY, Buffalo), Frank Kozusko (Old Dominion University), and Michele Macaraeg (NASA Langley Research Center).

S. Girimaji examined a diffusive-reactive model system and demonstrated that, under certain conditions, it could be considered a reasonable model of a non-premixed, chemically reacting turbulent flow. The model assumes that the scalar field evolves under the influence of molecular diffusion alone, with the diffusion coefficient being a function of time. The numerical solution of the model system was compared against other existing models. Insight into several important aspects of turbulent-combustion modeling was brought out. The strengths and limitations of the model was discussed.

The work of F. Jaberi and P. Givi focused on developing a phenomenological model for describing the evolution of single-point scalar probability density function (PDF) in incompressible turbulent flows. The model is based on the "Lamellar Description" and has proven promising in stochastic modeling of turbulent mixing phenomenon. In this work, Jaberi and Givi develop a closure which they term "the Inter-Layer Diffusion Model." They show that this model provides a unique feature in that it recovers all the family of PDF's generated by other existing closures.

Kozusko, Macaraeg and Jackson examined turbulence from a different perspective. Experiments indicate that combustion takes place only within small scale eddies associated with the diffusion of fuel and oxidizer into each other provided the value of the Damkohler number is large. Assuming, therefore, that turbulence can be viewed as a collection of small scale eddies, a mathematical model was constructed which examined the ignition dynamics within a small scale eddy. In particular, the model was that of a constant-density, one-step, irreversible Arrhenius reaction between initially unmixed species of fuel and oxidizer occupying adjacent half-planes which were then allowed

M. Y. Hussaini et al. (eds.), Transition, Turbulence and Combustion, Vol. II, 251–252.

to mix and react in the presence of a vortex pair. The evolution in time of the temperature and mass-fraction fields was followed with an emphasis on the ignition time as a function of the vortex Reynolds number, location of the vortex centers relative to the fuel/oxidizer interface and initial temperature differences of the reacting species.

DIFFUSION-REACTION SYSTEM MODEL FOR TURBULENT COMBUSTION

Sharath S. Girimaji

Institute for Computer Applications in Science and Engineering
NASA Langley Research Center
Hampton, Virginia 23681-0001

ABSTRACT

The diffusion-reaction system, under certain conditions, is shown to be a reasonable model of non-premixed, chemically reacting turbulent flow. Evolution of multiple scalar fields undergoing molecular diffusion and chemical reaction from non-premixed random initial conditions is studied numerically. The study provides insight into several important aspects of turbulent-combustion modeling.

1. Introduction

A turbulent combustion model is a composite of several closure models representing different processes. Often, when calculations from the composite model are in poor agreement with either experimental or direct numerical simulation (DNS) data, it is difficult to isolate the underlying cause (the model of a specific physical process) of the disagreement. For this reason, it is useful to construct sample test problems where only one or few of the physical processes are important (other processes are either negligible or completely absent). Data from the numerical simulations of such a test problem can be used to construct and validate the individual models.

The effect of velocity field on scalar mixing is studied, without the complexities of spatial inhomogeneities or chemical reaction, using direct numerical simulations (DNS) by Eswaran and Pope (1988) and others. One of the chief findings of DNS studies is that (for the range of Reynolds number investigated, $R_\lambda \approx 30$ - 90) the velocity field only affects the timescale of the scalar pdf evolution. In appropriately normalized time, the scalar pdf evolution is independent of the velocity field. This insight gained from the DNS studies spawned reasonably successful modeling approaches where the scalar evolution is considered only in the normalized time, thus decoupling the

M. Y. Hussaini et al. (eds.), Transition, Turbulence and Combustion, Vol. II, 253–264.

velocity field from the scalar field. Two such models are: the mapping closure model (Chen, Chen & Kraichnan, 1989; Girimaji, 1992, 1993a), and the assumed β-pdf model (Girimaji, 1991).

Studying a test case which combines scalar mixing with chemical reaction in isotropic turbulence should be a useful second step towards the ultimate objective of understanding and modeling turbulent combustion. In this work, we first analyze the test-case equations and propose a statistical model for the scalar (species and temperature) evolution. The model equations are a set of diffusion-reaction system (DRS) of equations. The model accounts for the effect of the velocity field via a time-dependent diffusion enhancement factor. The model equations are close enough to the original turbulent-reaction equations that other models which decouple the velocity field and scalar-field mixing cannot distinguish between the two sets of equations. Numerical simulations of the model DRS equations are computationally much less intensive than DNS of the original equations and may serve nearly as useful a purpose in understanding several aspects of turbulent combustion and evaluating other models.

2. Diffusion-Reaction System Model

Consider the mixing and reaction of n scalars (of mass fractions ϕ_α, $\alpha = 1,\cdots n$) in a turbulent velocity field. Subject to standard assumptions (Williams, 1988; Girimaji, 1993b), the scalar and temperature (T) evolution equations are

$$\rho[\frac{\partial \phi_\alpha}{\partial t} + u_i \frac{\partial \phi_\alpha}{\partial X_i}] = \frac{\partial}{\partial X_i}[\rho D_\alpha \frac{\partial \phi_\alpha}{\partial X_i}] + w_\alpha, \tag{1}$$

and

$$\rho[\frac{\partial h^T}{\partial t} + u_i \frac{\partial h^T}{\partial X_i}] = \frac{\partial}{\partial X_i}[\lambda \frac{\partial T}{\partial X_i}] + \sum_{\alpha=1}^{n} \frac{\partial}{\partial X_i}[\rho h_\alpha^T D_\alpha \frac{\partial \phi_\alpha}{\partial X_i}] - \sum_{\alpha=1}^{n} h_\alpha^0 w_\alpha. \tag{2}$$

In the above equations, D_α and w_α are respectively the Fickian diffusion coefficient and the chemical production rate of species α. The heat conductivity of the mixture is λ. The enthalpy of formation and thermal enthalpy of species α are h_α^0 and h_α^T. The total thermal enthalpy of the mixture, (h^T), is given by

$$h^T = \sum_{\alpha=1}^{n} \phi_\alpha h_\alpha^T. \tag{3}$$

The pressure (p), density of the mixture (ρ) and the mass fractions are related through the equation of state which could be the ideal gas law. It is assumed that the temperature and the chemical source terms can be determined given the thermal enthalpy and mass fractions.

Subject to homogeneity of the turbulent velocity field, the joint pdf of scalar composition and thermal enthalpy, F(h^T, $\underline{\phi}$), evolves according to (Pope, 1985)

$$\rho(h^T, \underline{\phi})\frac{\partial F(h^T, \underline{\phi})}{\partial t} = - \frac{\partial}{\partial h^T}[F(T, \underline{\phi})\{\Theta_T - R_T\}] - \sum_{\alpha=1}^{n} \frac{\partial}{\partial \phi_\alpha}[F(h^T, \underline{\phi})\{\Theta_{\phi\alpha} - R_{\phi\alpha}\}]. \tag{4}$$

In the above equation the conditional diffusions of temperature (Θ_T) and scalar ($\Theta_{\phi\alpha}$) are given by

$$\Theta_T = \langle \frac{\partial}{\partial X_i}[\lambda \frac{\partial T}{\partial X_i} + \sum_{\alpha=1}^{n} \rho h_\alpha^T D_\alpha \frac{\partial \phi_\alpha}{\partial X_i}] | T, \underline{\phi} \rangle, \tag{5}$$

and

$$\Theta_{\phi\alpha} = \langle \frac{\partial}{\partial X_i}[D_\alpha \frac{\partial \phi_\alpha}{\partial X_i}] | T, \underline{\phi} \rangle, \tag{6}$$

where the notation $\langle b|r \rangle$ denotes the conditional mean of b with respect to r. The pdf source terms of temperature (R_T) and scalar ($R_{\phi\alpha}$) due to reaction are given by

$$R_T = - \sum_{\alpha=1}^{n} h_\alpha^0 w_\alpha, \tag{7}$$

and

$$R_{\phi\alpha} = w_\alpha. \tag{8}$$

While the source terms due to reaction are known functions of temperature and scalar composition, the conditional diffusion terms need closure modeling. The modeling of these terms requires that the effect of the velocity field on the scalar field be known, at least approximately.

The effect of the velocity field on the scalar field is examined in detail in Girimaji (1993b). The analysis is performed in a Lagrangian frame of reference ($\mathbf{x}$, t) which is cartesian. For the sake of this analysis, the velocity field is assumed to be isotropic. The Eulerian coordinate $\mathbf{X}$ is then nonstationary, curvilinear and nonorthogonal. Following Girimaji (1993b), the conditional scalar diffusion is modeled as

$$\begin{aligned}\Theta_{\phi\alpha} &= \langle \frac{\partial x_a}{\partial X_i}\frac{\partial x_b}{\partial X_i}\frac{\partial}{\partial x_a}[\rho D_\alpha \frac{\partial \phi_\alpha}{\partial x_b}]|h^T,\underline{\phi}\rangle \\ &\approx \langle \frac{\partial x_a}{\partial X_i}\frac{\partial x_b}{\partial X_i}\rangle\langle \frac{\partial}{\partial x_a}[\rho D_\alpha \frac{\partial \phi_\alpha}{\partial x_b}]|h^T,\underline{\phi}\rangle .\end{aligned} \tag{9}$$

The quantity $\langle \frac{\partial x_a}{\partial X_i}\frac{\partial x_b}{\partial X_i}\rangle$ is a second-order tensor which, if the turbulent field is isotropic, can be written as

$$\langle \frac{\partial x_a}{\partial X_i}\frac{\partial x_b}{\partial X_i}\rangle = \langle S(t)\rangle \delta_{ab} \tag{10}$$

where $S(t) = \frac{1}{3}\frac{\partial x_a}{\partial X_i}\frac{\partial x_a}{\partial X_i}$. Hence, the model for $\Theta_{\phi\alpha}$ is

$$\Theta_{\phi\alpha}(T,\underline{\phi}) = \langle S(t)\rangle\langle \frac{\partial}{\partial x_a}[D_\alpha \frac{\partial \phi_\alpha}{\partial x_a}]|h^T,\underline{\phi}\rangle . \tag{11}$$

Using similar arguments the model for Θ_T can be derived:

$$\Theta_T = \langle S(t)\rangle\langle \frac{\partial}{\partial x_a}[\lambda \frac{\partial T}{\partial x_a} + \sum_{\alpha=1}^{n} \rho h_\alpha^T D_\alpha \frac{\partial \phi_\alpha}{\partial x_a}]|h^T,\underline{\phi}\rangle . \tag{12}$$

The model. Based on the Lagrangian frame analysis, it is proposed that conditional diffusion be modeled by

$$\Theta_T = \langle S(t)\rangle\langle \frac{\partial}{\partial x_a}[\lambda \frac{\partial T^m}{\partial x_a} + \sum_{\alpha=1}^{n} \rho h_\alpha^{Tm} D_\alpha \frac{\partial \phi_\alpha^m}{\partial x_a}]|h^{Tm},\underline{\phi}^m\rangle , \tag{13}$$

and

$$\Theta_{\phi\alpha} = \langle S(t)\rangle\langle \frac{\partial}{\partial x_a}[\rho D_\alpha \frac{\partial \phi_\alpha}{\partial x_a}]|T^m,\underline{\phi}^m\rangle . \tag{14}$$

In the above equation, the superscript m refers to the model variables.

The effect of the velocity field on the scalar evolution in the Lagrangian reference frame is much smaller than the effect in the Eulerian reference frame. Hence, the model variables (h^{Tm} and $\underline{\phi}^m$) can be considered to evolve according to

$$\rho \frac{d\phi_\alpha^m}{dt} = \langle S(t)\rangle \frac{\partial}{\partial x_a}[\rho D_\alpha \frac{\partial \phi_\alpha^m}{\partial x_a}] + w_\alpha(\underline{\phi}^m), \tag{15}$$

and

$$\rho\frac{dh^{Tm}}{dt} = \langle S(t)\rangle\frac{\partial}{\partial x_a}[\lambda\frac{\partial T}{\partial x_a} + \sum_{\alpha=1}^{n}\rho h_\alpha^{Tm}D_\alpha\frac{\partial\phi_\alpha^m}{\partial x_a}] - \sum_{\alpha=1}^{n}h_\alpha^0 w_\alpha(\underline{\phi}^m). \tag{16}$$

The above equations form a statistical model of the chemically reacting flow equations (1) and (2). The model equations are similar enough to the original equations that many of the other scalar mixing models – multiscalar mapping closure models (Pope, 1991; Gao, 1991; Girimaji, 1993a), multivariate β-pdf model, JET model (Madnia et al., 1992) – cannot distinguish the difference between the pdf's of the two sets of instantaneous equations. Hence, for validating many statistical aspects of the other models, DNS of equations (15) and (16) can be used rather than the more expensive DNS of turbulent reacting flow equations (1) and (2).

The quantity $\langle S(t)\rangle$ is a measure of the mean deformation of the Eulerian coordinate frame caused by the velocity field and can be interpreted as the diffusion enhancement factor. As explained in detail in Girimaji (1993b), $\langle S(t)\rangle$ can be modeled as

$$p(t) \equiv \frac{d\ln\langle S\rangle}{dt} \approx \frac{0.8}{\tau_\eta}, \tag{17}$$

where τ_η is the Kolmogorov time scale of the turbulence. The diffusion enhancement factor $\langle S(t)\rangle$ reflects the cascading effect of the velocity field on the scalar field.

3. Further Simplification and Simulation

For the sake of simulations, the simplest non-trivial case of multiscalar mixing and combustion where reaction affects the evolution of scalar pdf is considered (Pope, 1991):

$$\frac{\partial\phi_\alpha}{\partial t} + U_i\frac{\partial\phi_\alpha}{\partial X_i} = D_\alpha\frac{\partial^2\phi_\alpha}{\partial X_i\partial X_i} + w_\alpha. \tag{18}$$

The above equation represents constant-density, isothermal chemical reactions between scalars of different, but constant, diffusivities. If the velocity field $\mathbf{U}$ is isotropic, the enhanced-diffusion reaction model corresponding to equation (18) is

$$\frac{d\phi_\alpha^m}{dt} = \langle S(t)\rangle D_\alpha\frac{\partial^2\phi_\alpha^m}{\partial x_a\partial x_a} + w_\alpha(\underline{\phi}^m). \tag{19}$$

The time-dependent diffusion enhancement factor is calculated from equation (17).

We consider reversible reactions of the type

$$A + B \leftrightarrow 2P, \tag{20}$$

where scalars A and B are reactants and scalar P is the product. The factor two is included in equation (20) to conserve mole fractions. The chemical production rates of the three species are

$$\begin{aligned} w_A &= -k_f \phi_A \phi_B + k_b \phi_P^2 \\ w_B &= -k_f \phi_A \phi_B + k_b \phi_P^2 \\ w_P &= 2k_f \phi_A \phi_B - 2k_b \phi_P^2, \end{aligned} \tag{21}$$

where k_f and k_b are the forward and the backward reaction-rate coefficients. The backward coefficient is usually expressed in terms of the forward rate: $k_b = k_e k_f$, where k_e is the equilibrium constant.

Simulations. The problem of non-premixed, randomly and isotropically distributed scalars undergoing enhanced mixing and chemical reaction is considered. Numerical simulations of equation (19) are performed in a cubical domain with periodic boundary conditions. A pseudo-spectral method is used. The initial scalar field is specified in the manner described in Girimaji (1993a). The initial scalar field has a prescribed length scale l_α for each scalar, and the initial joint pdf, $F(\phi_A, \phi_B)$, is given by

$$\begin{aligned} F(\phi_A, \phi_B; 0) \;\; = \;\; & \mu_A(0)\delta(\phi_A - 1)\delta(\phi_B) + \\ & \mu_B(0)\delta(\phi_A)\delta(\phi_B - 1) + \mu_P(0)\delta(\phi_A)\delta(\phi_B). \end{aligned} \tag{22}$$

In the above equation, μ_A, μ_B and μ_P represent the means of species A, B and P, respectively. The mass fraction of P is given by

$$\phi_P = 1 - \phi_A - \phi_B. \tag{23}$$

The values of the parameters for the various simulations are shown in Table 1.

4. Results

The results of the simulations are discussed in detail in Girimaji (1993b). Here we present only the results of model evaluations performed using the simulation data. We seek to answer two specific questions: i) How good is the assumed multivariate β-pdf model (Girimaji 1991) for calculating reacting flows? ii) How does the computationally simple assumed-pdf model compare with mapping-closure model for multiscalar mixing?

Assumed multivariate β-pdf model. In the assumed-pdf approach the scalar joint pdf is prescribed knowing the first few moments of the scalar field. It is computationally far less intensive than the mapping closure model and is ideally suited for engineering computations. The assumed multivariate β pdf model for an N-scalar mixing process is given by (Girimaji, 1991)

$$F(\underline{\phi}) = \frac{\Gamma(\beta_1 + \cdots + \beta_N)}{\Gamma(\beta_1) \cdots \Gamma(\beta_N)} \phi_1^{\beta_1 - 1} \phi_2^{\beta_2 - 1} \cdots \phi_N^{\beta_N - 1} \delta(1 - \phi_1 - \ldots - \phi_N). \tag{24}$$

The parameters of the model $(\beta_1, \cdots, \beta_N)$ are functions of the mean mass fractions μ_α and turbulent scalar energy $Q = \sum_{\alpha=1}^{N} \sigma_\alpha^2$, where σ_α^2 is the variance of mass fraction of scalar α. The above model has been tested for two-scalar inert mixing process with reasonable success, but is yet to be validated for multiscalar mixing and reaction.

In Figure 1, the individual variances and some cross moments of species A and B, calculated using the model, are compared against corresponding simulation data for case C4b. The means and the turbulent scalar energy required as inputs to the model are taken from DRS data. The model predicts the variances reasonably well. The magnitude of the cross covariance is underpredicted by the model, but the difference is not too much. For the two higher cross moments compared ($\langle \phi_A'^2 \phi_B' \rangle$ and $\langle \phi_A' \phi_B'^2 \rangle$) the model does surprisingly well. Overall, the performance of the model is quite satisfactory, given the simplicity of the model. More detailed comparisons are performed in Girimaji (1993b).

Multivariate β-pdf model vs. Mapping closure model. Now the assumed β-pdf model is compared against the more detailed and computationally intensive mapping closure model of Girimaji (1993a). The two models are equally good for the case of inert nonpremixed two-scalar mixing (Girimaji, 1992). Comparison of the two models for multiscalar mixing has not been performed before, and is attempted presently for inert mixing (case C3). In Figure 2, various normalized cross moments (of species A and B) obtained from the simulations are plotted as a function of the variance of scalar A. (The variance decreases monotonically in time and, hence, can be used in lieu of time as the abscissa.) Also shown in the Figure are the cross moments calculated using the two models. The moments of the normalized variables (θ_α) are presented:

$$\theta_\alpha = \frac{\phi_\alpha - \langle \mu_\alpha \rangle}{\sigma_\alpha}. \tag{25}$$

For all the moments considered, both models agree equally well with the data (case C3). For the case when the initial length scales of the scalar are different (case C4), the performance of the mapping closure model is clearly superior (figure not shown) to that of the assumed-pdf model. These findings are in keeping with the arguments presented in Girimaji (1991) that the assumed β-pdf model is likely to be more accurate when the length scale and diffusivities of the scalars are similar (case C3) than when they are widely disparate (case C4). For case C4, even the mapping closure model is not very accurate. This leads to the question, for engineering calculations, is the increase in accuracy achieved using the mapping closure model worth the extra effort involved in computing? The answer will depend on the application for which the model is being used. In combustion calculations it is the unnormalized moments that appear in the scalar moment equations and need closure modeling, not the normalized moments. Even for case C4, despite the relatively poor agreement of the normalized moments, the unnormalized moments calculated using the β-pdf model are quite close to the data as seen in Figure 1 for the corresponding reacting case C4b. This appears to suggest that the assumed-pdf approach, although simplistic, may be quite adequate for modeling of turbulent combustion in engineering calculations.

5. Summary

It is shown that, in isotropic turbulence, the (enhanced) diffusion-reaction system (DRS) can be considered a statistical model of turbulent combustion equations (1) and (2) with the initial field and the diffusion enhancement factor ($\langle S(t) \rangle$) as inputs. The DRS model is close enough to the turbulent combustion equations that, in normalized time, many of the other models (that decouple the velocity and scalar fields) cannot distinguish between DRS equations and the original equations. Simulations of DRS equations have been performed and some of the other models have been evaluated against DRS data.

References

Chen, H., Chen, S., and Kraichnan, R. H., 1989. "Probability distribution of a stochastically advected scalar field," *Physical Review Letters* **63**(24), pp. 2657 - 2660.

Eswaran, V. and Pope, S. B., 1988. "Direct numerical simulations of the turbulent mixing of a passive scalar," *Phys. Fluids* **31**(3), pp. 506 - 520.

Gao, F., 1991. "Mapping closure for multispecies Fickian diffusion," *Phys. Fluids A* **3**(10), pp. 2438 - 2444.

Girimaji, S. S., 1991. "Assumed β-pdf model for turbulent mixing: validation and extension to multiple scalar mixing," *Combust. Sci. and Tech.* **78**, pp. 4 - 6, pp. 177 - 196.

Girimaji, S. S., 1992. "A mapping closure for turbulent scalar mixing using time-evolving reference field," *Phys. Fluids A* **4**(12), pp. 2875 - 2886.

Girimaji, S. S., 1993a. "A study of multiscalar mixing," *Phys. Fluids A* **5**(7), pp. 1802 - 1809.

Girimaji, S. S., 1993b. "Simulations of diffusion reaction equations with implications to turbulent combustion modeling," ICASE Report No. 93-69, NASA CR-191536.

Madnia, C. K., Frankel, S. H., and Givi, P., 1992. "Reactant conversion in homogeneous turbulence: mathematical modeling, computational validations and practical applications," *Theoret. Comput. Fluid Dynamics* **4**, pp. 79 - 93.

Pope, S. B., 1985. "PDF methods for turbulent reacting flows," *Prog. Energy Combust. Sci.* **11**, pp. 119 - 192.

Pope, S. B., 1991. "Mapping closures for turbulent mixing and reaction," *Theo. and Comp. Fluid Dynamics* **2**, pp. 255 - 270.

Williams, F. A., 1988. *Combustion Theory*, Addison - Wesley Publishing Company, Inc., New York.

Table: Parameters of the Simulations

Group	Case	K_f	K_e	μ_A	μ_B	D_A	D_B	$\kappa_A = \frac{2\pi}{l_A}$	$\kappa_B = \frac{2\pi}{l_B}$
1	C1	0	0	0.5	0.5	1	1	4	4
	C1a	20	0	0.5	0.5	1	1	4	4
	C1b	100	0	0.5	0.5	1	1	4	4
	C1c	100	1	0.5	0.5	1	1	4	4
2	C2	0	0	0.25	0.75	1	1	4	4
	C2a	20	0	0.25	0.75	1	1	4	4
	C2b	100	0	0.25	0.75	1	1	4	4
	C2c	100	0	0.25	0.75	1	1	4	4
3	C3	0	0	0.25	0.5	1	1	8	8
	C3b	100	0	0.25	0.5	1	1	8	8
4	C4	0	0	0.25	0.25	1	1	16	4
	C4b	100	0	0.25	0.25	1	1	16	4

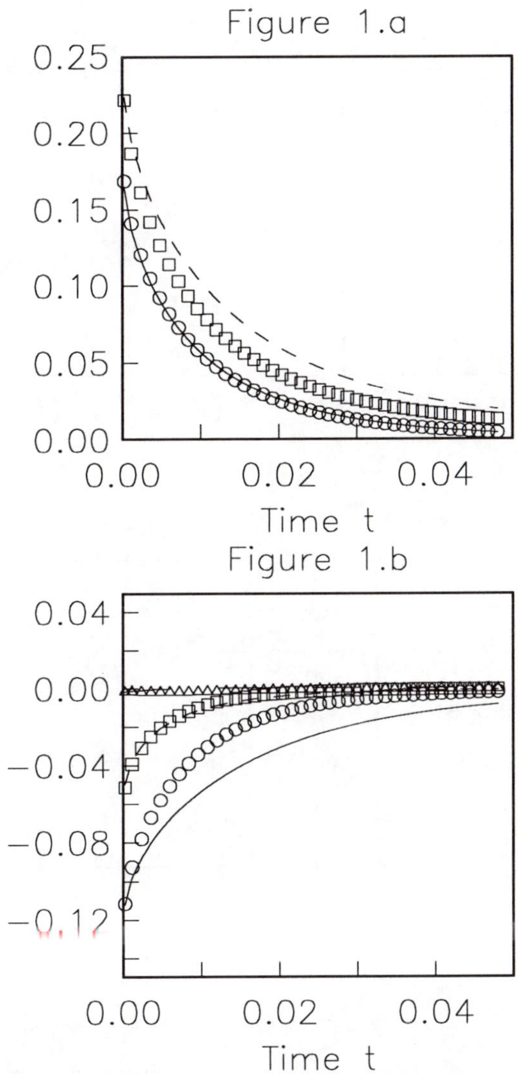

Figure 1: DRS data vs. assumed β-pdf model. Comparison of various unnormalized moments for case C4b. (a) σ_1^2: data (solid line), model (circle); σ_2^2 – data (- -), model (square). (b) $\langle \phi_1' \phi_2' \rangle$ – data (solid line), model (circle); $\langle \phi_1'^2 \phi_2' \rangle$: data (- -), model (square); and, $\langle \phi_1' \phi_2'^2 \rangle$ – data (— —), model (triangle).

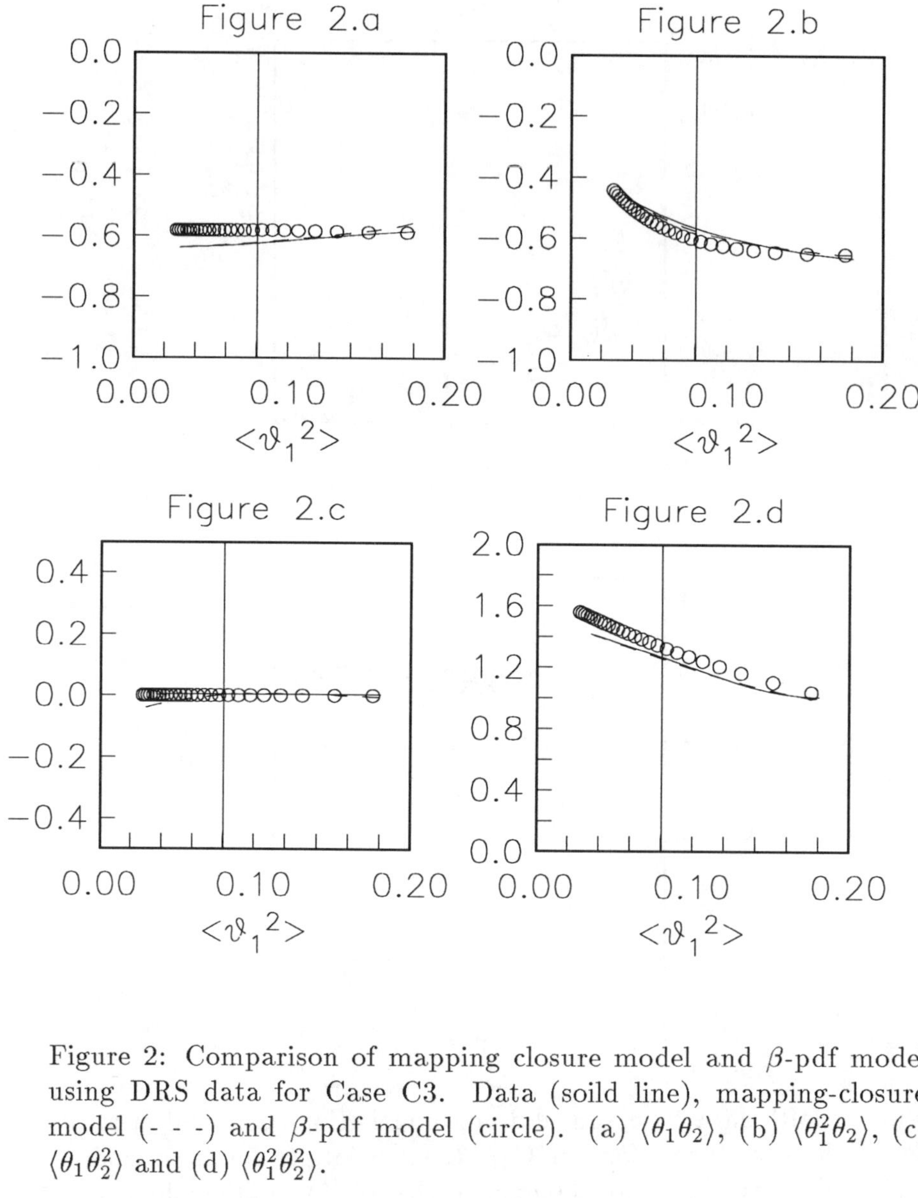

Figure 2: Comparison of mapping closure model and β-pdf model using DRS data for Case C3. Data (soild line), mapping-closure model (- - -) and β-pdf model (circle). (a) $\langle\theta_1\theta_2\rangle$, (b) $\langle\theta_1^2\theta_2\rangle$, (c) $\langle\theta_1\theta_2^2\rangle$ and (d) $\langle\theta_1^2\theta_2^2\rangle$.

THE LAMELLAR DESCRIPTION OF MIXING IN TURBULENT FLOWS

Farhad A. Jaberi and Peyman Givi[1]

State University of New York
Buffalo, New York 14260

ABSTRACT

Work is underway in developing a model based on the "lamellar theory" for statistical description of scalar mixing phenomenon in turbulent flows. A model termed the *Inter-Layer Diffusion Model* (ILDM) is proposed. This model suggests that there are two distinct but coupled mechanisms by which the mixing process is described. These mechanisms are due to: (1) local events and (2) integrated global events. It is shown that the conditional expected diffusion of the scalar field depicted by the model depends more directly on local events. The global events manifest themselves by the evolution of the conditional expected dissipation, and also the boundedness of the composition domain. With the manipulations of each of these two mechanisms, several families of scalar probability density functions (PDF's) can be generated.

1. Introduction

The description of scalar mixing continues to be a challenging task in mathematical modeling of turbulent flows (Kollmann, 1990, Pope, 1990; Kerstein, 1991b; Ottino, 1989). Within the past decade several strategies have been devised and implemented for a variety of applications. A standard test case which has proven very useful is the problem of passive scalar mixing from an initial binary state in homogeneous-isotropic turbulence (Pope, 1982; Kosaly and Givi, 1987; McMurtry and Givi, 1989; Eswaran and Pope, 1988; Pope, 1991; Madnia *et al.*, 1992; Frankel *et al.*, 1992; Miller *et al.*, 1993; Frankel *et al.*, 1993; Jiang *et al.*, 1992; Givi, 1989). In this setting, the problem of mixing is isolated from other competing physical mechanisms. It also provides a relatively simple condition to configure in

[1]This work is part of an effort sponsored by the Office of Naval Research under Grant N00014-90-J-4013 and by the National Science Foundation under Grant CTS-9253488.

M. Y. Hussaini et al. (eds.), Transition, Turbulence and Combustion, Vol. II, 265–274.

both laboratory and numerical experiments. In this context, turbulent mixing involves two physical mechanisms: turbulent convection and molecular diffusion. The first is a mechanical "stirring" process which results in the stretching of the intermaterial area of the scalar. Mixing induced by such stirring is completed at small scales by the molecular action through diffusion of substances across intermaterial surface areas. The results of recent laboratory experiments (Dahm *et al.*, 1990) and numerical simulations (Ruetsch and Maxey, 1992) suggest that a major part of scalar diffusion occurs in a system of *layer-like striation*, or *lamellae.* Each of these lamellae have a thickness (2ζ) which in general is a function of space and time, *i.e.* $\zeta = \zeta(\vec{x}, t)$. The rate of stretching and folding of the lamellar structures is governed by turbulent eddies. Development of a physical model which can accurately portray the mechanisms of lamellar stretching, as well as the distribution of these lamellae, is the subject of current research (Ottino, 1989; Muzzio and Ottino, 1989b; Muzzio and Ottino, 1989a; Sokolov and Blumen, 1991a; Sokolov and Blumen, 1991b). Our purpose in this work is to further examine the lamellar theory and to investigate some of its salient features for statistical modeling of mixing in turbulent flows.

2. Probability Modeling and the Lamellar Description

With the assumption of Fickian diffusion, the transport equation for a passive scalar variable, $\phi(\vec{x}, t)$, in a homogeneous-isotropic incompressible turbulent flow is expressed by:

$$\frac{\partial \phi}{\partial t} + \vec{U} \cdot \nabla \phi = \mathcal{D}_{\mathcal{M}} \nabla^2 \phi, \tag{1}$$

where $\vec{U}(\vec{x}, t)$ and $\mathcal{D}_M$ denote the velocity vector and the molecular diffusion coefficient, respectively. Using standard methods (Pope, 1985) the evolution equation for the single-point PDF, $P(\phi(\vec{x}, t))$, of the scalar is expressed by:

$$\frac{\partial P}{\partial t} + \frac{\partial (DP)}{\partial \phi} = 0, \quad \phi_\ell \le \phi \le \phi_u, \tag{2}$$

or alternatively by:

$$\frac{\partial P}{\partial t} + \frac{\partial^2 (EP)}{\partial \phi^2} = 0. \tag{3}$$

In these equations, ϕ_ℓ and ϕ_u denote the scalar bounds, D represents the expected value of the scalar diffusion conditioned on the scalar value, ϕ, and E denotes the expected value of scalar dissipation conditioned on ϕ. At the single-point level, D and E are unknown and must be provided externally.

Equation (1) indicates that mixing evolution is governed by two mechanisms: molecular diffusion and turbulent advection. The former occurs primarily at small scales, whereas the latter is a multi-scale phenomenon. For thin diffusion zones and sufficiently steep scalar gradients across the scalar interfaces, the diffusion process takes place in a direction approximately normal to the interface. We represent this direction by y which is bounded by *striation thickness*, ζ, that is $-\zeta \leq y \leq \zeta$. This thickness is defined as a positive random variable which has a frequency denoted by the *Striation Thickness Distribution* (STD) (Muzzio and Ottino, 1989b; Ottino, 1989) $f(\zeta)$. The PDF of the scalar, conditioned on the lamellae thickness is denoted by $g(\phi|\zeta)$ and is related to physical space (y) by:

$$g(\phi|\zeta) = \begin{cases} \frac{1}{2\zeta}\left|\frac{dy}{d\phi}\right| & \text{if } -\zeta \leq y \leq \zeta \\ 0 & \text{elsewhere} \end{cases} . \tag{4}$$

The unconditional PDF of the scalar is, therefore, determined by the spatial weighted average of the conditional PDF:

$$P(\phi, t) = \frac{\int_{|y|}^{\infty} g(\phi|\zeta)\zeta f(\zeta) d\zeta}{\int_0^{\infty} \zeta f(\zeta) d\zeta}. \tag{5}$$

Embedded in Eq. (5) is the assumption that $|\frac{dy}{d\phi}|$ is the same for all lamellae with the same ζ, and is therefore independent of the physical space. This implies that the scalar PDF is a function of the lamellae thickness and the gradient of the scalar in the lamellae. This equation can be simplified further by assuming independence of $\frac{dy}{d\phi}$ from striation thickness (Kerstein, 1991a),

$$P(\phi, t) = \left|\frac{dy}{d\phi}\right| \frac{\int_{|y|}^{\infty} f(\zeta) d\zeta}{2\int_0^{\infty} \zeta f(\zeta) d\zeta}. \tag{6}$$

Equation (5) indicates that with the knowledge of the STD and the scalar profile within lamellae, the PDF of the scalar field is determined. The approximate form of this relation given by Eq. (6), as first derived by Kerstein (1991a) is much simpler in that it can be evaluated analytically. This equation can be alternatively expressed in the form:

$$P(y(\phi, t), t) = |\frac{dy}{d\phi}| G(|y|), \tag{7}$$

where

$$G(|y|) = \frac{\int_{|y|}^{\infty} f(\zeta) d\zeta}{2 \int_0^{\infty} \zeta f(\zeta) d\zeta} = \frac{1 - C(|y|)}{2 < \zeta >}, \tag{8}$$

with $C(\zeta)$ and $< \zeta >$ denote the cumulative distribution function (CDF), and the probability average of ζ. In the form presented, Eq. (7) shows that the evolution of the PDF is determined through the combined influences of two terms. The first term on the RHS denotes the scalar distribution within inter-layer regions. The second term on the RHS represents the role of the STD and manifests the influences of stirring. It is useful to think of the first term as a representation of local events, and consider the second term as the influence of global events. Of course, these two terms are not independent and each of them is influenced by the other.

3. Conditional Statistics

Given the scalar PDF, Eqs. (2)-(3) can be used to determine the conditional statistics of the scalar field. With the PDF specified, the conditional expected diffusion is given by:

$$D(\phi, t) = -\frac{1}{P(\phi, t)} \frac{\partial F}{\partial t}, \tag{9}$$

where F denotes the CDF of the scalar. By relating ϕ to y by a function of the form $\phi = \phi(y, t)$, which indicates the local behavior within the lamellae, the conditional diffusion of ϕ can be expressed in terms of y:

$$D(y) = -(\frac{\partial y}{\partial \phi})^{-1} (\frac{\partial y}{\partial t}). \tag{10}$$

In deriving this relation, the function $\phi(y, t)$ must satisfy certain restricting conditions and it is further assumed that the magnitude of y_ℓ (the value of y at ϕ_ℓ) is fixed. Equation (10) exhibits an important feature. It indicates that D is not "directly" dependent on the STD and is governed by the transient scalar evolution within the lamellae.

Therefore, if the scalar distribution within the lamellar structures is the same - even approximately - the conditional expected diffusion portrays the same behavior. With this view, it is not surprising that the majority of PDF closures currently in use, as well as numerical and experimental data available to-date, portray similar conditional diffusion fields (Miller *et al.*, 1993; Pope and Ching, 1993).

The conditional expected dissipation is determined by using Eqs. (3) and (7):

$$E(y) = -\frac{1}{|\frac{\partial y}{\partial \phi}|G(|y|)}\int_{y_\ell}^{y} G(|y|)|\frac{\partial \phi}{\partial y}|\frac{\partial y}{\partial t}dy. \tag{11}$$

If $y = y(\phi, t)$ and $G = G(|y|)$ are known, Eqs. (7), (10), and (11) determine the PDF, the CDF, D, and E.

4. ILDM

The specification of a self-similar scalar distribution within the inter-material domain alleviates the formulation in that the exact "spatial" variation of the scalar field is not considered. Instead, this variation is governed by the STD and its temporal-spatial evolution. Thus, two closure problems are involved: (1) specification of the time-variant domains and accounting for their translation, rotation, and stretching, and (2) determination of the scalar transport in each of these domains. Clearly, the first closure is to determine $f(\zeta)$, and the second one deals with the specification of scalar distribution within the lamellae.

In a turbulent flow the STD is influenced by turbulent stirring, and the scalar distribution is governed by a convection-diffusion transport equation expressed in a frame of reference attached to the moving material element. This transport is governed by Eq. (1) with the velocity vector replaced by a relative velocity vector ($\vec{U} = \vec{U}_{rel}$). In the context of the lamellar theory, it is assumed that the transport occurs in one principal direction. By expressing Eq. (1) along the direction aligned with the local gradient of scalar, we have:

$$\frac{\partial \phi}{\partial t} + V_{rel}\frac{\partial \phi}{\partial y} = \mathcal{D_M}\frac{\partial^2 \phi}{\partial y^2}. \tag{12}$$

Here, V_{rel} is a time and space dependent function and denotes the effects of the relative velocity field $\vec{U}_{rel}$ in the direction of principal scalar transport. Equation (12), together with a striation thickness

PDF, constitutes the basis of what we term the *Inter-Layer Diffusion Model* (ILDM). With this model, turbulent mixing is modeled in terms of two distinct but coupled mechanisms: (1) those associated with local events, and (2) those of the summation of these events. The local events are described by the modeled form of the transport equation within the inter-layer material (Eq. (12)), and the global influences are exhibited through the specification of $f(\zeta)$. In the form presented, a "specific" STD cannot be suggested, nor is such a suggestion physically meaningful. However, it is argued that different physical scenarios can be expressed in terms of the model. The other modeling assumption involves the behavior of the advection term in Eq. (12).

As an example, for an initial binary state, one solution of Eq. (12) is of the form:

$$\phi = \mathrm{erf}\left(\frac{y}{\sqrt{2}\tau} + c\right), \tag{13}$$

where τ and c are functions of the physical time t. This distribution yields the PDF,

$$P(y) = \sqrt{\frac{\pi}{2}}\,\tau \exp\left[\left(\frac{y}{\sqrt{2}\tau} + c\right)^2\right] G(|y|), \tag{14}$$

the conditional expected diffusion,

$$D(y) = \frac{-2}{\sqrt{\pi}\tau}\left(\frac{y}{\sqrt{2}\tau}\frac{d\tau}{dt} - \tau\frac{dc}{dt}\right) \exp\left[-\left(\frac{y}{\sqrt{2}\tau} + c\right)^2\right], \tag{15}$$

and the conditional expected dissipation,

$$\begin{aligned} E(y) \;=\; & \frac{-2\sqrt{2}}{\pi\tau^2}\frac{\exp\left[-\left(\frac{y}{\sqrt{2}\tau}+c\right)^2\right]}{G(|y|)}\left(\frac{1}{\sqrt{2}\tau}\frac{d\tau}{dt}\int_{y_l}^{y} yG(|y|)\right. \\ & \exp\left[-\left(\frac{y}{\sqrt{2}\tau}+c\right)^2\right] dy \\ & \left.-\tau\frac{dc}{dt}\int_{y_l}^{y} G(|y|)\exp\left[-(\frac{y}{\sqrt{2}\tau}+c)^2\right] dy\right). \end{aligned} \tag{16}$$

An important feature of Eq. (15) is that regardless of the choice of a STD, the conditional expected diffusion obtained here is similar to that generated by the Amplitude Mapping Closure (AMC) of

Kraichnan (Chen *et al.*, 1989), as first derived by Miller *et al.* (1993). The exact forms of the PDF and the conditional expected dissipation are dependent on the choice of a STD. For example, the exponential STD has been proposed by Kerstein (1991a):

$$f(\zeta) = \beta(n+1)\zeta^n \exp(-\beta\zeta^{n+1}), \tag{17}$$

where $\beta = \left[\sqrt{\frac{2}{\pi}}\Gamma(\frac{n+2}{n+1})\right]^{n+1}$ and Γ denotes the gamma function. Using this STD, the scalar PDF adopts the form:

$$P(y) = \frac{\tau}{2}\exp\left[(\frac{y}{\sqrt{2}\tau} + c)^2 - \beta|y|^{n+1}\right]. \tag{18}$$

For the special case of the Rayleigh distribution ($n = 1$), the generated PDF is the same as that obtained by the AMC. In general, the conditional expected dissipation does not yield an analytical solution. However, for the Rayleigh distribution, Eq. (16) can be analytically integrated:

$$E(y) = \frac{\frac{d\tau}{dt}}{\pi\tau^3 A}\exp(-\left[(A+D)y^2 + 2By + 2c^2\right]) + \sqrt{\frac{1}{\pi A}\frac{1}{\tau^3}}$$

$$\left[\frac{d\tau}{dt}(\frac{B}{2A} + \frac{dc}{dt})\right]\left[1 - \text{erf}(\frac{2Ay+B}{2\sqrt{A}})\right]\exp\left[\frac{B^2}{4A} - (Dy^2 + By + c^2)\right]. \tag{19}$$

Here, $A = \frac{1}{2}(\frac{1}{\tau^2}+1)$, $B = \frac{\sqrt{2}c}{\tau}$, and $D = \frac{1}{2}(\frac{1}{\tau^2}-1)$. As expected, Eq. (19) is identical to the form derived by Miller *et al.* (1993).

By following the format suggested here, many different families of PDF's can be constructed. Also, the conditional statistics of the scalar field can be evaluated in a straightforward manner. The results of our current work in progress indicate that the advantage of the ILDM is due to its capability of generating PDF's which are in accord with those based on the Linear Eddy Model (Kerstein, 1988; Kerstein, 1991b; Kerstein, 1991c; Cremer *et al.*, 1992), the AMC (Gao, 1991; O'Brien and Jiang, 1991; Miller *et al.*, 1993), Johnson-Edgeworth Translation (Miller *et al.*, 1993), numerical simulations (Eswaran and Pope, 1988; Miller *et al.*, 1993) and other closures (Yakhot, 1989; Yakhot *et al.*, 1990; Pope and Ching, 1993; Ching, 1993). This issue will be discussed in detail in an upcoming paper.

References

Chen, H., Chen, S., and Kraichnan, R. H., 1989, Probability distribution of a stochastically advected scalar field, *Phys. Rev. Lett.* **63**, 2657–2660.

Ching, E. S. C., 1993, Probability densities of turbulent temperature fluctuations, *Phys. Rev. Lett.* **70**, 283–286.

Cremer, M. A., McMurtry, P. A., and Kerstein, A. R., 1992, Effects of turbulence length-scale distribution on scalar mixing in homogeneous turbulent flow, *Bull. Amer. Phys. Soc.* **37**, 1805.

Dahm, W. J. A., Southerland, K. B., and Buch, K. A., 1990, Direct, high resolution, four-dimensional measurements of fine scale structure of $Sc \gg 1$ molecular mixing in turbulent flows, Technical Report AFOSR-89-0541-1, Department of Aerospace Engineering, University of Michigan, Ann Arbor, MI.

Eswaran, V. and Pope, S. B., 1988, Direct numerical simulations of the turbulent mixing of a passive scalar, *Phys. Fluids* **31**, 506–520.

Frankel, S. H., Jiang, T.-L., and Givi, P., 1992, Modeling of isotropic reacting turbulence by a hybrid mapping-EDQNM closure, *AIChE J.* **38**, 535–543.

Frankel, S. H., Madnia, C. K., and Givi, P., 1993, Comparative assessment of closures for turbulent reacting flows, *AIChE J.* **39**, 899–903.

Gao, F., 1991, Mapping closure and non-Gaussianity of the scalar probability density functions in isotropic turbulence, *Phys. Fluids A* **3**, 2438–2444.

Givi, P., 1989, Model free simulations of turbulent reactive flows, *Prog. Energy Combust. Sci.* **15**, 1–107.

Jiang, T.-L., Gao, F., and Givi, P., 1992, Binary and trinary scalar mixing by Fickian diffusion-Some mapping closure results, *Phys. Fluids A* **4**, 1028–1035.

Kerstein, A. R., 1988, A linear eddy model of turbulent scalar transport and mixing, *Combust. Sci. and Tech.* **60**, 391–421.

Kerstein, A. R., 1991a, Flapping model of scalar mixing in turbulence, *Phys. Fluids A* **27**, 2819–2827.

Kerstein, A. R., 1991b, Linear-eddy modeling of turbulent transport. Part 6. Microstructure of diffusive scalar mixing fields, *J. Fluid Mech.* **231**, 361–394.

Kerstein, A. R., 1991c, Linear-eddy modeling of turbulent transport. Part V. Geometry of scalar interfaces, *Phys. Fluids A* **3**, 1110–1114.

Kollmann, W., 1990, The pdf approach to turbulent flow, *Theoret. Comput. Fluid Dynamics* **1**, 249–285.

Kosaly, G. and Givi, P., 1987, Modeling of turbulent molecular mixing, *Combust. Flame* **70**, 101–118.

Madnia, C. K., Frankel, S. H., and Givi, P., 1992, Reactant conversion in homogeneous turbulence: Mathematical modeling, computational validations and practical applications, *Theoret. Comput. Fluid Dynamics* **4**, 79–93.

McMurtry, P. A. and Givi, P., 1989, Direct numerical simulations of mixing and reaction in a nonpremixed homogeneous turbulent flow, *Combust. Flame* **77**, 171–185.

Miller, R. S., Frankel, S. H., Madnia, C. K., and Givi, P., 1993, Johnson-Edgeworth translation for probability modeling of binary mixing in turbulent flows, *Combust. Sci. and Tech.* **91**, 21–52.

Muzzio, F. J. and Ottino, J. M., 1989a, Dynamics of a lamellar system with diffusion and reaction: Scaling analysis and global kinetics, *Phys. Rev. A* **40**, 7182–7192.

Muzzio, F. J. and Ottino, J. M., 1989b, Evolution of a lamellar system with diffusion and reaction: A scaling approach, *Phys. Rev. Lett.* **63**, 47–50.

O'Brien, E. E. and Jiang, T.-L., 1991, The conditional dissipation rate of an initially binary scalar in homogeneous turbulence, *Phys. Fluids A* **3**, 3121–3123.

Ottino, J. M., 1989, *The Kinematics of Mixing: Stretching, Chaos, and Transport*, Cambridge University Press, Cambridge, U.K.

Pope, S. B. and Ching, E. S. C., 1993, Stationary probability density functions: An exact result, *Phys. Fluids A* **5**, 1529–1531.

Pope, S. B., 1982, An improved turbulent mixing model, *Combust. Sci. and Tech.* **28**, 131–145.

Pope, S. B., 1985, PDF methods for turbulent reacting flows, *Prog. Energy Combust. Sci.* **11**, 119–192.

Pope, S. B., 1990, Computations of turbulent combustion: Progress and challenges, in *Proceedings of 23rd Symp. (Int.) on Combustion*, pages 591–612, The Combustion Institute, Pittsburgh, PA.

Pope, S. B., 1991, Mapping closures for turbulent mixing and reaction, *Theoret. Comput. Fluid Dynamics* **2**, 255–270.

Ruetsch, G. R. and Maxey, M. R., 1992, The evolution of small-scale structures in homogeneous-isotropic turbulence, *Phys. Fluids A* **4**, 2747–2760.

Sokolov, I. M. and Blumen, A., 1991a, Diffusion-controlled reactions in lamellar systems, *Phys. Rev. A* **43**, 2714–2719.

Sokolov, I. M. and Blumen, A., 1991b, Distribution of striation thicknesses in reacting lamellar systems, *Phys. Rev. A* **43**, 6546–6549.

Yakhot, V., Orszag, S. A., Balachandar, S., Jackson, E., She, Z.-S., and Sirovich, L., 1990, Phenomenological theory of probability distributions in turbulence, *J. Sci. Comput.* **5**, 199–221.

Yakhot, V., 1989, Probability distributions in high Rayleigh number Bernard convection, *Phys. Rev. Lett.* **63**, 1965–1967.

STUDY OF IGNITION OF A DIFFUSION FLAME IN THE FIELD OF A VORTEX PAIR[1]

F. Kozusko

Department of Mathematics and Statistics
Old Dominion University
Norfolk, Virginia 23529

Michéle G. Macaraeg

NASA Langley Research Center
Hampton, Virginia 23681-0001

and

T. L. Jackson

Institute for Computer Applications in Science and Engineering
NASA Langley Research
Hampton, Virginia 23681-0001

ABSTRACT

The distortion of flames in flows with vortical motion is examined via asymptotic analysis and numerical simulation. The model consists of a constant-density, one-step, irreversible Arrhenius reaction between initially unmixed species occupying adjacent half-planes which are then allowed to mix and react in the presence of a vortex pair. The evolution in time of the temperature and mass-fraction fields is followed. Emphasis is placed on the ignition time and location as a function of vortex Reynolds number, spacing between the centers of the vortex pair and initial temperature differences of the reacting species. The study brings out the influence of the vortex pair on the chemical reaction.

[1]This work was supported by the National Aeronautics and Space Administration under NASA Contract No. NAS1-19480 while in residence at the Institute for Computer Applications in Science and Engineering (ICASE), NASA Langley Research, Hampton, VA 23681-0001 and while attending the ICASE-NASA Langley Research Center workshop on Transition, Turbulence and Combustion held during June 7 - July 2, 1993.

M. Y. Hussaini et al. (eds.), Transition, Turbulence and Combustion, Vol. II, 275–284.

1. Introduction

An important problem in the general area of non-premixed combustion is the theoretical study of chemical reactions in turbulent reacting flows. However, such flows are difficult to analyze because they are highly nonlinear, transient and involve a large number of parameters as well as a wide range of length and time scales. When the Reynolds number of the flow is large, experiments indicate that the length and time scales can be separated into large and small scales. The large scales are essentially inviscid and are related to the geometric configuration of the flow, while combustion takes place on the smaller scales associated with the diffusion of fuel and oxidizer into each other. It is on these smaller scales that theoretical work can be established, in the hope of gaining insight into the more complex larger problems. One such theoretical investigation is that of Marble (1985), who developed a simple model problem which locally describes an established thin flame that is being wrapped up by a small scale eddy. However, there are certain important aspects, such as ignition and flame structure, which were not included within the framework of the Marble problem.

We have recently completed a systematic analysis of the ignition process and the subsequent flame structure of diffusion flames in vortex dominated flows using asymptotics and direct numerical simulation (Macaraeg, Jackson and Hussaini, 1992). We considered a constant-density, one-step, irreversible Arrhenius reaction between initially unmixed species of fuel and oxidizer occupying adjacent half-planes, which were then allowed to mix and react in the presence of a single point vortex. Using large activation energy asymptotics, particular attention was paid to the ignition regime as a function of the vortex Reynolds number. For near equal initial temperatures of the fuel and oxidizer and expanding the temperature about its inert value, it was shown that the overall ignition times decreases slightly and then asymptotes to a finite value as the vortex Reynolds number is increased from zero. For finite values of the activation energy, a direct numerical solution was used to study post-ignition events. First, a hot spot develops within the viscous core of the vortex (ignition) and develops rapidly into a circular flame which grows in time according to a similarity rule. Next, a tribrachial flame configuration evolves at infinite distance to the right and left of the origin. The diffusion flames emanating from this configuration then moves toward and finally merges with the flame in the reacted core region.

The above mentioned problem of a flame interacting with a vortex is supposed to model certain fundamental mechanisms of turbulent diffusion flames which form the basis of many propulsion devices. Since acoustic waves are always present in these combustion systems, the role of acoustics in such flame/vortex interactions was investigated (Jackson, Macaraeg and Hussaini, 1993). The acoustic field was considered to be a spatially uniform but time dependent pressure wave of small amplitude. This pressure wave was then taken to have the sinusoidal form $P(t) = P_A sin(\omega t + \Phi)$, where P_A is the amplitude, ω is the frequency and Φ is the phase, and was regarded as a single Fourier mode of a more general pressure wave. It was concluded that for low frequencies, ignition could be accelerated or significantly delayed depending on the phase, while for high frequencies, ignition was always enhanced independent of phase.

The above ignition studies have been restricted to reaction in flows with a single point vortex. However, in turbulent flows, vortices are also likely to appear in pairs and so studies using counter rotating vortices are also important. The influence of a vortex pair on certain flame properties have been investigated in the past by modelling (e.g., Karagozian and Manda (1986)) and experimental (e.g., Roberts and Driscoll (1991)) techniques. However, the dynamics of ignition has not been previously investigated. Therefore, in this paper we extend our earlier ignition studies to flows with counter rotating vortices. In the next section we state the problem. Section 3 presents an analysis of the ignition regime using a combination of large activation energy (and hence Zeldovich number) asymptotics and numerics. Finally, results are presented in Section 4.

2. Problem Formulation

In this section the problem for the time evolution of initially unmixed species occupying adjacent half-planes which are then allowed to mix and react in the presence of a vortex pair is stated. The nondimensional equations governing this field in cylindrical coordinates are

$$U = \frac{1}{r}\frac{\partial \Psi}{\partial \theta}, \qquad V = -\frac{\partial \Psi}{\partial r} \tag{1}$$

$$T_t + UT_r + \frac{V}{r}T_\theta = \nabla^2 T + \beta\Omega \tag{2}$$

$$F_{j,t} + UF_{j,r} + \frac{V}{r}F_{j,\theta} = \nabla^2 F_j - \Omega, \qquad j = 1,2 \tag{3}$$

$$\Omega = DaF_1F_2e^{-Ze/T} \tag{4}$$

$$\Psi = -\frac{RSc}{2}\ln\left[\frac{(x-d)^2+y^2}{(x+d)^2+y^2}\right] \tag{5}$$

where (U, V) are the nondimensional radial and angular velocity components, respectively; Ψ the streamfunction with the Cartesian coordinates (x, y) being related to the cylindrical coordinates by the transformation $x = r\cos\theta$ and $y = r\sin\theta$; ∇^2 is the two-dimensional Laplacian operator in cylindrical coordinates; T is the temperature; and F_1 and F_2 the mass fractions of the fuel and oxidizer, respectively. The chemical model is a one-step irreversible Arrhenius reaction. The vortex pair is assumed potential with $2d$ being the distance between the vortex centers. The nondimensional parameters appearing above are the vortex Reynolds number R, defined as $\Gamma/2\pi\nu$ where ν is the kinematic viscosity; the Schmidt number $Sc = \nu/D$ assumed equal for both species and D is the species diffusion coefficient; the Zeldovich number $Ze = E/(R^oT_\infty)$ with E the dimensional activation energy and R^o the universal gas constant; the Damkohler number Da defined as the ratio of the characteristic diffusion time scale to the characteristic chemical time scale; and finally β the heat release per unit mass of F_1. The temperature and mass fractions were nondimensionalized by the values T_∞ and $F_{1,\infty}$, respectively. The velocities were nondimensionalized by U_o, some characteristic speed. Lengths and times are referred to the relevant diffusion characteristic scales $l_d = \lambda/\rho C_p U_o$ and l_d/U_o of the flow, respectively, where ρ is the density assumed constant, λ is the thermal conductivity, and C_p is the specific heat at constant pressure. Finally, the assumption of unit Lewis number is made, so that $Sc = Pr$, where Pr is the Prandtl number. These equations are to be solved subject to appropriate initial and boundary conditions.

3. Ignition Regime

At time $t = 0$, the reaction rate is exactly zero owing to the product $F_1F_2 = 0$. For $t > 0$, the fuel and oxidizer begin to mix by diffusion, as well as by convection due to the presence of the vortex pair, and the reaction rate is no longer zero. For small time, it can thus reasonably be assumed that the effect of the reaction on the overall flow field is small. The solution for small time in which the reaction rate term is neglected is known as the inert or chemically frozen solution, and will be denoted by the superscript I. In what follows, we only consider the case of near equal initial temperatures. This assumption leads to the inert solutions $T^I = 1+O(Ze^{-1})$, where the $O(Ze^{-1})$ term is included to allow for small initial temperature differences between the fuel and the oxidizer. As time increases, more of the combustible mixes until, at some finite time, a thermal explosion occurs characterized by significant departure from the inert. To analyze the ignition process, we determine the effect of the growing reaction rate by expanding about the inert solution as

$$T = 1 + Ze^{-1}T_1 + O(Ze^{-2}), \qquad F_j = F_j{}^I + O(Ze^{-1}) \tag{6}$$

and take the asymptotic limit $Ze \to \infty$. The leading-order equations are given by

$$T_{1,t} + UT_{1,r} + \frac{V}{r}T_{1,\theta} = \nabla^2 T_1 + F_1^I F_2^I e^{T_1} \tag{7}$$

$$F_{j,t}^I + UF_{j,r}^I + \frac{V}{r}F_{j,\theta}^I = \nabla^2 F_j^I, \qquad j = 1,2 \tag{8}$$

where we have chosen the Damkohler number Da to be

$$Da = \frac{e^{Ze}}{\beta Ze}. \tag{9}$$

This particular choice of the Damkohler number ensures that a distinguished limit exists, in that the reaction rate term is of the same order in a Zeldovich number expansion as the time derivative terms. The appropriate boundary and initial conditions are given by

$$FUEL \quad SIDE: \qquad T_1 = 0, F_1^I = 1, F_2^I = 0 \tag{10}$$

$$OXIDIZER \quad SIDE: \qquad T_1 = \beta_T, F_1^I = 0, F_2^I = \phi^{-1} \tag{11}$$

where ϕ is the equivalence ratio defined as the ratio of the mass fraction of the fuel (species 1) to the mass fraction of the oxidizer (species 2) divided by the ratio of their molecular weights times their stoichiometric coefficients, and β_T is the parameter which allows for small initial temperature differences. If $\phi = 1$, the mixture is said to be stoichiometric, if $\phi > 1$ it is fuel rich, and if $\phi < 1$, it is fuel lean. Also, if β_T is less than zero, the oxidizer is relatively cold compared to the fuel, and if β_T is greater than zero, it is relatively hot.

The system (7-8) was solved numerically for a range of R with $\beta_T = 0$ and $Sc = 1$. Implementation of boundary conditions are facilitated if the system is recast in Cartesian coordinates (x, y). The solution technique is a 2nd-order finite difference scheme on a nonuniform mesh. Grid resolution studies which at least doubled the computational mesh were carried out to ensure that structures were well resolved. The time-stepping scheme is a four-stage Runge-Kutta which is formally 2nd-order but has an extended stability region making it accurate and robust for moderately stiff problems. All runs were performed interactively on a Cray YMP.

4. Results

As t increases, the solution for T_1 becomes unbounded at some finite time (t_{ig}) and location (x_{ig}, y_{ig}). This characterizes the ignition regime. In Figure 1 we plot the ignition times as a function of the vortex Reynolds number R for $\beta_T = 0$, $\phi = 1$ and for a vortex spacing of $d = 4$. Note that the ignition time corresponding to the vortex pair lies between that of no vortex and that of a single vortex. We remark here that as d increases, the ignition time for a double vortex approaches that of a single vortex since the velocity fields associated with each vortex have little effect on each other. Figure 2 is a plot of the ignition times for a single point vortex $(*)$ and for a double vortex $(\square)$ of spacing $d = 4$ as a function of the equivalence ratio ϕ with $R = 20$. Note that for rich mixtures the ignition times are delayed significantly, while for lean mixtures the ignition times decreases and asymptotes to a constant. Also, as ϕ decreases from one, the ignition times for the point vortex and the double vortex are essentially indistinguishable. Finally, Figures 3 and 4 are contours of the temperature perturbation at ignition for the double vortex with centers at $d = 2$ and $d = 4$, respectively, and for various values of R with $\phi = 1$. Note that when the vortex centers are close (Figure 3), ignition initially occurs away from the vortex center which then moves towards the center as R increases. At $R = 100$ ignition is seen to occur at the centers (Fig. 3d). The same is true for centers at $d = 4$ (Figure 4), but with the ignition location moving into the centers at a smaller value of the vortex Reynolds number, at about $R = 10$.

References

Jackson, T. L., Macaraeg, M. G. and Hussaini, M. Y., 1993. "Role of acoustics in flame/vortex interactions," *Journal of Fluid Mechanics*, **254**, pp. 579-603.

Karagozian, A. R. and Manda, B. V. S., 1986. "Flame structure and fuel consumption in the field of a vortex pair," *Combustion Science and Technology* **49**, pp. 185-200.

Macaraeg, M. G., Jackson, T. L. and Hussaini, M. Y., 1992. "Ignition and structure of a laminar diffusion flame in the field of a vortex," *Combustion Science and Technology* **87**, pp. 363-387.

Marble, F. E., 1985. "Growth of a diffusion flame in the field of a vortex," in *Recent Advances in Aerospace Sciences*, ed. by C. Cassci, pp. 395-413.

Roberts, W. L. and Driscoll, J. F., 1991. "A laminar vortex interacting with a premixed flame: Measured formation of pockets of reactants," *Combustion and Flame* **87**, pp. 245-256.

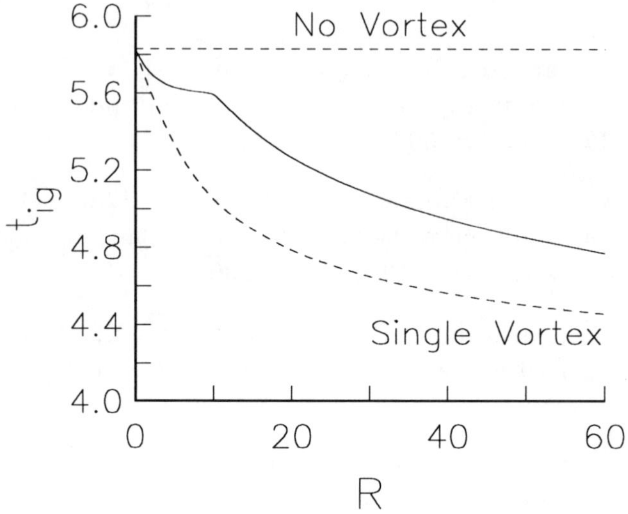

Figure 1. Plot the ignition times as a function of the vortex Reynolds number R for $\beta_T = 0$, $\phi = 1$ and a vortex spacing of $d = 4$.

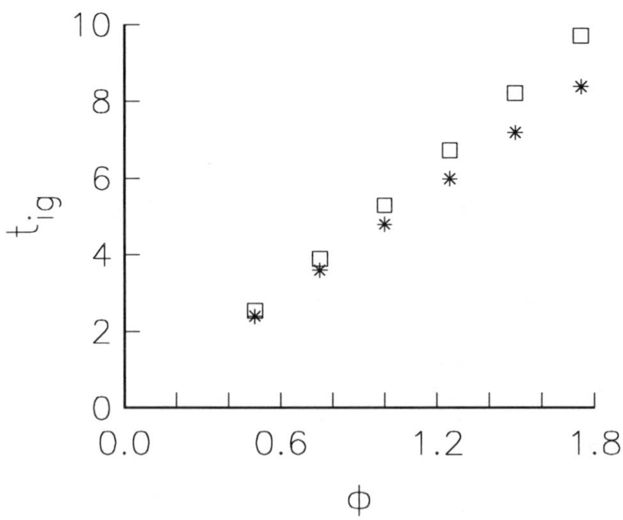

Figure 2. Plot of the ignition times for a single point vortex ($*$) and for a double vortex ($\square$) of spacing $d = 4$ as a function of the equivalence ratio ϕ with $R = 20$, $\beta_T = 0$.

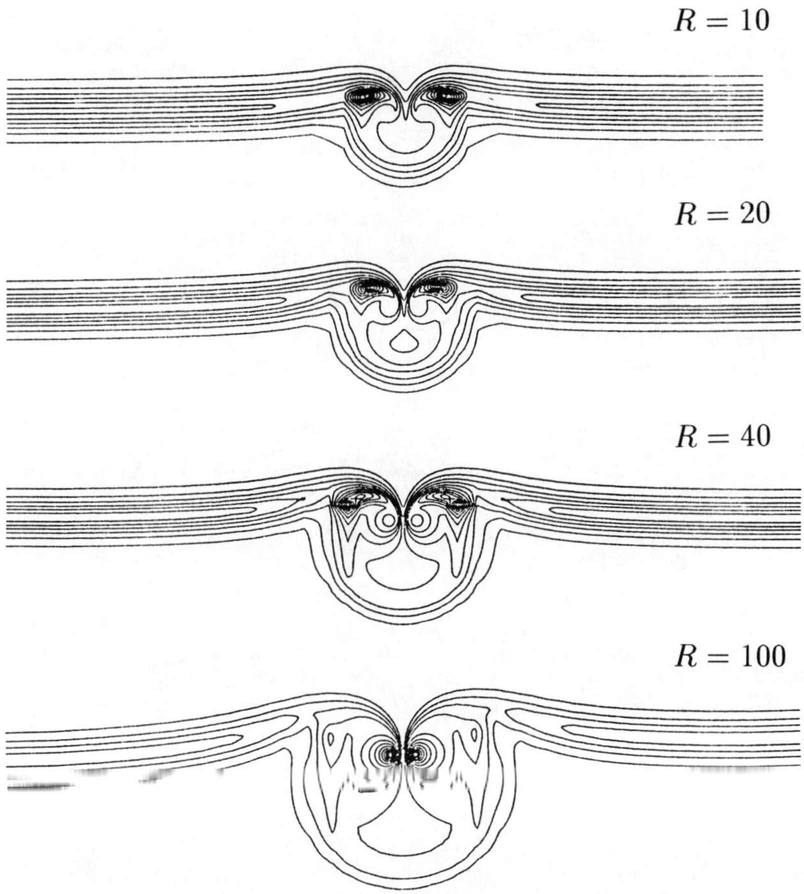

Figure 3. Contours of the temperature perturbation at ignition for the double vortex with centers $d = 2$ are plotted for various values of R and for $\phi = 1$, $\beta_T = 0$.

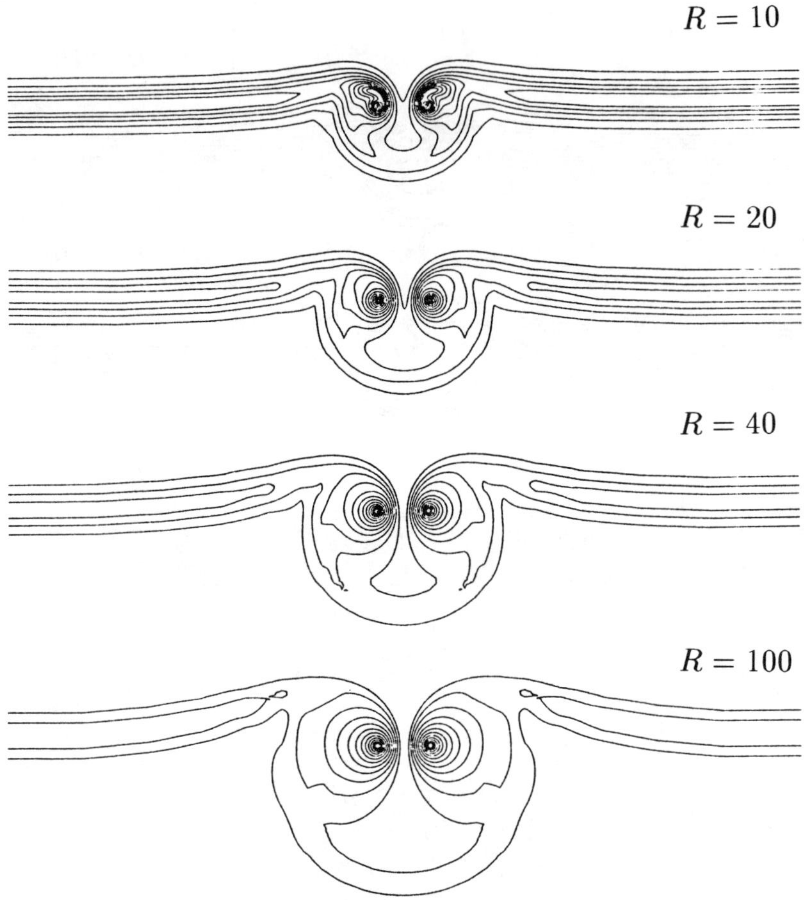

Figure 4. Contours of the temperature perturbation at ignition for the double vortex with centers $d = 4$ are plotted for various values of R and for $\phi = 1$, $\beta_T = 0$.

SHOCK-INDUCED COMBUSTION/DETONATION

GROUP SUMMARY: SHOCK-INDUCED COMBUSTION/DETONATION

Ajay Kumar
NASA Langley Research Center
Hampton, Virginia 23681-0001

The focus of the work in this group was on numerical investigation of shock-induced combustion and associated numerical issues. This topic is of particular interest to NASA Langley due to its relevance to high Mach number airbreathing propulsion. There are four papers from this group in the proceedings.

In the first paper, Quirk investigated the reflection process for a detonation front impinging on a ramp. In this investigation, he employed a high-order Gudonov scheme with a sophisticated adaptive mesh refinement technique that dynamically matches the resolution of the computational grid to the requirements of the evolving flow solution. Using the numerical data base, he generated soot-trace records, similar to those used by experimentalists to understand detonation wave structure, to demonstrate their use in understanding detonation phenomena by numerical simulations.

The second paper by Quirk, Jackson, and Kapila also employed the same numerical scheme as used above to understand the mechanism involved in the evolution of a pulse disturbance to detonation in a reactive atmosphere. The numerical results confirmed asymptotic studies in the limit of large activation energy.

The third paper by Carpenter, Atkins, and Singh develops two new techniques for shock-fitted boundary conditions by using a conventional characteristic treatment and a finite-wave approach. It is shown that both boundary condition approaches result in stable and accurate solutions; however, the finite-wave approach is simpler to implement and is suggested as the method of choice.

In the final paper, Kumar and Singh investigated the stability of shock-induced combustion past blunt projectiles. Axisymmetric Euler equations with a simple irreversible Arrhenius chemical model of (A $\rightarrow$ B) were solved using a shock-fitting technique. It was observed that the unsteadiness of the reaction front is triggered along the stagnation streamline. Following this observation, a special set of equations were obtained using the local similarity assumptions for the stagnation region. This set of equations will be used in further parametric studies and stability analysis of the reaction front.

M. Y. Hussaini et al. (eds.), Transition, Turbulence and Combustion, Vol. II, 287.

UNSTEADY SHOCK-INDUCED COMBUSTION PAST BLUNT BODIES

Ajay Kumar

NASA Langley Research Center
Hampton, Virginia 23681-0001

D. J. Singh[1]

Analytical Services & Materials Inc.
Hampton, Virginia 23666

ABSTRACT

A shock-fitting method is used to investigate unsteady shock-induced combustion past blunt projectiles. The flow is modeled by the inviscid axisymmetric Euler equations and a species continuity equation. A simple one-step irreversible chemistry model (A $\rightarrow$ B) is used to represent the chemical reaction system. The effects of projectile velocity and nose radius, reaction rate constant, activation temperature, and heat release on the stability of the reaction front are discussed. A simpler set of equations is developed that is valid in the stagnation region. It is shown that this set of equations predicts the behavior of the reaction front which is qualitatively similar to that predicted by the full axisymmetric analysis.

1. Introduction

Numerical simulation of Ballistic range experiments of shock-induced combustion in premixed fuel/air mixtures past blunt projectiles is a very active area of research. Ballistic data provide excellent opportunity to study complex flow physics that exist in supersonic reacting flows. The dominant features of these flows are controlled by chemical kinetics and inviscid flow physics. Hence, the uncertainties introduced by diffusion, mixing, and turbulence are eliminated, making it a very clean problem to study numerically.

Zeldovich (Zeldovich and Shlyapintokh, 1960) suggested that combustion can be stabilized by the shock waves produced by bodies

[1]Research supported by the National Aeronautics and Space Administration through NASA Contract No. NAS1-19320.

M. Y. Hussaini et al. (eds.), Transition, Turbulence and Combustion, Vol. II, 289–300.

moving at supersonic speeds in combustible mixtures. This technique was extensively used by various researchers (Alpert and Toong, 1972; Lehr, 1972; and McVey and Toong, 1972) to study the supersonic combustion/detonation. These studies involved firing projectiles into combustible mixtures at various levels of overdrive, and recording the detonation structures around the projectiles. The projectile is said to be overdriven if its velocity exceeds the Chapman-Jouget (C-J) velocity of the mixture in which it is traveling.

Figure 1 shows a typical ballistic range shadowgraph from experiments conducted by Lehr (1972). It shows a 15 mm diameter spherical-nosed projectile traveling at Mach 4.79 into a premixed hydrogen/air mixture at 0.421 atm. The C-J velocity of the mixture under these conditions is Mach 5.11; the projectile, therefore, is underdriven. A highly unsteady flow field can be seen in the figure. It shows two discontinuities. The outer, nearly smooth discontinuity is the bow shock and the inner, periodically corrugated discontinuity is the reaction front. The separation between the two (induction zone) is minimum near the stagnation line and increases away from it due to an increase in induction time as the shock becomes weak away from the stagnation line. The reaction front shows very clear periodic instabilities originating from the stagnation line and traveling downstream. The frequency of this instabilities was measured to be 0.72 MHz.

As mentioned earlier, a number of researchers (Matsuo et al., 1992; Matsuo and Fujiwara, 1991; Singh et al., 1992; Wang et al., 1988; Wilson and MacCormack, 1992; and Youngster et al., 1991) have simulated these flows numerically using a wide range of numerical techniques and various levels of complexity in chemical kinetics models for hydrogen/air ranging from a simple two-step model (Matsuo and Fujiwara, 1991; Wang et al., 1988) to a detailed hydrogen-air model (Matsuo et al., 1992; Singh et al., 1992; Wilson and MacCormack, 1992; and Yungster et al.) It is shown that the simple chemical reaction models are capable of reproducing all the flow features qualitatively, but detailed models are needed for quantitative prediction of features such as the frequency of unsteadiness.

In all the preceeding numerical studies, the bow shock was captured; although fitting this shock is the better choice for such problems. In the present study, a shock-fitted axisymmetric blunt-body Euler code is used with a one-step irreversible chemistry model ($A \rightarrow B$). A parametric study is conducted over a wide range of parameters to identify their effects on the stability of the reaction front. It is observed that irrespective of physical parameters causing the instability of the reaction front, the instability originates in the stagnation region and propagates downstream. To simplify the problem and to make it amenable to formal stability analysis, a locally similar set of equations is obtained which is valid in the stagnation region. This set of equations is solved along the stagnation line, and a comparison is made with the full axisymmetric solution to asses whether or not the instability behavior of the reaction front agrees qualitatively in the stagnation region with the full axisymmetric predictions. In this paper, the full axisymmetric approach is discussed first, followed by the locally similar approach in the stagnation region.

2. Governing Equations and Computational Procedure

The two-dimensional/axisymmetric Euler equations and species continuity equation in the body-oriented coordinate system shown in Figure 2 can be written as

$$\frac{\partial \boldsymbol{U}}{\partial t} + \frac{\partial \boldsymbol{M}}{\partial s} + \frac{\partial \boldsymbol{N}}{\partial n} + Q = 0 \tag{1}$$

where

$$U = \lambda\beta \begin{bmatrix} \rho \\ \rho u \\ \rho v \\ \rho H - p \\ \rho Z \end{bmatrix}$$

$$M = \beta \begin{bmatrix} \rho u \\ \rho u^2 + p \\ \rho uv \\ \rho u H \\ \rho u Z \end{bmatrix}, \quad N = \lambda\beta \begin{bmatrix} \rho v \\ \rho vu \\ \rho v^2 + p \\ \rho v H \\ \rho v Z \end{bmatrix}$$

$$Q = \begin{bmatrix} 0 \\ \beta\kappa\rho uv - p\lambda sin\theta \\ -\beta\kappa\rho u^2 - p(\lambda cos\theta + \beta\kappa) \\ 0 \\ \lambda\beta\dot{w} \end{bmatrix}$$

$$\lambda = 1 + n\kappa, \quad \beta = r + ncos\theta$$

$$H = h + \frac{u^2 + v^2}{2} + q_o Z \tag{2}$$

$$\dot{w} = K \rho Z e^{-\frac{E^+}{T}} \tag{3}$$

Here, u and v are the velocities along and normal to the body surface; ρ, p, h, and H are the density, pressure, static enthalpy, and total enthalpy, respectively; Z is the mass fraction of species A; $\dot{w}$ is the production rate; K is the reaction rate constant; E^+ is the activation temperature and q_o is the heat release due to combustion; s and n represent distance along and normal to the body, r is the distance normal to the body axis; and t represents time. κ is the body curvature which is equal to 1 for a spherical body. All the quantities have been nondimensionalized with respect to freestream conditions and nose radius, $\overline{r_n}$. Since the above set of equations becomes singular as r $\rightarrow$ 0, a limiting form of the equation is obtained along the stagnation line. The details of this limiting form of equations are given in Kumar and Graves (1972) and are not repeated here.

A one-step irreversible chemical model, A $\rightarrow$ B, is used in this analysis. The production rate of species B is given by Arrehenius law. The governing equations are solved using MacCormack's predictor-corrector method (MacCormack, 1969). Inviscid flow boundary conditions are applied on the body, and the bow shock is fitted.

3. Results and Discussion

The numerical simulations are carried out to study the effects of projectile velocity and nose radius, reaction rate constant, activation temperature, and the heat release. It is assumed that the projectile is moving in a uniformly-mixed mixture of species A. A set of baseline conditions are assumed as follows:

$$\overline{R_{gas}} = 300 \quad m^2/sec^2K \qquad \gamma = 1.2$$
$$\overline{T_\infty} = 300 \quad K \qquad \overline{K} = 2.3e+08 \quad /sec$$
$$\overline{E^+} = 15000 \quad K \qquad \overline{q_o} = 3.969e+06 \quad m^2/sec^2.$$

For these conditions, the C-J velocity is 1972 m/sec. A projectile nose radius of 0.01m is assumed for the baseline case. All the results presented here have been obtained using a uniform grid of 41x101 points.

Figure 3 shows the temperature contours for a case with 10 percent overdrive. The outer boundary is the bow shock, and the reaction front lies between the shock and the body. The reaction front is separated from the bow shock by an induction distance. The wave pattern between the shock and the body consists of compression waves and contact discontinuities. The compression waves originate from the unsteadiness in the reaction front and travel towards the body and the bow shock. Contact discontinuities are generated at the bow shock when a compression wave (originated at the reaction front) interacts with the bow shock. The unsteadiness in the reaction front originates at the stagnation line and is convected downstream.

To help understand the generation of instabilities, attention is now focussed on the temporal behavior of temperature along the stagnation line. Figure 4 shows the contour plot of temperature along the stagnation line as a function of time. It shows the bow shock, reaction front, compression waves, and the contact discontinuities. The unsteadiness in the reaction front discussed earlier can now be seen clearly. McVey and Toong (1971) used a wave interaction model to explain the mechanism of instability. According to this model, the instabilities are generated in the subsonic region near the nose of the projectile. A comparison of Figure 4 with the proposed model revealed that all the pertinent processes are captured in the numerical simulation. The generation of unsteadiness in the reaction front is identical to what is predicted by the model. A detailed discussion on the mechanism of instability is given in Singh et al. (1992) and will not be repeated here.

Figure 5 shows the effect of projectile velocity on the stability of reacting flow. Shown in the figure are the plots of stagnation point shock-standoff distance as a function of time for five cases ranging from 5 percent underdrive to 30 percent overdrive. It is seen that the flow remains steady for 5 percent overdrive primarily because the shock is not strong enough to cause significant combustion. But, as the projectile speed is increased, the flow becomes unsteady until the projectile is moving at sufficient overdrive (30 percent in the present case) where the flow again appears to be steady. At this high value of overdrive, the shock is quite strong, causing combustion close behind the bow shock. It is difficult to say whether the flow is truely steady or it simply appears steady but has very high frequency, low amplitude unsteadiness. A further grid refinement study is required to resolve this issue.

A similar kind of study was done for other flow parameters, namely the projectile nose radius, reaction rate constant, activation temperature, and heat release, by varying them one at a time. It was observed that increasing nose radius, reaction rate constant, or heat release, or reducing activation temperature caused the reaction front to go unsteady. It also appears that frequency of unsteadiness is dependent on how fast the reaction takes place. The results of the present study, as well as earlier studies, clearly point to the fact that the unsteadiness of the shock-induced combustion front is triggered in the stagnation region and propagates downstream.

It is, therefore, possible to focus on the stagnation region to better understand the mechanisms causing unsteadiness and to develop any relationship between frequency and amplitude of unsteadiness and the length of induction and reaction zones. Using the local similarity approach, the flow variables are expanded in the stagnation region as a function of angle θ as follows:

$$\begin{array}{llcll} v = & v_1 cos\theta & \mid + & v_2 cos\theta \sin^2\theta & + - - - - \\ u = & u_1 sin\theta & \mid + & u_2 sin^3\theta & + - - - - \\ \rho = & \rho_1 & \mid + & \rho_2 sin^2\theta & + - - - - \\ p = & p_1 + p_2 sin^2\theta & \mid + & p_3 sin^4\theta & + - - - - \\ Z = & Z_1 & \mid + & Z_2 sin^2\theta & + - - - - \end{array} \tag{4}$$

where v_1, v_2, u_1, u_2, etc. are functions of n and t. Retaining terms up to the truncation shown above by the dashed line and substituting in eq. (1), one obtains

$$\frac{\partial \boldsymbol{U}}{\partial t} + \frac{\partial \boldsymbol{N}}{\partial n} + Q = 0 \tag{5}$$

where

$$U = \begin{bmatrix} \rho_1 \\ \rho_1 u_1 \\ \rho_1 v_1 \\ \rho_1 H_1 - p_1 \\ \rho_1 Z_1 \end{bmatrix} ; N = \begin{bmatrix} \rho_1 v_1 \\ \rho_1 v_1 u_1 \\ \rho_1 v_1^2 + p_1 \\ \rho_1 v_1 H_1 \\ \rho_1 v_1 Z_1 \end{bmatrix}$$

$$Q = \frac{(u_1 + v_1)}{(1+n)} \begin{bmatrix} 2\rho_1 \\ 3\rho_1 u_1 \\ 2\rho_1 v_1 \\ 2\rho_1 H_1 \\ 2\rho_1 Z_1 \end{bmatrix} + \begin{bmatrix} 0 \\ 2p_2/(1+n) \\ 0 \\ 0 \\ \dot{w} \end{bmatrix}$$

and p_2 can be obtained from the following equation

$$\frac{\partial p_2}{\partial n} = -\frac{\partial p_1}{\partial n} + \frac{\rho_1 u_1}{(1+n)}(u_1 + v_1). \qquad (6)$$

Equation (5) is again solved by the same procedure as used for Eq. (1).

Figure 6 shows the contour plot of temperature along the stagnation line as a function of time for the flow conditions corresponding to the case shown in Figure 4. It is very obvious from this figure that the analysis of shock-induced combustion in the stagnation region only predicts qualitatively similar behavior as observed in the full axisymmetric analysis. This simpler approach, therefore, can be used to conduct a detailed parametric study to determine the dependence of the flow field instability on the induction distance and reaction strength.

4. Conclusions

Unsteady shock-induced combustion past blunt projectiles is studied by solving the inviscid Euler equations and a simple chemistry model, (A $\rightarrow$ B). The bow shock due to the blunt body is fitted. The analysis predicts all the mechanisms causing instabilities of the reaction front even with this simple chemistry model. It is shown that overdrive stabilized the reaction front, whereas increasing the nose radius, reaction rate constant, or heat release, or reducing activation temperature caused the reaction front to go unsteady. Since the instabilities of the reaction front develop in the stagnation region, a set of equations is developed that is valid in the stagnation region. It is found that this simple approach is able to predict qualitatively similar behavior of the reaction front as predicted by the full analysis. It, therefore, can be used to conduct a detailed parametric study and to establish relationships between various parameters and reaction front instabilities.

References

Alpert, L. R. and Toong, T. Y., 1972. "Periodicity in exothermic hypersonic flows about blunt projectiles," *Astronautica Acta* **17**, pp. 539-560.

Kumar, Ajay and Graves, R. A., Jr., 1972. "Numerical solution of the viscous hypersonic flow past blunted cones at angle of attack," AIAA Paper 77-172.

Lehr, H. F., 1972. "Experiments on shock-induced combustion," *Astronautica Acta* **17**, pp. 589-597.

MacCormack, R. W., 1969. "The effect of viscosity in hypervelocity impact cratering," AIAA Paper 69-354.

Matsuo, A. and Fujiwara, T., 1991. "Numerical simulations of shock-induced combustion around an axisymmetric blunt body," AIAA Paper 91-1414.

Matsuo, A., Fujiwara T., and Fujii, K., 1993. "Flow features of shock-induced combustion around projectiles traveling at hypervelocities," AIAA Paper 93-0451.

McVey, J. B. and Toong, T. Y., 1971. "Mechanism of instabilities of exothermic hypersonic blunt-body flows," *Combustion Science and Technology* **3**, pp. 63-76.

Singh, D. J., Ahuja, J. K., and Carpenter, M. H., 1992. "Numerical simulation of shock induced combustion/detonation," *Computing Systems in Engineering* **3**(1-4), pp. 201-215.

Wang, Y., Fujiwara, T., Aoki, T., Arakawa, H., and Ishiguro, T., 1988. "Three-dimensional standing oblique detonation wave in a hypersonic flow," AIAA Paper 88-0478.

Wilson, G. J. and MacCormack R. W., 1992. "Modeling supersonic combustion using a fully-implicit numerical method," *AIAA Journal* **30**(4), pp. 1008-1015.

Yungster, S., Eberhardt, S., and Bruckner, A. P., 1991. "Numerical simulation of hypervelocity projectiles in detonable gases," *AIAA Journal* **29**(2), pp. 187-199.

Zeldovich, Y. B. and Shlyapintokh, I. Y., 1960. "Ignition of explosive gaseous mixtures in shock waves," Office of Technical Services, U.S. Dept. of Commerce, Washington, D.C., OTS:60-41, 533.

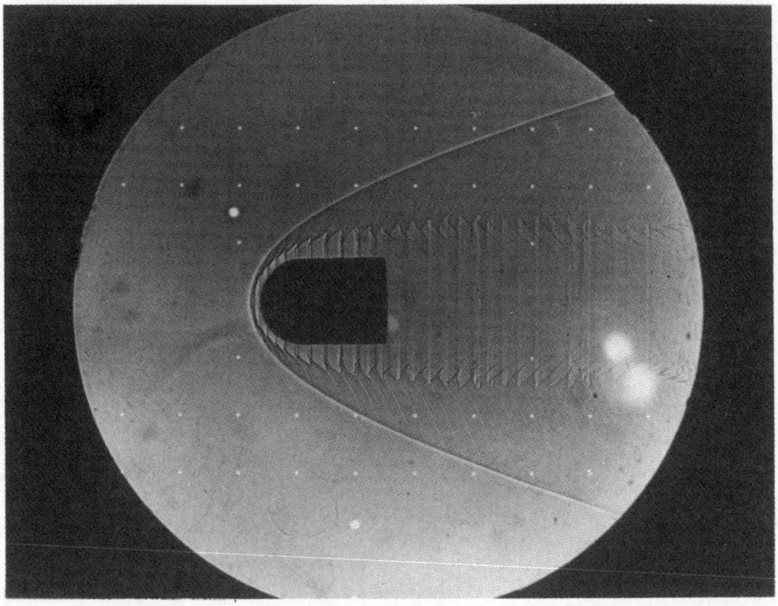

Figure 1: Shadowgraph of a spherical projectile moving at Mach 4.79 into a premixed stoichiometric hydrogen-air mixture. (Courtesy of H.F. Lehr)

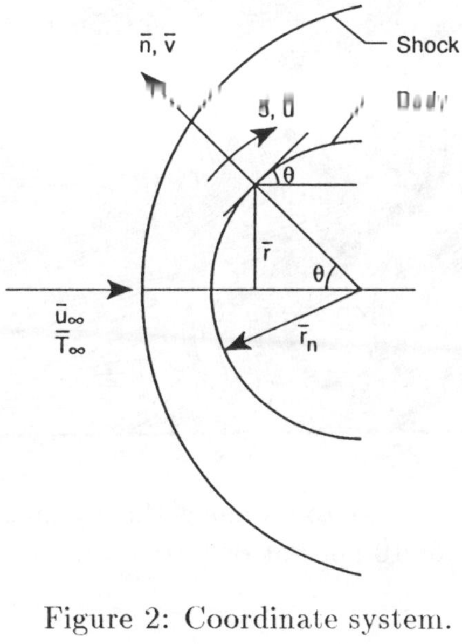

Figure 2: Coordinate system.

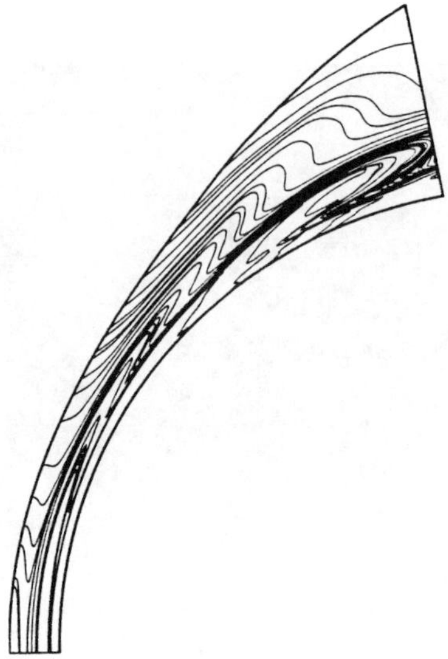

Figure 3: Temperature contours for 10 percent overdrive case.

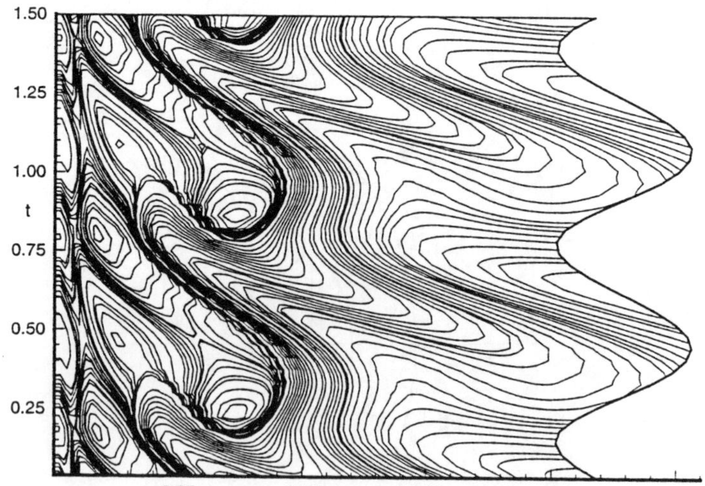

Figure 4: Temperature contours along the stagnation streamline as a function of time for 10 percent overdrive case.

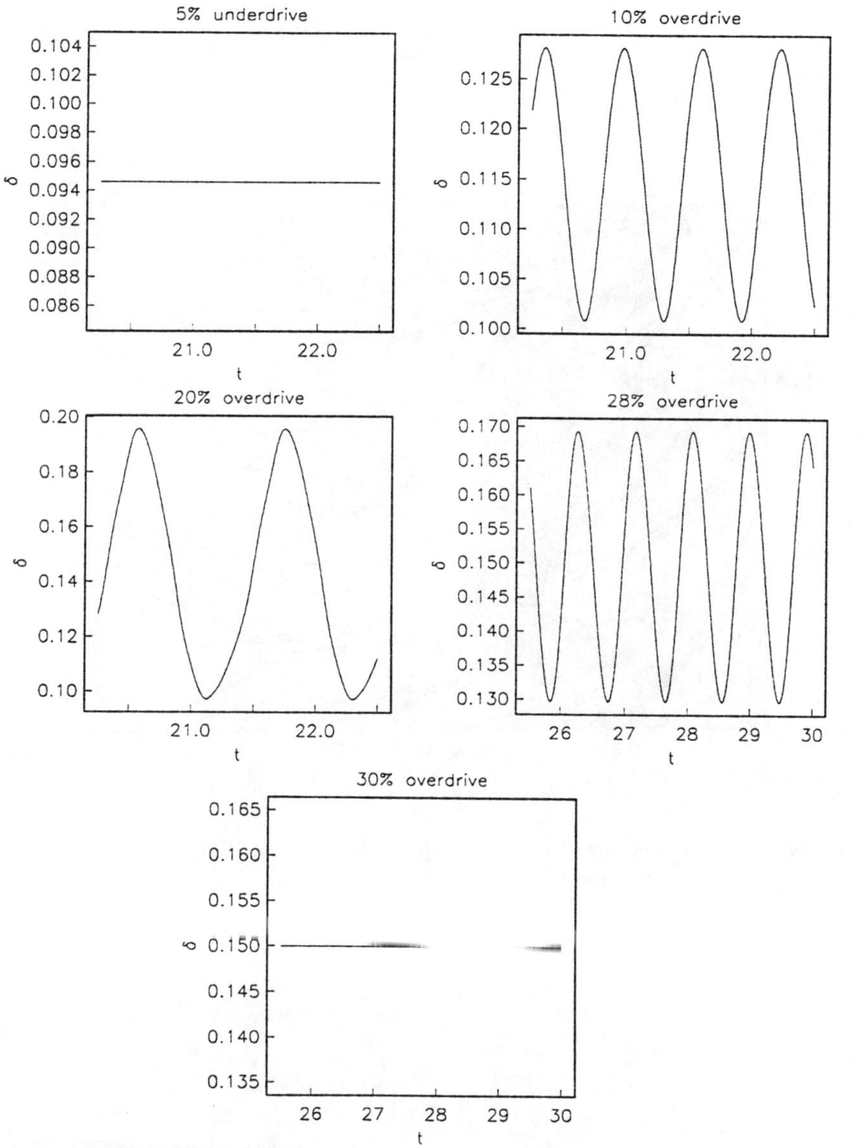

Figure 5: Shock standoff distance at the stagnation point as a function of time.

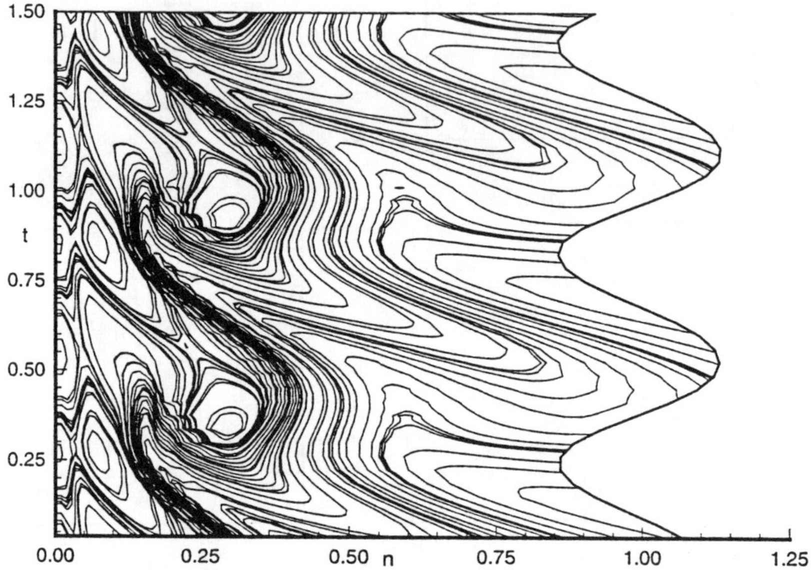

Figure 6: Temperature contours along the stagnation streamline as a function of time for 10 percent overdrive case with reduced equations.

CHARACTERISTIC AND FINITE-WAVE SHOCK-FITTING BOUNDARY CONDITIONS FOR CHEBYSHEV METHODS

Mark H. Carpenter

NASA Langley Research Center
Hampton, Virginia 23681-0001

Harold L. Atkins

NASA Langley Research Center
Hampton, Virginia 23681-0001

D. J. Singh

Analytical Services and Materials Inc.
107 Research Drive
Hampton, Virginia 23666

ABSTRACT

We use shock-fitting techniques to solve the discontinuous flow through a quasi 1-D nozzle with a Chebyshev spectral collocation method. Special attention is given to the boundary and shock velocity equations. We show that stable and accurate solutions can be obtained by either of two new techniques for the shock fitted boundary condition: 1) a conventional characteristic treatment, or 2) a finite-wave approach. Comparisons are made between the new methods, and the method proposed by Kopriva et al. (1991).

1. Introduction

Shock-fitting techniques have been widely used in computational fluid dynamics for over twenty years (see e.g. Moretti (1968) and Moretti (1972)). By fitting the shock, one knows the precise location of the discontinuity, thereby eliminating the need for interpolation or differentiation across the shock. The solution on either side of the interface is smooth, and can be calculated to high order without oscillations. This does introduce additional complexity into the problem, since the shock position, velocity, and curvature must be accounted

M. Y. Hussaini et al. (eds.), Transition, Turbulence and Combustion, Vol. II, 301–312.

for in multiple dimensions. In spite of the complexity, shock-fitting provides a means of achieving spectral accuracy throughout the entire domain (Kopriva et al., 1991).

The Rankine-Hugoniot (R-H) equations alone are not sufficient to determine the shock jump conditions and the shock velocity, simultaneously. An additional equation must be supplied to close the system. Ideally, the outward running compatibility equation along the characteristic reaching the shock provides an independent equation to close the system. Moretti (1972) provides an elaborate iterative method for simultaneously satisfying the R-H and compatibility relations, within the context of a finite-difference technique. Alternatively, Kentzer (1970) and Barnwell (1971), independently devised methods for closing the system. Both derived a shock acceleration equation from the R-H relations, using the compatibility relation on the high pressure side of the shock. Integrating the shock acceleration equation and using the R-H relations provides closure for the shock-fitting problem.

It should be noted that during the ensuing years many different techniques have been proposed to deal with closing the R-H, shock velocity system. Some have abandoned the notion of using the outward running compatibility condition, as described by Moretti or Kentzer. Remarkably, (possibly due to the dissipative nature of the numerical methods being used) these techniques have proven to be satisfactory for certain applications.

For numerical methods with less dissipation (high-order or spectral) the precise treatment of the shock jump equations is more essential (Hussaini et al., 1985, and Kopriva et al., 1991). The work of Kopriva et al. (1991) demonstrated that to ensure time asymptotic stability for a Chebyshev spatial operator (with no filter), the outward running compatibility relationship must be used for the shock acceleration procedure. Following the work of Kentzer (1970), Kopriva solved a shock acceleration equation for the shock velocity, derived from the R-H relations and the outward facing compatibility relation. The R-H relations were then used to specify the flow behind the shock. Various other techniques not based on characteristic information, were shown to diverge exponentially for long times.

As is apparent from Kopriva's work, the Chebyshev spatial operator is extremely sensitive to incorrect imposition of boundary information; being stable only for those techniques which faithfully approximate the physical waves at the boundary. Thus, the Chebyshev spatial operator is an obvious choice for testing new boundary procedures. In general, procedures which work for the Chebyshev spatial operator will work for conventional low order finite-differences.

The present paper deals with two new methods of imposing the shock jump conditions for the time dependent shock-fitting problem. The first technique, referred to as the 1) characteristic approach, is an extension of the conventional characteristic boundary decomposition. The second technique, referred to as the 2) finite-wave approach, is an extension of the finite-wave model for boundary closure, developed by Atkins et al. (1993). Both methods are shown to be stable for the classical four stage Runge-Kutta (RK) scheme with a Chebyshev spatial operator. In addition, both methods yield temporal accuracies which are consistent with the formal accuracy of the RK method on coarse grids. (On fine grids, the linearization in the characteristic approach yields a first order solution.)

In Section 2 we introduce the governing equations for the test problem, in a coordinate frame which accounts for the movement of a discontinuity. Section 3 introduces the R-H conditions and derives an equation for the shock velocity. Section 4 describes the new characteristic approach to boundary closure, while Section 5 details the finite-wave approach. Section 6 presents a comparison of the two new methods and the method proposed by Kopriva, on the one-dimensional nozzle problem.

2. Equations

To establish the stability and robustness of the new shock-fitting procedures, it is desirable to have a test problem which has a moving shock that asymptotically converges to a stable shock location. It is necessary that the new procedures track the movement of the shock correctly from the initial location of the shock to its final steady-state position. Supersonic flow into a diverging nozzle with a specified back pressure provides such a test problem. Figure (1) shows a schematic of a 1-D nozzle, for which it is assumed that the Euler equations are valid. The initial pressure at the exit of the nozzle, is chosen such that a shock exists in the nozzle. The domain of interest in the nozzles is the subsonic portion between the shock and the exit plane. The Chebyshev method is used on the subsonic portion of the nozzle, while the new shock-fitting procedures are tested on the shock.

The governing equations are the quasi-one-dimensional Euler equations $\vec{U}_t + \vec{F}_x = \vec{R}$ and can be found in any text on fluid dynamics. It is convenient to solve the equations in a coordinate frame that explicitly accounts for the movement of the shock. Specifically, the coordinates (x,t) are mapped to (ξ,τ) according to the transformations $\xi = \frac{x-x_a(t)}{x_b(t)-x_a(t)}$ and $\tau = t$. In the new coordinate frame, the quasi 1-D equations (in nonconservation form) become

$$\frac{\partial \vec{U}}{\partial \tau} + (I\xi_t + B\xi_x)\frac{\partial \vec{U}}{\partial \xi} = \vec{R} \tag{1}$$

where $\vec{U} = [\rho A, \rho u A, \rho e_t A]^T$, $\vec{R} = [0, PA_\xi \xi_x, 0]^T$, and $B = \frac{\partial F}{\partial U}$. The area $A = A(x)$, and $e_t = C_v T + \frac{u^2}{2}$. The analytic forms of the metrics become $\xi_t = \frac{-(x_b-x_a)x_a-(x-x_a)(x_b-x_a)}{(x_b-x_a)^2}$ and $\xi_x = \frac{1}{(x_b-x_a)}$. Equation (1) can be rearranged into the convenient form $\vec{U}_\tau + \bar{B}\,\vec{U}_\xi = \vec{R}$ where $\bar{B}$ is defined as

$$\bar{B} = \xi_x \begin{bmatrix} \frac{\xi_t}{\xi_x} & 1 & 0 \\ \frac{u^2\,(\gamma-3)}{2} & u\,(3-\gamma) + \frac{\xi_t}{\xi_x} & \gamma - 1 \\ -\gamma\, u\, e + u^3\,\frac{\gamma-2}{2} & \gamma\, e - u^2\,\frac{2\,\gamma-3}{2} & u\,\gamma + \frac{\xi_t}{\xi_x} \end{bmatrix}$$

with $e = C_v\, T = \frac{a^2}{\gamma(\gamma-1)}$. Note that the term $\frac{\xi_t}{\xi_x}$ can be interpreted as the velocity of the grid in physical space.

3. Boundary Conditions

The physical boundary conditions for the shock interface located at x_a, are the R-H equations relating the states on either side of a moving shock. They are described in detail in many other works [e.g. Moretti (1972)], as well as in the finite-wave section of this paper. Both new shock-fitting procedures rely on the R-H equations. In addition, the characteristic approach relies on determining the shock velocity from the pressure ratio across the shock. Rearranging the R-H equations, an expression for the shock velocity becomes

$$w = u_1 - a_1\sqrt{\frac{\gamma+1}{2\gamma}\frac{P_2}{P_1} + \frac{\gamma-1}{2\gamma}} \tag{2}$$

In a quasi-one-dimensional nozzle, the free-stream information needed by the R-H equations as well as the shock velocity equation (2), is a function of the spatial position of the shock x_a. The following assumptions are made to ensure a consistent set of boundary conditions: 1) From the nozzle throat to the shock, [$x_{th} \leq 0 \leq x \leq x_a$ in figure (1) with x_{th} the throat position], the flow is supersonic and has a constant mass flow rate. 2) At $x = 0$, the Mach number is chosen to be $M_0 = 1.5$, and expands isentropically from there to the shock. The local upstream Mach number at x_a is thus determined from the area ratio $\frac{A}{A^*} = \frac{1}{M}[\frac{2}{\gamma+1}(1+\frac{\gamma-1}{2}M^2)]^{\frac{\gamma+1}{2(\gamma-1)}}$ where $\frac{A}{A^*}$ is the ratio of the area at the point where the Mach number is M to the throat area (Ames Research Staff, 1953). Given the Mach number, and the total quantities T_o and ρ_o, the local properties assuming isentropic flow are $\frac{T_o}{T} = 1 + \frac{\gamma-1}{2}M^2$; and $\frac{\rho_o}{\rho} = [1+\frac{\gamma-1}{2}M^2]^{\frac{1}{\gamma-1}}$. The pressure, speed of sound and flow velocity were then calculated from $P = \rho RT$, $a^2 = \gamma RT$, and $u = Ma$.

The subsonic outflow boundary conditions at $x = x_b$ are treated by characteristic decomposition. (See e.g. Don et al. (1990).) The incoming characteristic is specified from the infinity values at the nozzle exit, while the out-running characteristics are calculated numerically. Since the position of the outflow boundary was fixed at $x_b = 10$, the flow quantities at x_b were specified such that a stationary shock exists at the position $x = 5$.

4. Characteristic Decomposition

For a stationary shock, characteristic decomposition of the Euler equations on the boundary is a well documented means of imposing stable and accurate physical boundary conditions for subsonic flows. (See Don and Gottlieb (1990), Oliger and Sundstrom (1978), or Gottlieb et al. (1982).) For a moving shock, the flow on the high pressure side of the shock is subsonic relative to the moving shock. In the transformed coordinate frame (relative to the shock), the shock is stationary and located at $\xi = 0$. Clearly, imposing conventional characteristic decomposition in the computational coordinate frame is necessary to ensure stability of the boundary procedure. Unfortunately, the shock velocity is not known a priori from the R-H relations, and must be determined by another means.

The following has proven to be a stable and accurate procedure for updating the shock velocity. Specifically, two temporary states are formed at the boundary, a) the "inner state" from the time advancement of the physical equations on the numerical boundary, b) the "outer state" from the R-H relations corresponding to the instantaneous shock velocity and position. In-running characteristics are then formed from the outer state, while out-running characteristics are formed from the inner state. The primitive variables are then solved from the correct characteristic information, followed by an evaluation of the shock velocity from equation (2).

To illustrate the procedure, we begin with equation (1). A similarity transformation exists which rotates the matrix $\bar{B}$ into the form, $P^{-1}\,\bar{B}\,P = \Lambda$, where Λ is a diagonal matrix. Assuming locally frozen coefficients in the matrix P the characteristic (canonical) form of the equations become $\frac{\partial \vec{V}}{\partial \tau} + (\Lambda)\frac{\partial \vec{V}}{\partial \xi} = P^{-1}\vec{R}$ where $P^{-1}\,\vec{U} = \vec{V}$. The precise forms of P, P^{-1}, and λ are easily determined, and are not presented here. Note that the eigenvalues are $\bar{u} \pm a$, and $\bar{u}$, scaled by the metric ξ_x, where $\bar{u}$ is the sum of the Eulerian fluid velocity and the grid velocity.

The eigenvalues Λ_j are the wave speeds corresponding to the characteristics V_j, and determine whether a physical boundary condition is appropriate. For well-posedness, if $\Lambda_j > 0$ the value of V_j must be imposed on the boundary. For $\Lambda_j < 0$ the value of V_j must be calculated from a numerical procedure within the domain. Given the physically correct information in $\vec{V}$, the primitive variables can be obtained by rotating $\vec{V}$ back to $\vec{U}$ with the P matrix as $P\,\vec{V} = \vec{U}$.

The entire process can be simplified to a correction procedure written as $\vec{U} = \vec{U}^{rh} + P\,I^{-}\,P^{-1}\,(\vec{U}^{n} - \vec{U}^{rh})$ where I^{-} is the reduced identity matrix, having a nonzero entry only in the (1,1) position. Following the correction, the shock velocity w is updated from equation (2). Given the new w, a new $\vec{U}^{rh}$ could be formed, and the entire procedure repeated until convergence is reached.

The characteristic decomposition is equally valid in multi-dimensions and can be generalized to other hyperbolic problems. Several notes are appropriate at this point. No increased robustness or accuracy is achieved by iterating the procedure to convergence. (The method is no better than first order on fine grids, regardless of the number of iterations performed). In addition, updating the shock velocity before or after the characteristic decomposition does not affect the robustness of the procedure. For these reasons, one pass through the characteristic procedure is sufficient. Similarly, the procedure is insensitive to whether the matrix P was formed using R-H information or numerical information.

5. Finite-Wave Model

The finite-wave model was originally developed as a non-reflective boundary conditions for acoustic problems that would preserve the order properties of a high-order numerical flow solver. In the work of Atkins et al. (1993), this was achieved by using the exact non-linear relations of finite acoustic waves and entropy waves to couple the numerical boundary to a prescribed ambient condition. This approach eliminates the linearization inherent to characteristic methods and is exact for 1-D inviscid flows. The accuracy depends primarily on the validity of the choice of finite waves incorporated in the model. For acoustic flows, the model was composed of left- and right-traveling acoustic waves and a convecting entropy wave. The equations that govern this wave structure are

$$u_2 - u_1 = \frac{-2\,n}{\gamma-1}(a_2 - a_1) \quad ; \quad \frac{a_2}{a_1} = \left(\frac{P_2}{P_1}\right)^{(\gamma-1)/2\gamma} \qquad (3)$$

$$u_3 = u_2 \quad ; \quad P_3 = P_2 \qquad (4)$$

$$u_4 - u_3 = \frac{2\,n}{\gamma-1}(a_4 - a_3) \quad ; \quad \frac{a_4}{a_3} = \left(\frac{P_4}{P_3}\right)^{(\gamma-1)/2\gamma} \qquad (5)$$

where the subscript denotes the fluid state, u is the flow speed, P is the pressure, and n is a sign parameter that is introduced for convenience and depends only on whether the boundary condition is applied to the left side (n=+1) or the right side (n=-1) of the domain. Equation (3) describes a left-traveling acoustic wave; equation (4) describes a convective entropy wave; and equation (5) describes a right-traveling acoustic wave. Equations (3) - (5) can be rearranged to give the properties at states *2* and *3* in terms of the properties at states *1* and *4* and can be found in the work of Atkins et al. (1993).

For the present problem, the finite-wave model is modified by replacing equation (3): the left-traveling acoustic wave, with the R-H conditions for a moving shock

$$\frac{u_2 - u_1}{a_1} = \frac{-2}{\gamma+1}\left(M_x - \frac{1}{M_x}\right) \qquad (6)$$

$$\frac{\rho_2}{\rho_1} = \frac{1}{1 - \frac{2}{\gamma+1}(1 - \frac{1}{M_x^2})} \tag{7}$$

$$\frac{P_2}{P_1} = 1 + \frac{2\gamma}{\gamma+1}(M_x^2 - 1) \tag{8}$$

where $M_x = \frac{w}{a_1}$ and $a^2 = \gamma P/\rho$. Note that the number of equations exactly equals the number of unknowns; however, unlike the acoustic case, the new set of equations cannot be solved in closed form. Instead, the equations must be solved by an iterative procedure. In the current implementation, the equations (4), (5) and (6) - (8) are simplified to give a single equation involving M_x and conditions at states *1* and *4*, which is solved by Newton iteration. Once the shock speed is known, everything else is easily determined. In the present test case, the left boundary is moving with the shock: thus the flow direction is always into the domain and state *2* is always the desired boundary state.

6. Results

The standard Chebyshev collocation technique was used as the spatial operator for all calculations, with between 16 and 40 collocation points. With 40 points, the error produced by the spatial operator was below machine precision. Time was advanced with the classical fourth-order Runge-Kutta method. Both of these methods are well documented elsewhere (Canuto et al., 1987) and will not be discussed here.

Three shock-fitting procedures: 1) the characteristic approach, 2) the finite-wave approach, and 3) the shock acceleration equation of Kopriva et al. (1991), were tested in terms of stability and accuracy. In all cases physical time was used as the relaxation technique to drive the equations to steady-state. To test stability, the initial shock position was perturbed to a position $2 \leq x \leq 8$, and allowed to relax back to its equilibrium value of $x = 5$. In the cases involving characteristic decomposition, the matrix P was formed with the R-H values $\vec{U}^{rh}$. For the finite-wave approach, three iterations of the solution were performed at each stage of the R-K procedure. All three of the procedures converged to the exact analytic solution, and converged at the same physical rate (governed by the physics). The solutions remained stable indefinitely after steady state was reached for all three procedures. This established the stability of the two new shock-fitting procedures.

A case which diverged exponentially was 4) non-characteristic treatment of the inflow boundary for which the shock velocity was calculated from equation (2) based on a numerically determined downstream pressure, followed by imposition of all the flow variables from the R-H relations.

The temporal accuracy of the two new methods was tested with the following transient problem. The exit pressure P_{inf} was specified to establish the steady-state shock position $x_a = 5$. The flow was then initialized with the steady state solution throughout the subsonic portion of the domain. A pressure perturbation of the form $P_b = P_{\text{inf}}[1 + \epsilon sin^4(\omega t)]$, $t > 0$, was then imposed at the exit plane. The left traveling acoustic wave eventually reaches the stationary shock, and establishes a sinusoidal movement of the shock. Since an analytic solution is not known for this problem, the exact solution at a time T was taken to be the limiting solution as the time step $\delta t \rightarrow 0$. (Again note that temporal error dominates the error in these calculations, and should decrease at a rate of $(\delta t)^4$ for the fourth-order RK method.)

Table (I) shows the results from a temporal refinement study for the characteristic, and finite-wave approach, respectively. The time T, was taken such that the shock had moved through several periods of oscillation. For CFL's ≤ 0.1, there was no change in the solution for the finite-wave approach, and was used as the exact solution. The Chebyshev operator with 16 collocation points was used to allow a large time step.

	Characteristic	Method	Finite	Wave
CFL	Log(Error)	Slope	Log(Error)	Slope
2.5	-6.37	NA	-6.70	NA
2.0	-6.60	2.37	-7.07	3.81
1.25	-6.93	1.62	-7.86	3.87
1.0	-7.06	1.06	-8.24	3.92
0.5	-7.38	1.00	-9.42	3.92
0.25	-7.68	1.00	-10.44	3.39
0.10	-8.08	1.00	Exact	NA
0.05	-8.39	1.03	Exact	NA

Table (I). Log of the solution error for various different CFL's, with the characteristic approach, (columns two and three) and the finite wave approach (columns four and five).

The characteristic approach converges initially at a rate of fourth order (clearly evident on finer meshes), but asymptotically converges at a rate of first order. This is consistent with the findings of Atkins et al. (1993), who showed that the characteristic decomposition on the boundary is inherently first order. The finite-wave model converges at the theoretical rate of fourth order to the exact solution. At least three iterations must be performed at each stage of the R-K procedure to ensure fourth-order accuracy in the finite-wave approach. Note that even the characteristic approach provides a solution accurate to seven significant digits before the first order error components become dominant.

By far, the simplest technique to implement was the finite-wave approach, followed by the characteristic approach. There was no significant difference in the execution time between the procedures, since most of the work was done outside the boundary routines. For time dependent calculations, the accuracy of the finite-wave approach is clearly superior to that of the characteristic approach.

7. Conclusions

Two new techniques have been developed for specifying the shock-fitting jump conditions. The first technique uses classical characteristic decomposition techniques with the R-H relations to provide a stable and accurate means of specifying the subsonic upstream boundary conditions. The second uses the notion of finite waves to relate the numerically determined inner state and the free stream conditions across the shock. The accuracy and robustness of each new method was demonstrated on a quasi one-dimensional nozzle problem for which an exact solution is known. The standard Chebyshev collocation method was used as the spatial operator and fourth-order Runge-Kutta was used in time. Comparisons with the shock acceleration techniques proposed by Kopriva are favorable, and are considerably simpler to implement. The overall simplicity of the finite-wave approach, as well as its formal accuracy, make it the method of choice for shock-fitting boundary conditions.

References

Ames Research Staff, 1953. "Equations, tables, and charts for compressible flow," NACA Rept. 1135.

Atkins, H. L. and Casper, J., 1993. "Nonreflective boundary conditions for high-order methods," AIAA-93-0152. Accepted for publication in the *AIAA Journal.*

Barnwell, R. W., 1971. "A time-dependent method for calculating supersonic angle-of-attack flows about axisymmetric blunt bodies with sharp shoulders and smooth non-axisymmetric blunt bodies," NASA TN D-6283.

Canuto, C., Hussaini, M. Y., Quarteroni, A., and Zang, T. A., 1987. *Spectral Methods in Fluid Mechanics*, Springer-Verlag, New York.

Don, W. S and Gottlieb, D., 1990. "Spectral simulation of unsteady compressible flow past a circular cylinder," NASA Contractor Report 182030, ICASE Report No. 90-29.

Gottlieb, D., Gunzburger, M., and Turkel, E., 1982. "On numerical boundary treatment of hyperbolic systems for finite difference and finite element methods," *SIAM J. Numer. Anal.* **19**(4).

Hussaini, M. Y., Kopriva, D. A., Salas, M. D., and Zang, T. A., 1985. "Spectral methods for the Euler equations: Part II-Chebyshev methods and shock fitting," *AIAA Journal* **23**(2).

Kentzer, C., 1970. "Discretization of boundary conditions on moving discontinuities," *Proceeding of the II International Conference on Numerical Methods in Fluid Dynamics.*

Kopriva, D. A., Zang, T. A., and Hussaini M. Y., 1991. "Spectral methods for the Euler equations: The blunt body problem revisited," *AIAA Journal* **29**(9).

Moretti, B., 1968. "Inviscid blunt body shock layers," Polytechnic Institute of Brooklyn, New York, PIBAL Rept. 68-15.

Moretti, B., 1972. "Thought and afterthoughts about shock computations," Polytechnic Institute of Brooklyn, New York, PIBAL Rept. 72-37.

Oliger, J. and Sundstrom, A., 1978. "Theoretical and practical aspects of some initial boundary value problems in fluid dynamics," *SIAM Journal of Applied Mathematics* **35**, pp. 419-446.

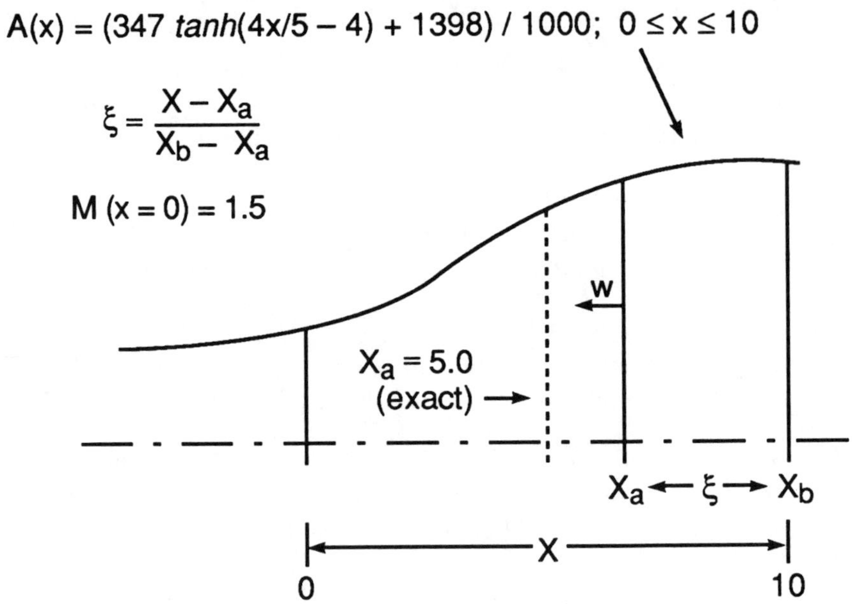

Figure 1: Schematic of the quasi-one-dimension nozzle.

EVOLUTION OF A COMPRESSIVE PULSE IN A REACTIVE ATMOSPHERE: TRANSITION TO DETONATION[1]

James J. Quirk and T. L. Jackson

Institute for Computer Applications in Science and Engineering
NASA Langley Research
Hampton, Virginia 23681-0001

and

A.K. Kapila

Rensselaer Polytechnic Institute
Troy, New York 12180-3590

ABSTRACT

The evolution of a pulse disturbance in a reactive atmosphere is examined, asymptotically and numerically, with a view towards describing the phenomenon of transition to detonation.

1. Introduction

This paper is a further step in the study of the manner in which introduction of a small-amplitude gasdynamic disturbance in an otherwise uniform, reactive environment can provoke a detonation. Attention is restricted to planar waves in a premixed gas that reacts in accordance with one-step Arrhenius kinetics with a large activation energy.

Spatial gradients are essential if a detonation, or indeed, any sort of wave motion, is to occur. Otherwise, each element of the medium would react in exactly the same way as its neighbor, resulting in a constant-volume explosion at the end of an induction delay. The

[1]This work was supported by the National Aeronautics and Space Administration under NASA Contract No. NAS1-19480 while in residence at the Institute for Computer Applications in Science and Engineering (ICASE), NASA Langley Research, Hampton, VA 23681-0001 and while attending the ICASE-NASA Langley Research Center workshop on Transition, Turbulence and Combustion held during June 7 - July 2, 1993. AKK also acknowledges the support from the National Science Foundation.

M. Y. Hussaini et al. (eds.), Transition, Turbulence and Combustion, Vol. II, 313–329.

presence of a spatial gradient allows neighboring elements to expand differentially as they release chemical energy, leading to the birth of pressure waves which, under appropriate conditions, may cause the medium to detonate.

Different evolution scenarios are possible depending on the type and length scale of the disturbance. For long-wavelength disturbances, large-activation-energy asymptotics has been remarkably successful in elucidating the details of the process. For shorter disturbances, asymptotics had provided only partial answers todate. The present study takes the asymptotic analysis further, and guided by the asymptotics, employs highly accurate numerics to provide a complete picture of the sequence to detonation.

In the general area of finite-amplitude waves in reactive gases, the pioneering effort is due to Clarke; see, especially, Clarke (1978) and Clarke (1979) and the review articles Clarke (1981, 1985, and 1989). Most of the subsequent studies have built on the foundation laid by Clarke, and the present work is no exception.

2. The Model

On the time and length scales of interest, the role of diffusion is minimal, outside shock waves. Therefore, it suffices to treat the medium as an ideal gas, modelled by the reactive Euler equations:

$$\rho_t + (\rho u)_x = 0, \quad (\rho u)_t + (p + \rho u^2)_x = 0, \tag{1}$$

$$(\rho E)_t + (\rho E u + pu)_x = 0, \quad (\rho Y)_t + (\rho Y u)_x = -\frac{\epsilon}{\beta}\rho W, \tag{2}$$

with

$$p = \rho T, \quad a^2 = \gamma p/\rho, \quad E = \frac{p}{(\gamma - 1)\rho} + \frac{1}{2}u^2 + \frac{\beta Y}{\gamma - 1}, \tag{3}$$

and

$$W = Y \exp\left[\frac{1}{\epsilon} - \frac{1}{\epsilon T}\right]. \tag{4}$$

Here, p, ρ, T, u and Y are, respectively, the gas pressure, density, temperature, velocity and reactant mass fraction. The variables have been made dimensionless with respect to the initial uniform reference state p_0, ρ_0, T_0 and Y_0. Velocity is referred to u_0, defined by

$$u_0 = \sqrt{(p_0/\rho_0)} = a_0/\sqrt{\gamma}$$

where a_0 is the frozen acoustic speed at the reference state. Time is referred to the constant-volume induction time t_0 (derived from classical explosion theory in the limit of large activation energy), and length to $u_0 t_0$. The dimensionless parameters appearing above are the reciprocal activation temperature ϵ and the polytropic exponent γ, while β is the heat release per unit mass, made dimensionless with respect to $c_v T_0$, where c_v is the specific heat at constant volume.

The conservative form of the equations, given above, is particularly well-suited to a numerical treatment. For analytical purposes it is more convenient to turn to the Lagrangian version, written below in characteristic form:

$$p_t \pm \alpha p_\psi \pm \alpha(u_t \pm \alpha u_\psi) = \frac{\epsilon}{v} W, \tag{5}$$

$$p_t + \alpha^2 v_t = \frac{\epsilon}{v} W, \quad Y_t = -\frac{\epsilon}{\beta} W, \tag{6}$$

with

$$T = pv, \quad \alpha^2 = \gamma p / v, \tag{7}$$

and W defined as before. The density ρ has now been replaced by the specific volume $v = 1/\rho$ and x by a Lagrangian particle label ψ, defined by

$$\psi = \int_0^x \rho(\lambda, t) d\lambda.$$

At time $t = 0$, the medium is assumed to be at the uniform state

$$p = v = T = Y = 1, \quad u = 0,$$

occupying the half plane $\psi > 0$.

3. Uniform Evolution

Devoid of any spatial disturbances, the state of the ambient medium satisfies

$$u = 0, \quad v = 1, \quad p = 1 + \beta(1 - Y) = T.$$

In the limit $\epsilon \to 0$, the early evolution of temperature occurs according to the asymptotic expansion

$$T \sim 1 + \epsilon\phi(t)$$

with the perturbation ϕ satisfying

$$\phi_t = e^{\phi}, \quad \phi(0) = 0$$

whence

$$\phi = -\ln(1-t). \tag{8}$$

Thermal runaway displayed by the temperature perturbation identifies $t = 1$ as the induction delay, and signals the onset of vigorous reaction. Further analysis (see Kapila (1982), for example) reveals that following runaway, the constant-volume explosion proceeds with exponential rapidity, ending with $Y = 0$ and $T = 1 + \beta$.

This analysis suggests that gasdynamic disturbances of $O(\epsilon)$ are likely to play an important role in the evolutionary process, comparable as they are to the chemically-induced *spatially uniform* perturbations of the initial state. In particular, such disturbances will have the ability to alter local induction times significantly, thereby introducing new physical phenomena in the induction process. Henceforth, therefore, disturbance amplitude shall be taken to be $O(\epsilon)$. It now remains to specify the length of the disturbance.

4. Long Disturbances

A natural length scale in the problem is the acoustic length, $a_0 t_0$, whose dimensionless measure is of order unity; it is the distance an acoustic signal would travel in a unit (dimensionless) time. Such disturbances occur naturally in a variety of situations. One example is the hot, chemically unstable region created by the passage of a strong shock through a gas that is essentially inert ahead of the shock because it is too cold there. In the induction stage one again considers $O(\epsilon)$ departures from the initial state and seeks the asymptotic expansions

$$p \sim 1 + \epsilon P, \quad T \sim 1 + \epsilon\phi, \quad v \sim 1 + \epsilon V, \quad Y \sim 1 - \epsilon\beta Z, \quad u \sim \epsilon U. \tag{9}$$

The perturbations satisfy

$$P_t + V_t = P_t \pm P_\psi \pm (U_t \pm U_\psi) = Z_t = e^{\phi} \tag{10}$$

subject to appropriate initial and boundary data. The solution can be obtained via relatively simple numerics (see Clarke and Cant (1984), Jackson and Kapila (1985)).

The principal interaction in these equations is between linearized acoustics and weak, but exponentially nonlinear, chemistry. If all the other variables are eliminated, the temperature perturbation satisfies Clarke's wave equation (Clarke, 1981)

$$\left(\phi_t - \gamma e^{\phi}\right)_{tt} = \left(\phi_t - e^{\phi}\right)_{\psi\psi} . \tag{11}$$

This equation displays a heirarchy of processes characteristic of chemico-acoustic interaction, including constant-volume and constant-pressure heating, and waves travelling with the adiabatic and isothermal sound speeds. In the context of detonation initiation, the principal feature of solutions of this equation is finite-time blowup. For the right sort of initial disturbance (such as a temperature bump), blowup occurs first at a single location (Jackson, Kapila and Stewart, 1989; Dold, 1988), and then continues along a well-defined path $t = \tilde{t}(\psi)$.

Of course, blowup of perturbation quantities is simply an asymptotic manifestation of substantial, imminent variations in the state of the gas. Further analysis reveals the blowup locus to be the trajectory of the head of a reaction wave. Typically, the wave starts out at an infinite speed but decelerates as it advances, remaining supersonic and shockless, and possessing a quasisteady structure determined by the Rankine-Hugoniot conservation conditions across it. Quasisteadiness gives way to genuine unsteadiness when the reaction wave has slowed down sufficiently for a sonic point to appear behind it. A weak shock then emerges near the sonic point and advances rapidly towards the front of the wave, strengthening as it does so, and the shock-flame complex quickly transforms into the standard, ZND detonation travelling at the Chapman-Jouguet speed. Details of this evolution have appeared elsewhere (Kapila and Dold, 1989; Dold, 1991).

The evolution problem has also been studied numerically. For relatively low activation temperatures (of the order of 5), the numerical scenario departs significantly from the asymptotic one, the essential difference being the appearance of additional unsteady structures (Clarke and Singh, 1989). However, at moderate activation temperatures (of the order of 20), Dold, Clarke and Short (1993) report a remarkable agreement between the asymptotic results and the fully numerical solution.

5. Short Disturbances

This section is concerned with disturbances that are considerably narrower, of dimensionless length $O(\epsilon)$. These were first considered by Clarke (1978 and 1979) and Blythe (1989), and later by Majda and Rosales (1987) and Almgren (1991). Attention will be confined to the special case of a single compressive pulse input, imparted to the medium through the motion of a piston at $\psi = 0$, so that the boundary condition is, for example,

$$u(0,t) = f(t/\epsilon) \equiv \begin{cases} \epsilon a \sin[\pi t/(\epsilon b)] & \text{if } 0 \le t \le \epsilon b \\ 0 & \text{if } t < 0 \text{ or } t > \epsilon b \end{cases}$$

with scaled amplitude $a > 0$ and scaled pulsewidth $b > 0$.

As the wave propagates away from the piston, essentially along the forward acoustic characteristic, it is convenient to examine it in terms of the coordinates s and t, where

$$s = (t - \psi/\sqrt{\gamma})/\epsilon. \tag{12}$$

The backward acoustic characteristic and the particle path pass through the wave in a time of order ϵ, and therefore interact with it only weakly. Solution is sought in the form

$$p \sim 1 + \epsilon P, \quad T \sim 1 + \epsilon\phi, \quad v \sim 1 + \epsilon V, \quad u \sim \epsilon U, \quad Y \sim 1 - \epsilon\beta Z - \epsilon^2\beta\eta. \tag{13}$$

One finds that all hydrodynamic variables are linear functions of the velocity perturbation U, i.e.,

$$P = \sqrt{\gamma}U - \ln(1-t), \quad V = -U/\sqrt{\gamma}, \quad \phi = \frac{\gamma-1}{\sqrt{\gamma}}U - \ln(1-t) \tag{14}$$

while U itself satisfies a reactive Burgers equation

$$U_t - \frac{1}{2}\left\{\frac{\gamma+1}{\sqrt{\gamma}}U - \ln(1-t)\right\}U_s = \frac{e^{(\gamma-1)U/\sqrt{\gamma}} - 1}{2\sqrt{\gamma}(1-t)}. \tag{15}$$

This equation is the counterpart of Clarke's equation (11), with which it shares exponential chemical amplification, but from which it differs in two important aspects: acoustics is now nonlinear and propagation unidirectional. It is also noteworthy that reactant depletion is now affected by the disturbance only at $O(\epsilon^2)$. Thus, the $O(\epsilon)$ reactant mass-fraction perturbation Z retains its ambient value

$$Z = -\ln(1-t)$$

while the $O(\epsilon^2)$ perturbation, η, evolves according to

$$\eta_s = \frac{e^{(\gamma-1)U/\sqrt{\gamma}} - 1}{1-t}.$$

Solution to the Burgers equation can be obtained quite simply by introducing a charactertistic coordinate, θ, leading to

$$U(\theta,t) = -\frac{\sqrt{\gamma}}{\gamma-1}\ln\left[1 - g(\theta)(1-t)^{-\lambda}\right] \tag{16}$$

with $\theta(\hat{s},t)$ defined implicitly by the relationship

$$\hat{s} = \theta - \frac{\gamma+1}{2\sqrt{\gamma}}\int_0^t U(\theta,\tau)d\tau. \tag{17}$$

Here,

$$\lambda = \frac{\gamma-1}{2\gamma} \quad \text{and} \quad g(\theta) = 1 - e^{-(\gamma-1)f(\theta)/\sqrt{\gamma}} \tag{18}$$

while the new spatial variable $\hat{s}$, related to s by

$$\hat{s} = s - \frac{1}{2}\int_0^t \ln(1-\tau)d\tau \tag{19}$$

simply accounts for the distortion of the forward characteristic due to passage through the uniformly reacting atmosphere.

Besides blowup, one must now contend with shock formation as well. There are two distinct possibilities, discussed below.

5.1. Broad pulse

For a broad pulse blowup occurs in the interior, near the peak of the pulse and well behind the shock, if the shock forms at all. The solution to the Burgers equation exhibits a self-similar, asymptotic structure near blowup. If $\hat{s} = \hat{s}_e$ and $t = t_e$ are the coordinates of the blowup point, then, as $t \to t_e^-$ with the similarity coordinate

$\Psi = (\hat{s} - \hat{s}_e)/\sqrt{t_e - t}$ fixed, U is found to have the logarithmically singular structure

$$U \sim -\frac{\sqrt{\gamma}}{\gamma - 1}\left[\ln\{A(t_e - t)\} + \ln(B^2\Psi^2)\right]$$

where the positive constants A and B are known. Correspondingly, the $O(\epsilon^2)$ depletion term η displays a stronger, square-root singularity,

$$\eta \sim \frac{\pi}{AB(1 - t_e)\sqrt{t_e - t}}\left(\frac{\pi}{2} + \tan^{-1}(B\Psi)\right)$$

while Z, the $O(\epsilon)$ perturbation in Y, remains bounded.

The singular behavior demands a revised description, for which the appropriate variables are σ and χ, defined by

$$t = t_e - \epsilon^2\sigma, \quad \psi = t_e - \epsilon\hat{s}_e - \frac{1}{2}\epsilon\int_0^{t_e} \ln(1 - x)dx - \epsilon^2\chi.$$

These scales ensure that ϵZ and $\epsilon^2\eta$ are now comparable, and that the similarity variable Ψ continues to be of order unity in the blowup region. Guided by (9) and (14), we now set

$$p \sim 1 - \frac{2\epsilon\gamma}{\gamma - 1}\ln\epsilon + \epsilon\bar{P}, \quad T \sim 1 - 2\epsilon\ln\epsilon + \epsilon\bar{\phi}, \quad v \sim 1 + \frac{2\epsilon\gamma}{\gamma - 1}\ln\epsilon + \epsilon\bar{V}$$

and

$$u \sim \frac{2\epsilon\sqrt{\gamma}}{\gamma - 1}\ln\epsilon + \epsilon\bar{U}, \quad Y \sim 1 - \epsilon\beta\bar{Z}.$$

The perturbations satisfy the reduced set

$$\bar{P}_\sigma + \bar{V}_\sigma = \bar{P}_\sigma \pm \bar{P}_\chi \pm (\bar{U}_\sigma \pm \bar{U}_\chi) = \bar{Z}_\sigma = -e^{\bar{\phi}}.$$

These are precisely the long-disturbance equations (10), albeit written in the new variables; thus Clarke' equation emerges again. The Burgers blowup has therefore been moderated, and one has entered a *second* induction stage, in which the space-time scales are narrowed, the rectant depletion elevated to leading order, and multidirectional wave motion restored. Evolution to detonation will now follow the path outlined earlier for the long-wave case.

5.2. Narrow pulse

A sufficiently narrow pulse steepens rapidly, and the shock devours the peak of the pulse before reactivity has had a chance to assert itself. Thus, temperature maximum occurs at the shock which, in due course, also becomes the site of blowup. Asymptotic solutions of the Burgers equation can be developed as before, with three important distinctions. First, the blowup layer, adjacent to the shock, is now much thicker; the appropriate similarity coordinate turns out to be

$$\Psi = \frac{\hat{s} - \hat{s}_s}{(t_e - t) \mid \ln(t_e - t) \mid}.$$

Second, along with U, the shock speed perturbation also displays a logarithmic singularity. Third, the $O(\epsilon^2)$ reactant consumption η now remains *bounded* at the blowup site (as it must, since it is bounded in the ambient region ahead of the shock for $t < 1$, and is continuous across the shock). The spatial gradient of η does become singular, however, suggesting that reactant consumption becomes stronger as one moves farther away from the shock. Once again, the Burgers blowup merely provides a transition into a second induction region, in which Clarke's equation applies and the long-wave scenario reemerges.

6. Numerics

In order to test the predictions of the asymptotic theory, an accurate numerical solution of the Eulerian form (1) - (4) of the governing equations has been obtained. The numerical scheme combines time-operator splitting with the Godunov approach. As the solution proceeds, the computational mesh is refined adaptively to ensure that rapidly evolving thin structures are resolved with precision. Shocks are captured rather than fitted.

As in the asymptotic analysis, the gas is assumed to occupy the half space $x > 0$. Perturbation is now introduced not via the piston but as an initial bump in velocity, given by

$$u(x,0) = \begin{cases} \frac{2\epsilon}{\gamma-1} \sin(\pi x / w) & \text{for } 0 < x < w \\ \\ 0 & \text{for } x > w\,. \end{cases}$$

The other data pertaining to the gas are

$$\gamma = 1.4, \quad \epsilon = 0.05, \quad \beta = 2.8.$$

For these values, the steady Chapman-Jouguet detonation travelling in the undisturbed medium would have velocity $v_{CJ} = 3.809$, von-Neumann peak pressure $p_{CJ} = 11.921$ and maximum temperature $T_{CJ} = 4.029$.

Two pulse widths are considered: a broad pulse with $w = 0.209$, corresponding to the scaled value $w/\epsilon = 4.18$, and a narrow pulse with $w = 0.0158$, corresponding to the scaled value $w/\epsilon = 0.316$. Results for each case are presented for blowup as well as post-blowup events. For economy in presentation, the displays start not at $t = 0$ but at a later time, shortly before blowup.

6.1. Broad pulse

Figure 1 is an x, t diagram that illustrates the overall scale of events; it shows, in three panels representing contiguous time intervals, the locus of the temperature maximum. The corresponding temperature profiles, plotted relative to the shock position, are displayed in the panels of Figure 2. Panel (a) of each figure corresponds to the time interval $0.100 < t < 0.554$. At this stage evolution is rather benign, with temperature evolving slowly on either side of the shock. Panels (b) correspond to the interval $0.554 < t < 0.574$. Blowup occurs around $t = 0.568$, indicated by the close packing of the maximum-temperature locations in Figure 1(b) and by the dotted curves in Figure 2(b). During this period of utmost brevity, when temperature in the blowup region increases substantially, the ambient region ahead of the shock is essentially frozen. At the end of blowup, temperature has reached a maximum value of 3.8, agreeing precisely with the constant-volume explosion value $1 + \beta$, and thus confirming that in the core of the hot spot the gas is inertially confined.

Post-blowup events, involving the birth of a reaction wave and its movement away from the blowup site, are described in panels (c) which correspond to the time interval $0.574 < t < 0.608$. Restricting attention to the wave travelling towards the shock, Figure 1(c) indicates that the wave is slowing down, reaching a speed of 4 units at $t = 0.580$, and essentially maintaining this speed thereafter. This speed agrees favorably with the CJ-value of 3.809, indicating that the wave is a fully-developed CJ detonation. For the parameters chosen here, the CJ wave has a mild galloping instability (Lee and Stewart, 1990) that is plainly visible in Figure 2(c).

6.2. Narrow pulse

Results for the narrow pulse are displayed in Figures 3 and 4, again in a three-panel format. Panels (a), beginning at $t = 0.643$, show the temperature maximum to be at the shock, as predicted by the Burgers blowup. Passage of time then causes the temperature maximum to move away from the shock, and the eventual Clarke-type blowup occurs well behind the shock, at around $t = 0.6835$ (panels (b)). Thereafter, evolution follows the broad-pulse case described above, with one important distinction: the reaction wave runs into the shock before it has had a chance to mature fully into the CJ detonation. Panel 3(c) shows that the wave is still decelerating, while panel 4(c) indicates that the galloping instability is yet to develop.

7. Conclusions

A highly accurate numerical scheme, combining the Godunov approach with time-operator splitting and adaptive mesh refinement, has been used to study the evolution to detonation of a pulse disturbance in a reactive atmosphere. It is observed that the pulse has a tendency to steepen into a weak shock. Both for short and long pulses, an induction period characterized by mild chemical amplification culminates in a localized rapid reaction at a site well behind the shock. The resulting reaction wave transits into a detonation, through a mechanism that is now well-understood. The numerical results confirm, and are themselves better understood in view of the existence of, asymptotic studies in the limit of large activation energy.

References

Almgren, R. F., 1991. "High frequency acoustic waves in a reacting gas," *SIAM J. Appl. Math.* **51**, pp. 351-373.

Blythe, P. A., 1989. "Wave propagation and ignition in a combustible mixture," *17th Symposium (International) on Combustion,* The Combustion Institute, Pittsburgh, pp. 909-916.

Clarke, J. F., 1978. "Small-amplitude disturbances in an exploding atmosphere," *J. Fluid Mech.* **89**, pp. 343-356.

Clarke, J. F., 1979. "On the evolution of compressive pulses in an exploding atmosphere; initial behaviour," *J. Fluid Mech.* **94**, pp. 195-208.

Clarke, J. F., 1981. "On the propagation of gasdynamic disturbances in an explosive atmosphere," *Prog. Aeronautics and Astronautics* **76**, pp. 383-402.

Clarke, J. F., 1985. "Finite amplitude waves in combustible gases," in "The Mathematics of Combustion," J. D. Buckmaster, Ed., SIAM Publications, Philadelphia, pp. 183-245.

Clarke, J. F., 1989. "Fast flames, waves and detonation," *Prog. Energy Combust. Sci.* **15**, pp. 241-271.

Clarke, J. F. and Cant, R. S., 1984. "Nonsteady gasdynamic effects in the induction domain behind a strong shock wave," *Prog. Aeronautics and Astronautics* **95**, pp. 142-163.

Clarke, J. F. and Singh, G., 1989. "A numerical simulation of shock generation using the random choice method," in "Numerical Combustion," A. Dervieux and B. Larrouturou, Eds., *Lecture Notes in Physics* **351**, Springer-Verlag, New York, pp. 22-35.

Dold, J. W., 1988. "Dynamic transition of a self-igniting region," in "Mathematical Modelling of Combustion and Related Topics," C. M. Brauner and C. Schmidt-Lainé, Eds., Martinus Nijhoff, Dordrecht, pp. 461-470.

Dold, J. W., 1991. "Emergence of a detonation within a reacting medium," in "Fluid Dynamical Aspects of Combustion Theory," A. Tesei and M. Onofri, Eds., Longman, U.K.

Dold, J. W., Clarke, J. F. and Short, M., 1993. "Steady and unsteady aspects of detonation initiation," submitted.

Jackson, T. L. and Kapila, A. K., 1985. "Shock-induced thermal runaway," *SIAM J. Appl. Math.* **45**, pp. 130-137.

Jackson, T. L., Kapila, A. K. and Stewart, D. S., 1989. "Evolution of a reaction center in an explosive material," *SIAM J. Appl. Math.* **49**, pp. 432-458.

Kapila, A. K., 1982. *Asymptotic Treatment of Chemically Reacting Systems*, Pitman Advanced Publishing Program, Boston.

Kapila, A. K. and Dold, J. W., 1989. "A theoretical picture of shock-to-detonation transition in a homogeneous explosive," *9th Symposium (International) on Detonation,* pp. 111-118.

Lee, H. I. and Stewart, D. S., 1990. "Calculation of linear detonation instability: one-dimensional instability of plane detonation," *J. Fluid Mech.* **216**, pp. 103-132.

Majda, A. J. and Rosales, R. R., 1987. "Nonlinear mean field - high frequency wave interaction in the induction zone," *SIAM J. Appl. Math.* **47**, pp. 1017-1039.

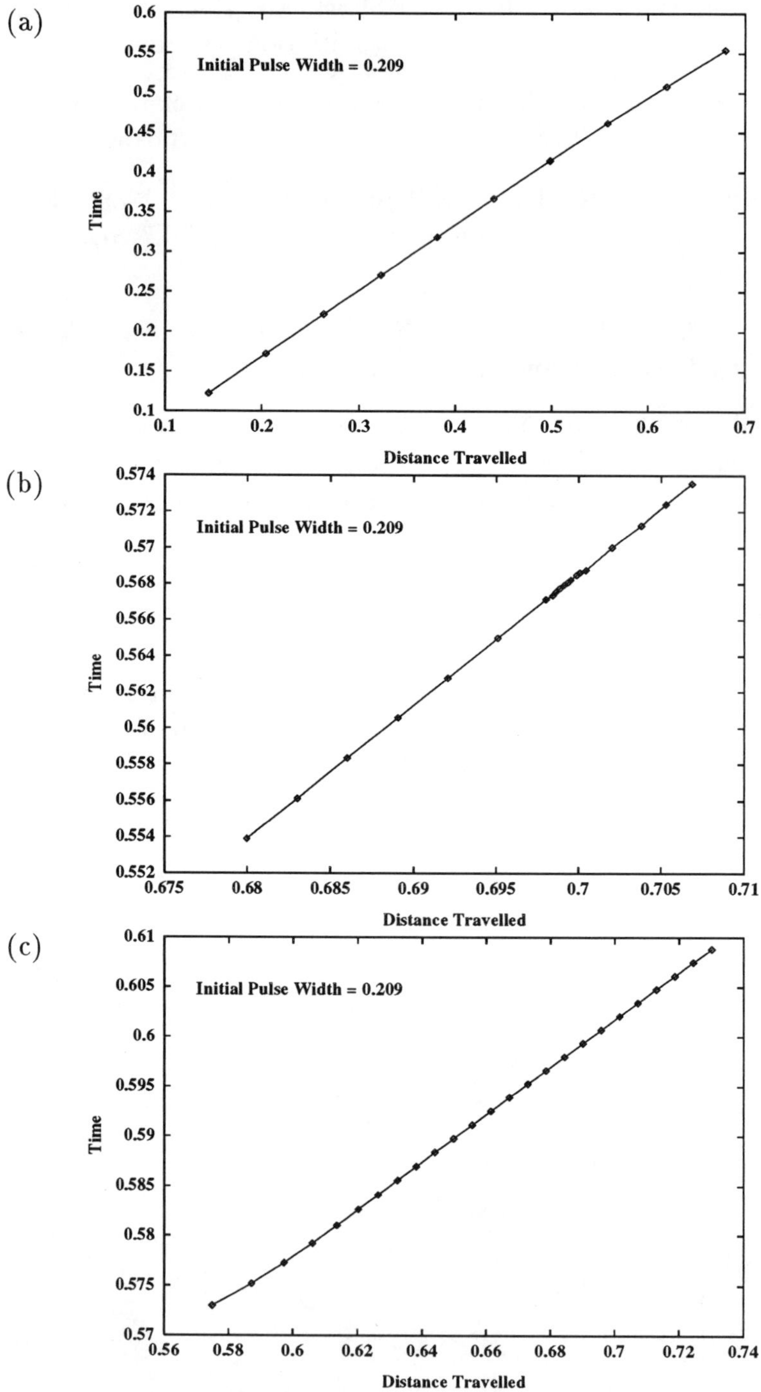

Figure 1: Locus of the temperature maximum for the broad pulse.

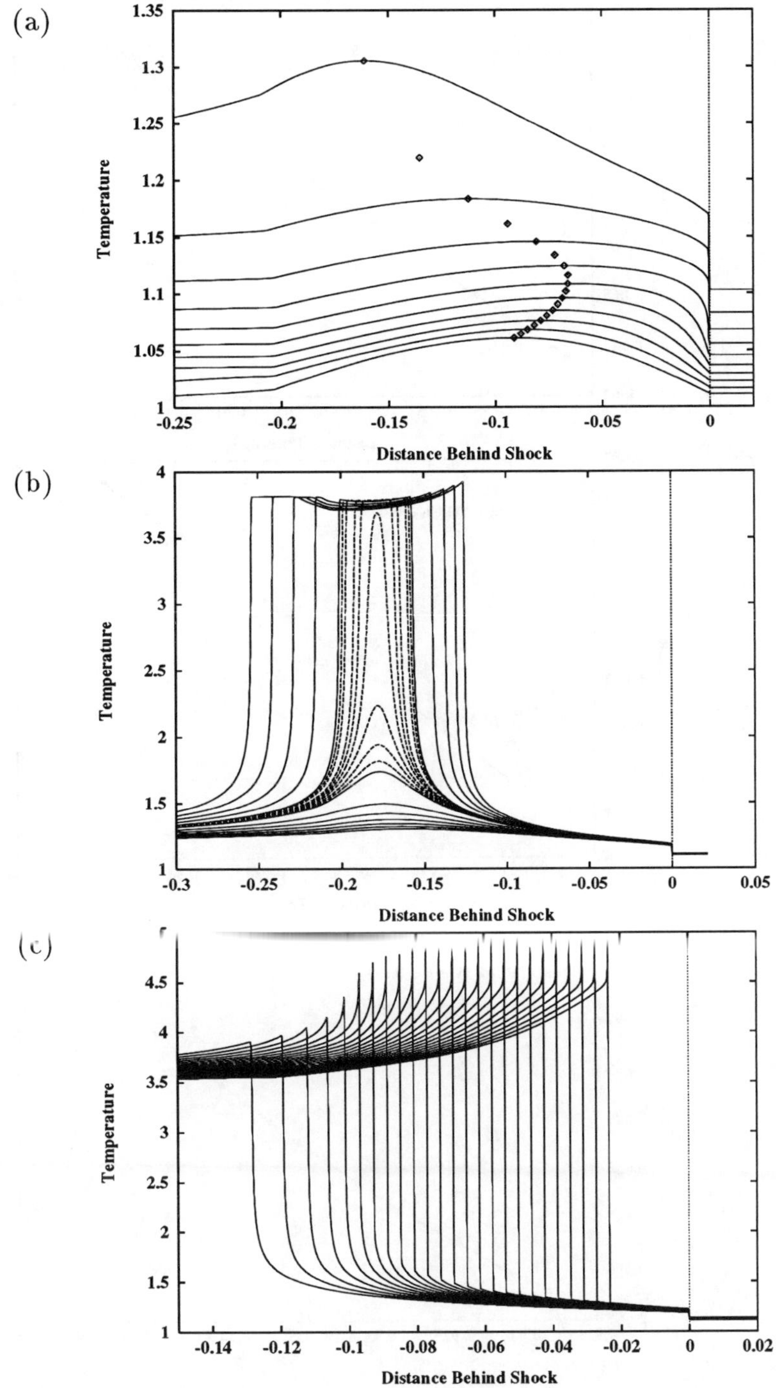

Figure 2: Temperature profiles for the broad pulse.

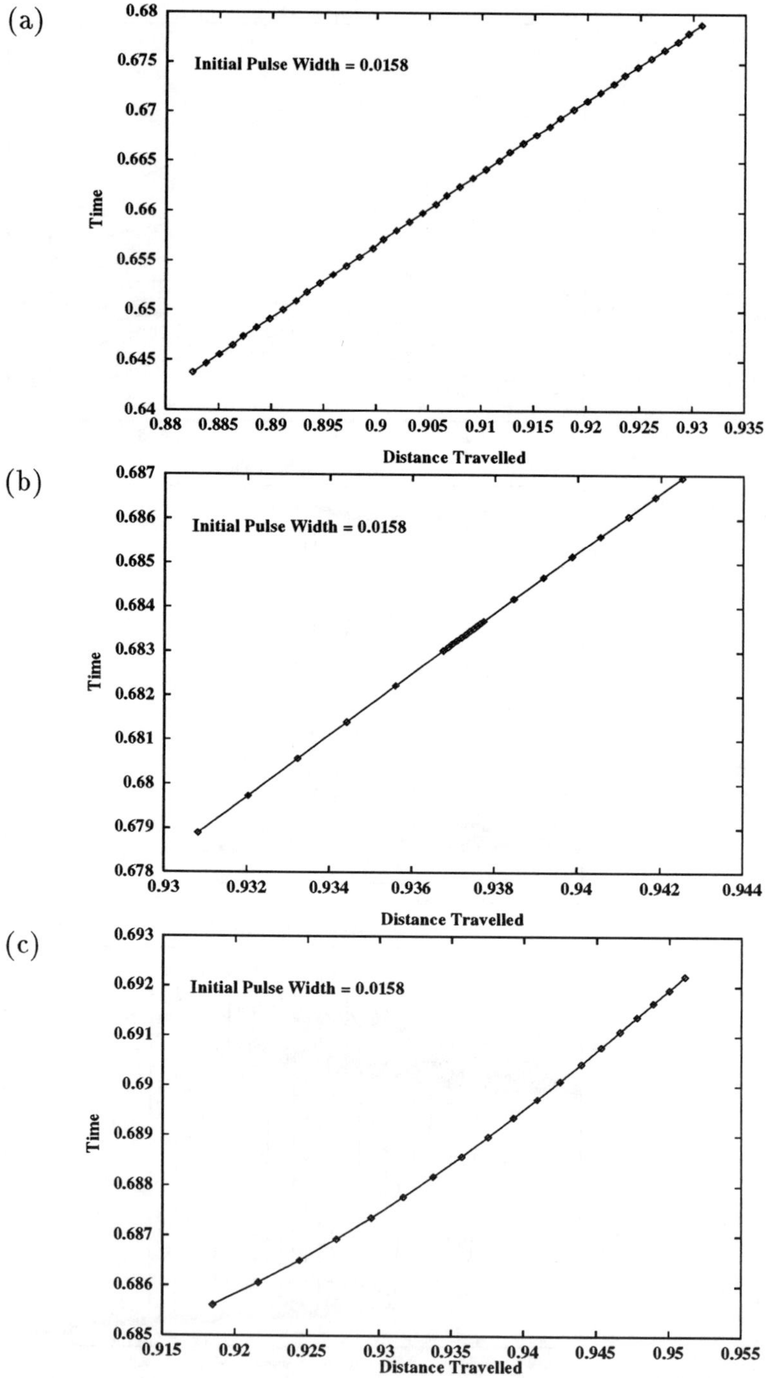

Figure 3: Locus of the temperature maximum for the narrow pulse.

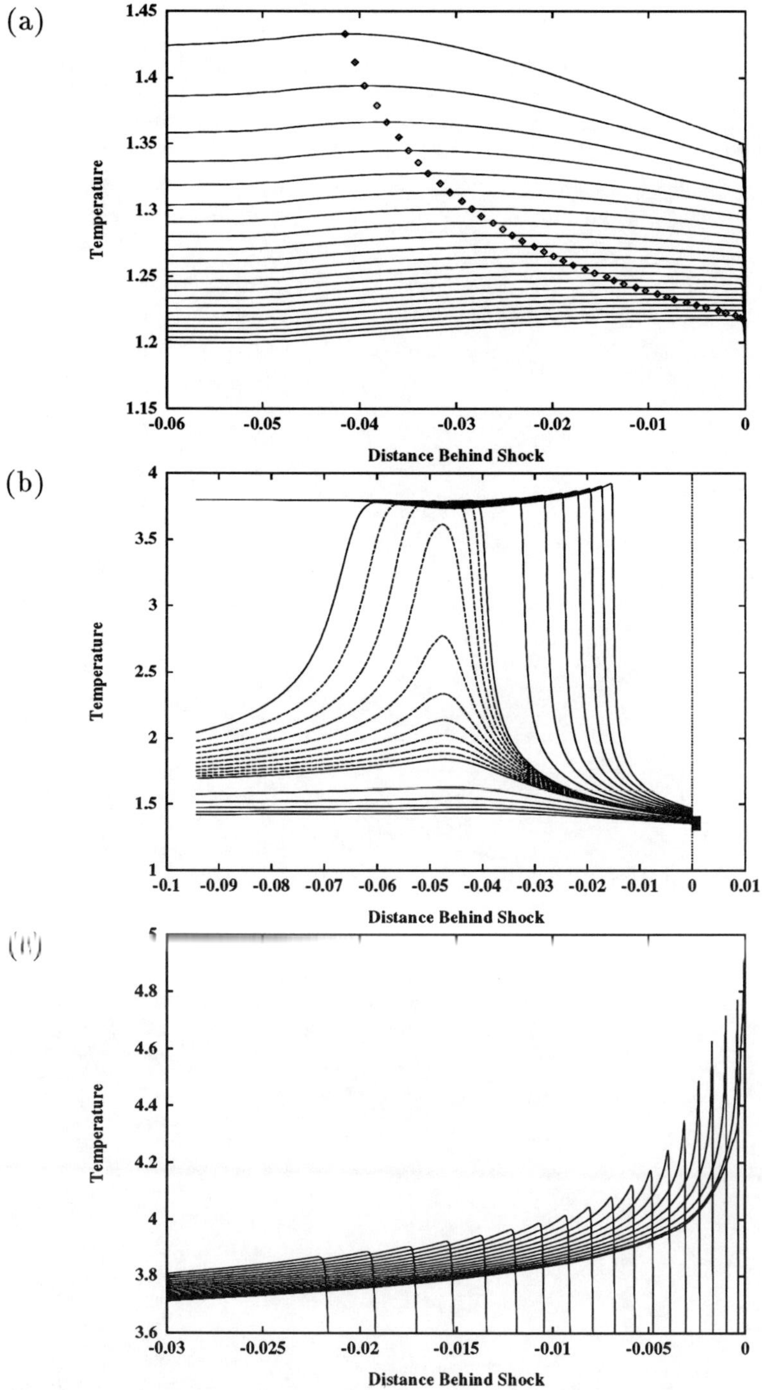

Figure 4: Temperature profiles for the narrow pulse.

NUMERICAL SOOT TRACES: "THE WRITING'S ON THE WALL"

James J. Quirk[1]

Institute for Computer Applications in Science and Engineering
Mail Stop 132C, NASA Langley Research Center
Hampton, Virginia 23681-0001

ABSTRACT

Until relatively recently, our knowledge of the structure of a detonation front rested almost entirely on experimental observation. In particular, much information had been gleaned using the so-called soot-film technique in which aluminium foils covered with lamp-black are used to line the walls of a shock tube; as a detonation wave passes down the tube, so it leaves imprints on the soot film that give vital clues as to the wave's structure. With the development of Computational Fluid Dynamics (CFD), it is now possible to perform numerical simulations that reveal a detonation front's structure in great detail. However, CFD simulations do not always improve our basic understanding of fluid phenomena, since it is all too easy to become overwhelmed by the large amounts of data that are generated. It has long been our experience that it is both useful and convenient to present computational results using traditional forms of communication. For example, we routinely make use of numerically generated schlieren-type images rather than making do with contour plots. In this paper we demonstrate how numerically generated soot-trace records may be used to good effect in understanding simulations of detonation phenomena.

1. Introduction

Since the early 1980s, impressive strides have been made in the development of numerical schemes that integrate the Euler equations for simulations of inert shock wave phenomena. More recently, algorithm developers have started to take interest in devising schemes for

[1]Research was supported by the National Aeronautics and Space Administration under NASA Contract No. NAS1-19480 while the author was in residence at the Institute for Computer Applications in Science and Engineering (ICASE), NASA Langley Research Center, Hampton, VA 23681.

M. Y. Hussaini et al. (eds.), Transition, Turbulence and Combustion, Vol. II, 331–337.

simulating reactive flows such as detonation phenomena (Bourlioux and Majda, 1992; Quirk, 1993). Although some progress has been made, it is clear that we are still far from being able to perform large scale simulations of reactive shock wave phenomena. The basic difficulty lies with the fact that the reaction process necessarily introduces a length scale into the problem, and unless this scale is adequately resolved the results from a simulation are largely worthless. Unfortunately, compared to the hydrodynamic length scale the reaction length scale is usually exceedingly small, thus reactive simulations are inherently expensive. Indeed, because of the disparateness of these two scales, one can easily show that something other than just a very large computer is required to perform multi-dimensional simulations (Quirk and Hanebutte, 1993).

For our simulations we employ a sophisticated adaptive mesh refinement algorithm that dynamically matches the resolution of the computational grid to the requirements of the evolving flow solution (Quirk, 1991). Thus it is possible to resolve the reaction zone without utilizing a prohibitively large number of mesh cells. For example, Figure 1 shows a schlieren-type snapshot from the simulation of a detonation wave impinging on a ramp. This calculation employed roughly 200 thousand cells, the equivalent uniform mesh calculation would require over 25 million cells! Even with the enormous savings gained via local mesh refinement, detonation simulations remain expensive to run. In view of this, we have invested considerable effort in parallelizing our mesh refinement scheme (Quirk and Hanebutte, 1993). Today, we can spread calculations across several workstations so as to reduce the turnaround time to acceptable levels. However, given the complex dynamical behaviour of detonation fronts it is very difficult to assimilate the information produced during a computation. Note that a typical simulation might take more than 60 thousand time steps, while the flow field solution at each time step might require tens of Mbytes of storage. Leaving aside the practical difficulties of storing all this data, it begs the question, what use would it be anyway?

Like it, or loathe it, the usefulness of a direct numerical simulation is often closely related to the extent to which computer graphics may be exploited in depicting its output. Unfortunately, however, there is a danger of placing aesthetics before scientific merit when plotting results. Now it has long been our experience that traditional forms of communication are more than adequate for displaying the results of most CFD applications. For example, numerically generated schlieren-type images are far superior to garishly coloured contour plots for understanding shock wave simulations. First, from a physiological viewpoint, the eye is more sensitive to varying shades of grey than it is towards varying colours. Second, from a psychological viewpoint, fluid dynamicists are conditioned to thinking in terms of schlieren images, shadowgraphs, interferograms etc. rather than contour plots. Of course, intelligent use of colour will always help highlight certain features and so improve our understanding, but it should be remembered that computer graphics are the medium of communication and not the message. In this vein, one obvious candidate for graphical output from a detonation simulation is that of the soot-film record.

2. Numerical Soot Traces

In contrast to a shock wave, a detonation front is often inherently unstable and exhibits complex dynamical behaviour. To quote from Oppenheim and Soloukhin (1973): "The placid image of a plane, steady flow detonation wave, whose progress and structure are virtually unencumbered by the kinetics of fast chemical reactions, was shattered in the nineteen sixties." One novel, yet simple, experimental technique helped this process. Aluminium foils coated in soot can be used to line a shock tube down which a detonation wave is propagated, the dynamics of the front will then be captured for posterity by the imprints left on the soot-film. Such experiments have shown that detonation fronts may support a transverse wave system that can trace out remarkably regular "fish-scale" patterns (Strehlow, 1969).

It has been argued that it is the scouring action of the vorticity field induced by the transverse waves that leaves the tell-tale imprints on the soot-film (Oppenheim and Soloukhin, 1973). To test this thesis we have produced smear images of the vorticity field predicted by our numerical simulations. In essence, the magnitude of the vorticity at each point in space is monitored and the peak value is saved and then plotted as a grey-scale image; the lighter the image, the higher the peak vorticity that has passed that point in space. The vorticity is computed in a straightforward manner using central differences. Note that no special treatment need be applied at the detonation front. After all, the computed value is only used to provide an indication for the strength of the vorticity; it is not used to evolve the flow, and so no cumulative error can ensue.

Figures 1 and 2 are taken from a study of the reflection process for a detonation front impinging on a ramp. The so-called reactive Euler equations were used to model the flow. For the non-dimensionalization used by Quirk (1993), the incident detonation wave is fixed by the following values: 50, 10, 1.2 and 1.2 which were used for the heat release, activation energy, ratio of specific heats and the overdrive parameter, respectively. From Figure 1, as is to be expected for low ramp angles, the schlieren image shows that the detonation wave undergoes single Mach reflection. Note that the schlieren image represents but a single snapshot from the simulation and so it fails to convey the dynamic nature of the detonation front. But something of this nature can be gleaned from the numerical soot-trace record. Firstly, the familiar cellular pattern, which is traced out by the numerous triple points that move along the length of the detonation front, is strongly reminiscent of those produced experimentally. Also note the two forward facing plumes that emanate from one side of each "fish-scale". Such plumes have been observed experimentally, for certain gaseous mixtures, and would appear to be caused by the vortical structure associated with a Rayleigh-Taylor instability that is born when two transverse waves collide. Interestingly, a smear trace of the pressure field also captures the gross features of the "fish-scale" pattern, albeit less sharply than the vorticity field, but it fails to detect these plume structures.

One other detail that can be seen from the soot trace is the fact that the detonation cell size behind the Mach stem is smaller than that behind the incident wave. Again this is in accordance with experimental observation. The degree of overdrive for the stem is necessarily higher than that for the incident wave, since it moves normal to the ramp's surface. Overdrive has a stabilizing affect on the detonation front; the higher the overdrive, the smaller the detonation cell spacing. From Figure 2 it can be seen that for large ramp angles the detonation wave undergoes regular reflection which again is consistent with experimental observation. The soot trace record confirms the absence of Mach reflection. Note that the extreme regularity of the detonation cells shown here is due to the fact that the simulation employed quite a low activation energy and, for the chosen parameters, the detonation wave is stable to one-dimensional disturbances.

Thus far all of our simulations have been for stylized reactions; the sheer expense of performing these simulations prevents us from using a more realistic reaction model. Nevertheless, it is gratifying that the simulations can faithfully reproduce, at least qualitatively, all the important features that have been observed experimentally.

3. Conclusions

It has long been argued that it is the scouring action of the vorticity field that leaves the "fish-scale" imprints which are observed experimentally for detonation waves when using the soot-film technique. Our computations are in complete agreement with this thesis. Numerically generated soot trace records provide a convenient means of assimilating the dynamics predicted via simulations of detonation phenomena, and they allow direct comparisons to be made with experimental results so as to gauge the reliability of a simulation.

References

Bourlioux, A. and Majda, A.J., 1992. "Theoretical and Numerical Structure for Unstable Two-Dimensional Detonations," *Combust. Flame* **90**, pp. 211–229.

Oppenheim, A.K. and Soloukhin, R.I., 1973. "Experiments in Gasdynamics of Explosions," *Ann. Rev. Fluid Mech.* **5**, pp. 31–58.

Quirk, J.J., 1991. "An Adaptive Grid Algorithm for Computational Shock Hydrodynamics," Ph.D. Thesis, College of Aeronautics, Cranfield Institute of Technology, U.K. .

Quirk, J.J., 1993. "Godunov-Type Schemes Applied to Detonation Flows," *ICASE Report No. 93-15.* To appear in the Proceedings of the 2nd ICASE/NASA LaRC Combustion Workshop, Hampton VA, 1992.

Quirk, J.J., and Hanebutte, U.R., 1993. "A Parallel Adaptive Mesh Refinement Algorithm," *ICASE Report No. 93-63.* Submitted to *J. Comput. Phys.*

Strehlow R.A., 1969. "The Nature of Transverse Waves in Detonations," *Astronaut. Acta* **14**, pp. 539–548.

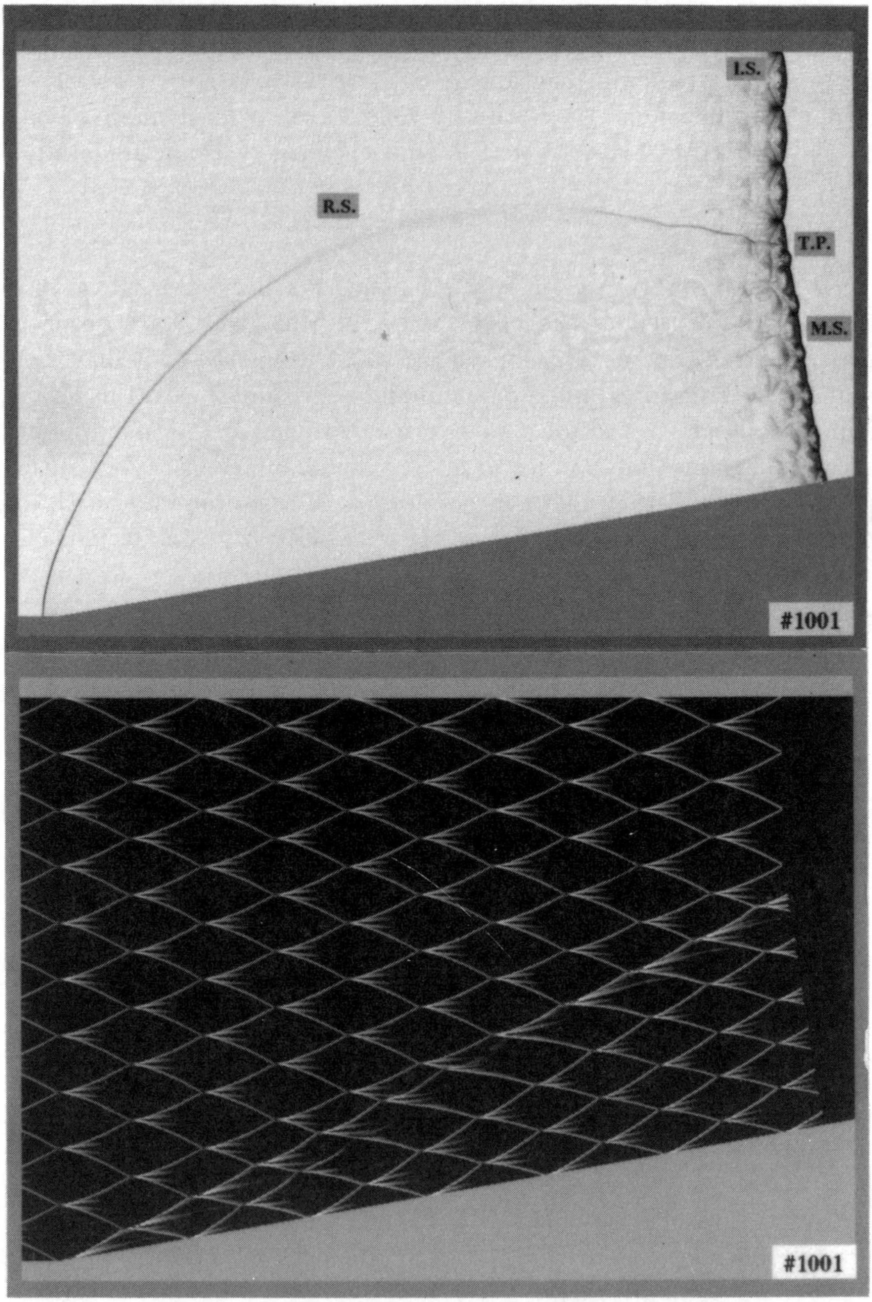

Figure 1: Schlieren-type image and numerical soot trace for a detonation wave reflecting from a 10° ramp. Key: I.S., Incident Shock; R.S., Reflected Shock; M.S., Mach Stem; T.P., Triple Point.

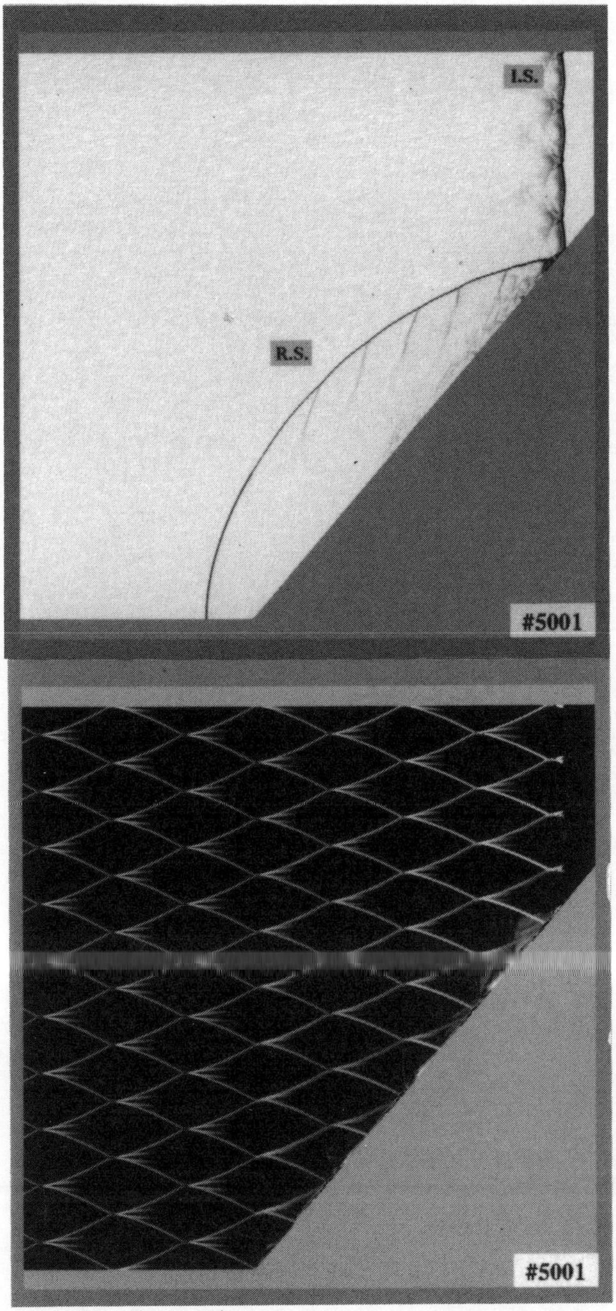

Figure 2: Schlieren-type image and numerical soot trace for a detonation wave reflecting from a 50° ramp. Key: I.S., Incident Shock; R.S., Reflected Shock.

CONTENTS

Volume I